Levinas's Philosophy of Time

Levinas's Philosophy of Time

Gift, Responsibility, Diachrony, Hope

Eric Severson

DUQUESNE UNIVERSITY PRESS
Pittsburgh, Pennsylvania

Published in the United States of America by
Duquesne University Press
600 Forbes Avenue
Pittsburgh, Pennsylvania 15282

Library of Congress Cataloging-in-Publication Data

Severson, Eric R.
Levinas's philosophy of time : gift, responsibility, diachrony, hope / Eric R. Severson.
pages cm
Includes bibliographical references and index.
Summary: "A chronological approach that examines the progression of Levinas's deliberations on time over six decades, thus providing new insights about aspects of Levinasian thought that have consistently troubled readers, including the differences between Levinas's early and later writings, his controversial invocation of the feminine, and the blurry line between philosophy and religion in his work"—Provided by publisher.
ISBN 978-0-8207-0462-3 (pbk. : alk. paper)
1. Levinas, Emmanuel. 2. Time—Philosophy—History—20th century.
I. Title.
B2430.L484S48 2013
115.092—dc23

2012044961

∞ Printed on acid-free paper.

For Misha,
as am I

Contents

Abbreviations ix

Acknowledgments xi

Introduction 1

One • Time, in the Beginning 5

Two • The Freedom and Horror of the Instant 39

Three • From Darkness to the Other 76

Four • The Recession of Time 108

Five • Between Four Walls 141

Six • Time in Transition 179

Seven • Diachrony and Narration 228

Eight • The Time of Restoration 267

Notes 303

Bibliography 341

Index 359

Abbreviations

All abbreviations are of primary works by Emmanuel Levinas unless otherwise noted. Full publication information can be found in the bibliography.

AE	*Autrement qu'être ou au-delà de l'essence*
AT	*Alterity and Transcendence*
BPW	*Basic Philosophical Writings*
BV	*Beyond the Verse: Talmudic Readings and Lectures*
CPP	*Collected Philosophical Papers*
DEH	*Discovering Existence with Husserl*
DEL	"Dialogue with Emmanuel Levinas" (with Richard Kearney)
DF	*Difficult Freedom: Essays on Judaism*
EE	*Existence and Existents*
EI	*Ethics and Infinity: Conversations with Philippe Nemo*
EN	*Entre Nous: Thinking of the Other*
GCM	*Of God Who Comes to Mind*
GDT	*God, Death, and Time*
IR	*Is It Righteous to Be?*
ITN	*In the Time of the Nations*
LR	*The Levinas Reader*
NT	*Nine Talmudic Readings*
OB	*Otherwise than Being, or Beyond Essence* (translation of *AE*)
OE	*On Escape*
OS	*Outside the Subject*
PN	*Proper Names*

RPH “Reflections on the Philosophy of Hitlerism”
TI *Totality and Infinity: An Essay on Exteriority*
TIH *The Theory of Intuition in Husserl’s Phenomenology*
TO *Time and the Other*
UH *Unforeseen History*

Other Abbreviations

BT *Being and Time* (Martin Heidegger)
CE *Creative Evolution* (Henri Bergson)
SR *The Star of Redemption* (Franz Rosenzweig)
VM “Violence and Metaphysics” (Jacques Derrida)

Acknowledgments

What is philosophy? The study of philosophy had little appeal to me until I read *Totality and Infinity* and was summoned by Levinas to consider the origins of philosophy in the face of the other person. In this sense, philosophy is merely fumbling for language to describe an encounter that is underway before I am even conscious of the other. This makes philosophy pure response and responsibility. One of Levinas's major philosophical innovations relates to the dynamics of language, particularly to the way that every word spoken already demonstrates a primordial debt. Each utterance should be prefaced with "acknowledgment" for the resources from which it springs, always from before and outside the subject. I have far more debts than I can acknowledge, to a host of friends, family, and students that made this effort possible.

I am profoundly grateful for guidance and mentorship provided by Shelly Rambo, Richard Kearney, and Jeffrey Bloechl, as well as my supportive and attentive colleagues at Eastern Nazarene College. I am appreciative for many suggestions offered by Boston University professors John Hart, Bryan Stone and John Berthrong. I thank my friends Christina Gschwandtner, Luke Cochran, and Andrew David for helpful feedback on the preparation of this volume, and especially Kurtis Jardim for his careful eye, thoughtful conversations, and deep generosity. My three children, Jasmine Marie, Ty, and Luke, have demonstrated supreme patience through the long hours of research, writing, and compiling this study. Most of all, I thank my wife Misha for love, patience, and support that defy articulation.

Introduction

Time has mesmerized and perplexed philosophers since prehistory. In the last century, the notion of time has received particularly intense reconsideration, and it plays a substantial role in the development of the philosophy of Emmanuel Levinas (1906–1995). However, in the scholarly field surrounding his work, it may be Tina Chanter who has pointed most directly to the need for such attention. In 2001 she wrote, "Despite the fact that Levinas's notion of time is central to his philosophy, it is singularly neglected by even self-proclaimed Levinasians."[1] The last decade, thanks in no small part to Chanter's own work, has featured a surge in interest regarding Levinas's discussions of time.

The philosophy of time plays a pivotal role in each of the developmental stages of Levinas's career, and by the end of his career, Levinas rests his resounding philosophy of radical responsibility on his understanding of time as diachrony. Though more famous for his direct discourse on ethics, it is his understanding of time that makes possible his radical claim that ethics is *first* philosophy. Unfortunately, and for a complex set of reasons, there have been very few extended treatments of Levinas's unique theory of time.[2] Many shorter examinations of this sort have been attempted, including thoughtful chapters and essays by Richard Cohen, Shmuel Wygoda, and Alfred Tauber, as well as French language studies by Rudolf Bernet, Robert Legros, and many others.[3] Wygoda notes, however, that the confines of a shorter study cannot do justice to this complicated theme in Levinas's work.[4] Tauber calls his sketch "preliminary,"[5] and Chanter has repeatedly demonstrated the need for further reconsideration of the connection between Levinas's philosophy of time and his more frequently referenced themes relating to responsibility and ethics.[6] Thus, the time is ripe for an extended exploration of this theme in Levinas's work.

Levinas delivers his understanding of time in pieces over the course of his long career. Though remarkably consistent in many respects, his overall philosophy moves into progressively more radical expressions of ethical responsibility. The notion of time is not an abstract philosophical puzzle in Levinas's philosophy. Even at the very end of his career, he expresses an ongoing desire, which is apparent from the beginning, to convey the meaning of time in less formalized ways. It is not the case that Levinas's considerations of time develop from a single, coherent seed into a predictable pattern of maturation. He leaves ideas aside, picks up new ones, takes a noticeable hiatus from developing his ideas about time, and surprises his readers with a number of twists and turns in this development. There is, however, an obvious continuity between Levinas's early and late writings on time. Early works like *Time and the Other* (1947) and late works like "Diachrony and Representation" (1982) are both deeply concerned with time. But Levinas's development of the concept of time leaves behind as much as it develops. Some concepts, such as Levinas's unique understanding of "the hypostasis," generate many pages of careful analysis in the early writings but disappear almost entirely from the later ones. At times, this transition is an abandonment and critique of prior ideas and themes. I do not wish to press any theory of development on Levinas's work. I subscribe to neither Gillian Rose's claim that Levinas's is an "authorship without itinerary,"[7] nor John Llewelyn's sense that Levinas operates with "foresight" of his future developments.[8] Attempts to simplify this development run the risk of obscuring the complicated way that Levinas procedes in his understanding of time.[9]

According to Alphonso Lingis, "Levinas's work contains not only a wholly new analysis of the forms of time—of the present, the past, the future—but also a new concept of the work of time."[10] Levinas does not arrive at his unique position on temporality from a vacuum, but relies heavily on the work of several key philosophers. The early work of Martin Heidegger provides a launching point for Levinas's creative treatment of time, though his unique position on time includes a definitive break from Heidegger's analysis of temporality. Levinas also understands time in ways that are at least partly informed by Henri Bergson, Franz Rosenzweig, and Edmund Husserl, and certainly the classical works in Western philosophy.

Levinas's philosophy intensifies over seven decades through a wide range of experiences and encounters. He consistently uses these interactions to demonstrate the universal reach of his central theme: "the absolute primacy of the ethical relation."[11] Levinas's emphasis on responsibility builds like a crescendo; with each phase of his thought, the voice and face of the other[12] becomes increasingly stringent, demanding, captivating, and philosophically significant. While this movement to increasingly radical articulations of responsibility is well documented, less attention has been paid to the critical role that *time* plays in the way Levinas progresses toward his mature position on ethics. As Richard Cohen writes in his introduction, "Because Levinas binds time to alterity, the theory of time articulated in *Time and the Other* marks but one moment in a progressively radicalized theory of time that unfolds in Levinas's work as a whole. Each of Levinas's works presents a distinct analysis of time, and each analysis is progressively more radical than the prior analysis, as the analysis of alterity is progressively radicalized" (*TO* 4).

Given the development of Levinas's thought on time over the course of his career, it seems fitting to analyze it in a chronological manner, focusing particularly on his philosophical writings without ignoring his other primary sources.[13] Considering his early work, chapter 1 outlines both the biographical and early philosophical influences on Levinas, investigating some of his first writings for indications about the influence and direction of his thought at the beginning of his career. Chapter 2 continues this investigation of Levinas's influences through an exegetical analysis of time as it is presented in *Existence and Existents* (1947). Then, chapter 3 investigates Levinas's increased emphasis on the other person in relationship to time as it appears in his "Time and the Other" lectures and other writings from the late 1940s.[14]

In somewhat of an interlude, chapter 4 explores the decade of the 1950s, in which Levinas writes very little about the concept of time, and suggests some reasons for and consequences of this waning emphasis. Chapters 5 and 6 then investigate Levinas's changing understanding of time across the 1960s, beginning with his first major work, *Totality and Infinity* (1961), and particularly the consequences of his invocation of the feminine. Chapter 6 also raises several possible explanations for the obvious differences between *Totality and Infinity* and his later work.

Moving to *Otherwise than Being*, chapter 7 proposes that Levinas's second major work is consistently and thoroughly dependent upon his understanding of time as diachrony. In conclusion, chapter 8 suggests directions for further development on the concept of time after Levinas based on the implications of his notion of time as diachrony and includes a critique of his spoken and unspoken invocations of the maternal. Not only are Levinas's claims about the transcendence of the face of the other funded and supported by his unique understanding of time, but some of the more severe problems in his philosophy relate to an unfortunate lapse in his own innovative development on time. His troubling use of the feminine, particularly in *Time and the Other*, *Totality and Infinity*, and less obviously in *Otherwise than Being*, relates directly to the concept of time.

Levinas repeatedly invokes concepts related to time throughout his career. Even having constricted this inquiry to Levinas's deliberations on time, the research here is far from comprehensive, as his work spans many decades, thousands of pages, and countless encounters and experiences. Levinas's thought is both serious and insistent, turning eventually against any idea that could mitigate the deepest conceivable obligation of the self to the other. The most mature and advanced expressions of this responsibility are also the most unrelenting and rely most obviously on Levinas's innovative notion of time. Levinas eventually intertwines his understanding of time with his articulations of radical responsibility, making the study of his unique use of time critical for interpreting his work overall.

ONE

Time, in the Beginning

> Time, the condition of human existence,
> is above all the condition of the irreparable.
> —Emmanuel Levinas, "Some Thoughts
> on the Philosophy of Hitlerism"

Introduction

The earliest of Levinas's writings indicate his awareness that the next step for philosophy involves a reconsideration of the concept of time. The briefest of glances at Levinas's biography and bibliography indicate that he was quite familiar with the work of Henri Bergson and Edmund Husserl, even before his famous encounter with Martin Heidegger. These thinkers each called, in vastly different ways, for a reconsideration of time in light of philosophy's historical negligence on the topic of temporality, and thus, influenced Levinas's own reflections on time. To this list, I will later add Franz Rosenzweig, who contributes vitally to a religious component in Levinas's reconsideration of time. Even in the initial parsing of influences on Levinas's philosophy, it is already clear that this philosopher is staking out a radically new path for the philosophy of time.

One can identify in Levinas's writings a consistent, escalating connection between time and intersubjectivity. Levinas's fixation on time is neither a passing stage in his development nor a topic secondary to his primary interests, but rather, he forges his moral philosophy alongside an original conception of time. Jacques Derrida compared this intensification to the progressive waves of surf on a shoreline. "It proceeds," writes Derrida, "with the infinite insistence of waves on a beach: return and repetition, always, of the same wave against the same shore, in which, however, as each return recapitulates itself, it also infinitely renews and enriches itself" (VM 312n7). This is an apt illustration for Levinas's work on time, which sometimes recedes with the tide from the foreground of his thinking, yet always remains an operative concept, even if this is less evident at certain points in his career. But on the concept of time, as with his famous emphasis on obligation to the other person, the waves wash high on the shoreline by the end of Levinas's career. In his final writings, time is *the* operational philosophical concept that Levinas uses to support his understanding of radical responsibility.

This becomes clear already from exploring several preliminary influences on Levinas's understanding of time, many of which are evident in his early and sometimes neglected prewar essays, particularly the three noteworthy works to be discussed later in this chapter: "Some Thoughts on the Philosophy of Hitlerism," *On Escape*, and "The Work of Edmund Husserl." However, an exegetical approach to Levinas's early influences is thorny on several levels. First, it is clear that Levinas has a host of reasons to raise his disagreements with Heidegger and Husserl gingerly. *On Escape*, for instance, is best understood as a prolonged engagement with Heidegger—Levinas asks a series of Heideggerian questions, uses language and analyses familiar to his writings, and raises some of his earliest polemics against Heidegger's methodology and ontology. Yet despite all these clear markings of a response to Heidegger in *On Escape*, Levinas does not even once mention Heidegger by name. Of course, there are obvious political, religious, historical, and personal reasons for this treatment. Two years before Levinas published *On Escape*, Heidegger had vocally and publicly aligned himself with Adolf Hitler's Third Reich. The complications of Levinas's relationship with Heidegger, and Heidegger's Nazi affiliation in particular, have a distinct influence on the development of Levinas's philosophy of time. The oblique but increasingly strident critiques of Heidegger certainly provide a puzzle for those who would understand Levinas's initial trajectory with regard to time from his first writings. It is not always clear, on first reading, where Levinas is making these breaks.

It is also a rather delicate matter to establish the connections between Levinas's rich and sometimes painful biography and his philosophical reflections. One cannot help but notice that one of his first independent writings is a 1934 essay identifying the roots and dangers of Hitlerism. Nor are we wise to ignore the fact that his first book, *Existence and Existents,* was mostly written in notebooks between Levinas's grueling workdays in a Nazi prisoner-of-war camp. All these factors make the early works of Levinas complex, foreboding, and intriguing. These works, then, must be navigated with a specific eye for the way that Levinas's philosophy begins with a steady suspicion that philosophy has serious work to do on the concept of time. Levinas suspects that the consequences of ongoing negligence on this topic are dire, relating directly to violence and totalization.

In sorting through the complexities of Levinas's personal and historical context, we are fortunate in the study of Levinas to have his own later reflections on his early work, including some of his notes and his endorsement of several of his interpreters. This is particularly helpful in studying *On Escape*, for which Jacques Rolland wrote an extended introduction that has been lauded by Levinas.[1] There are also numerous studies on Levinas's influences, both biographical and philosophical, which have meticulously explored the sources from which his unique philosophy springs. But while Husserl, Heidegger, Rosenzweig, Bergson, and others remain key interlocutors for Levinas throughout his career, and an exhaustive study of these influences is both laudatory and instrumental to this investigation, the purpose of this first chapter is simply to explore the influences on Levinas's early thoughts about time and to demonstrate that even his first writings are concerned with the connection between time and ethics. As "Some Thoughts on the Philosophy of Hitlerism" (1934) makes clear, the philosophical questions at stake in this investigation are neither trivial nor abstract, so it is important to keep historical and biographical factors in mind without allowing them to determine any conclusions about Levinas's philosophy.

Time, in the Beginning

Levinas was born in Lithuania in 1906, where he received a traditional Jewish education. Kovno, his Lithuanian hometown and the place he spent his first decade, was known as a center of talmudic scholarship in Europe,[2] but Levinas did not become familiar with talmudic studies until much later in life. Nevertheless, he certainly did carry into his philosophical studies a rigorous knowledge of Jewish Scripture and an appreciation for Jewish life, practices, and customs.

Levinas appeared on French Radio in 1937, giving an interview on the program *Voix d'Israël* that underscored a series of commitments to Jewish practices. It seems clear from his comments that Levinas's first 20 years fostered in him a strong appreciation for Jewish life. The relationship he sees between Judaism and time is already evident in this interview, where he claims that the rituals of Judaism cause an interruption in the flow of time. The seventh day does not "dawn like all the rest; it remains impervious to the concerns of the week." Rituals mark a pause, reminding Jews that they are strangers in the world, that the world is a gift and a miracle, and that "belief in creation—the basis of Judaism" is nothing other than this wonder. Jews cannot depend on "instinctual" responses, nor can they trust the "current" of time that "constantly connects us to things." At every "instant," the Jew experiences "the fact—so simple and yet so extraordinary—that the world is there." It is important to note that even in 1937, Levinas appreciates the way that ancient rituals set Jews apart, attuning them to a very different kind of time. For non-Jews the world is familiar and immediately comfortable; people and objects are encountered as "old acquaintances; they are familiar, everyday and profane."[3] Levinas, even at this early juncture, already considers Jewish time to be markedly different from the time of history and nature.

Prior to his graduate work, Levinas received an undergraduate education steeped in the philosophy of Bergson, mediated through psychologist Charles Blondel at Strasbourg University.[4] Indeed, Levinas calls Bergson his first "contemporary influence" (DEL 13). Levinas later credits Bergson for his contribution to the "destruction of the primacy of clock time," the understanding that time as it is offered by physics is derivative (*EI* 27). In Strasbourg in the 1920s, Levinas would later recount, "Bergson was being hailed as France's leading thinker" (DEL 13).

By 1928 Levinas was prepared by Bergson, through Blondel, to rethink time beyond the formal and traditional sense. Levinas happens to arrive at the university in Freiburg in time to encounter both Husserl and Heidegger even as Heidegger is sharply drawing away from Husserl on the very concept of time. In fact, Richard Cohen points out that Heidegger's own innovative thoughts on time can in some ways be seen not as opposed to Bergson's own work, but as a continuation of "a revolution that has already taken place."[5] Heidegger nevertheless repeatedly insists that Bergson has *not* managed to advance the philosophy of time past Aristotle's conception.[6] Levinas's studies of Bergson and the "great masters in the history of philosophy" may have uniquely disposed Levinas to detect the significance of Heidegger's critique, and to think creatively and innovatively on his own about the philosophy of time.[7] His awareness of Bergson indicates that Levinas would not have been caught off guard by the conversations about time already underway in Freiburg.

Once at Freiburg, it was through the mentorship and tutelage of Husserl that Levinas acquired a love for the phenomenological method, the methodology recently initiated by Husserl himself. Levinas returned the favor by translating and introducing Husserl to French philosophers. Levinas's exposure to Husserl's book on time, *Lectures on the Phenomenology of Internal Time-Consciousness* (1928) helps explain Levinas's own prolonged fascination in the topic of time. Husserl's interest in all corners of his phenomenological inquiry is to describe reality without presuppositions.[8] Describing *anything* without presuppositions is a daunting task, which partly explains the voluminous Husserl archives.[9] The topic of temporality was particularly interesting to Husserl precisely because so many presuppositions are loaded into the conception, experience, and measure of time. Husserl's methodology drove him to analyze the various forms of meaning that arise from the experience of time and to question the overly simplistic reduction of time to the ticking hands of the clock. Robert Sokolowski notes, "Once Husserl's thought on time had been developed, he often distinguished between three levels: (a) temporal objects in objective, worldly time, like melodies, races, local motions, or illnesses; (b) inner objects or immanent objects, which are sensations and acts and the combination of sensations animated by intentional acts; and (c) the absolute, time-constituting consciousness."[10] By peeling back the layers of meaning inherent in a concept, in this case the concept of time, Husserl sought to describe temporality in its richer and more primitive sense.

Levinas will eventually call this pursuit of a more original sense of time the "deformalization of time" (*EN* 175–77). The basic thesis that drives Husserlian phenomenology is that philosophy has distorted its analysis of reality by ignoring or overlooking presuppositions. These oversights are unintentional and quite natural, and many times, they may be innocuous. In some critical cases, however, these oversights may have dire consequences for the value and accuracy of philosophical inquiry. This makes Husserl's questions about time similar to Bergson's, who builds his unique understanding of duration on the accusation that philosophy is impoverished for failing to analyze layers of inherited assumptions about the concept of time. This similar effort to press back against the formal time of traditional philosophy was not lost on Levinas.

Levinas and Heidegger share with their common teacher, Husserl, an urgency to explore the gaps in traditional philosophy that phenomenology exposed. They each name different blind spots and suppositions in the history of philosophy, but they are bound by a common methodological aim: to interrogate philosophy for the sake of uncovering assumptions that have led it astray. Husserl set the stage for both Heidegger and Levinas by pointing to the *internal* experience of time as the more fundamental and original phenomenon of time. Clock time synchronizes trains and appointments, but it is an abstraction from the way time is experienced originally and internally. Though each of these three thinkers turns in different directions to describe the most fundamental sense of time, they share a basic mistrust of the history of philosophy when it comes to adequately considering temporality. Husserl published his lectures on internal time consciousness the same year he met Levinas, whom he called "a very gifted Lithuanian student."[11] It seems hardly coincidental, then, that Levinas showed such an early and sustained concern for the concept of time, for he encountered both Husserl and Heidegger when their attention was directed explicitly to the subject.

As much as Husserl enchanted Levinas, both personally and professionally, it was in Heidegger's lectures and his breathtaking *Being and Time* (1927) that Levinas believed himself to have found truly novel insight, and of which he steadfastly maintained his admiration for and often expressed debt to Heidegger's early work.[12] *Being and Time* played an unparalleled and profound role in the development of young Levinas, which can be observed throughout his career of exploring time and temporality. In fact, Levinas sustains some of the investigations initiated in *Being and Time* long after Heidegger appears to move on to other concerns.[13] His enduring admiration for *Being and Time,* if not its author, must however, be mentioned alongside Heidegger's alignment with National Socialism in the 1930s. Heidegger joined the Nazi party in 1933, a month after becoming the rector of the University of Freiburg. In an article he wrote shortly afterward for the university newspaper, Heidegger claimed, "The German people must choose its future, and this future is bound to the Führer."[14] Heidegger's participation in National Socialism certainly complicates the philosophical relationship between him and Levinas, but for our purposes here, it is most important to identify an early rift between Levinas and Heidegger on the concept of *time,* a rift that may even precede their more celebrated differences on the importance of intersubjectivity.

Encountering Heidegger

When Levinas came to Freiburg in 1928, he arrived on the scene in the midst of the transition in academic leadership between Husserl and his student and rising star, Heidegger. Levinas often said of his semesters at Freiburg, "I came to see Husserl, and what I saw was Heidegger."[15] He crowded into packed lecture halls to hear Heidegger unfold his broad and stunning insights into a lacuna in the history of Western philosophy regarding the nature of time.[16] The content of Heidegger's lectures closely matched the themes in the recently published *Being and Time*, which Heidegger rushed to publication to ensure his academic promotion to Husserl's position as chair at the University of Freiburg. The massive impact of these lectures on Levinas can be observed in his comment in 1931 that Heidegger's name "is now Germany's glory" (*UH* 64).

For all the disagreements that he will raise with Heidegger's philosophy throughout his career, some of which are already evident in his earliest works, Levinas will never waiver in his conviction that Heidegger has made a brilliant discovery. Clearly then, one must move through Heidegger to arrive at a reasonable understanding of Levinas's early texts and the motivations that drove him to write about time. Indeed, Levinas's own philosophy of time is tied closely to Heidegger's deconstruction of temporality. Tina Chanter claims that the difficulty of Heidegger's critique of Western philosophy prevents many readers of Levinas from seeing the brilliance of Levinas's own ideas about time.[17] It is therefore vital to engage in a summary of Heidegger's critique, with particular attention to the elements that relate to Levinas's early writings.

A fixation on the perplexing concept of time had been brewing for several centuries, perhaps most evident in the works of René Descartes, Immanuel Kant, G. W. F. Hegel, Husserl, and Bergson. In *Being and Time*, Heidegger utilizes his investigation of the concept of time to drive his sweeping critique of Western thought, and he points out a seemingly innocuous flaw in ancient philosophy that he believes previous generations have neglected. This neglect has devastating results, though Levinas and Heidegger disagree sharply on the form this devastation takes. For Heidegger, the inability to think properly about time leads to inauthenticity. Heidegger accuses Western philosophy of turning a blind eye to the temporality of being, of ignoring the primordial temporality that became obscured as philosophy consistently overlooked its temporal roots. This move, claims Heidegger, happens when Plato and Aristotle screen out essential elements of ontology.[18] Western philosophy, Heidegger surmises, continues to suffer the consequences of their unintentional censorship of Presocratic thought. Heidegger proposes, then, that both being and time be rethought in light of his discovery.

"Ever since Aristotle," writes Heidegger in *Being and Time*, "all discussions of the concept of time have clung *in principle* to the Aristotelian definition" (*BT* 473). Heidegger identifies the following statement of Aristotle's as his definition of time: "This is time: that which is counted in the movement which we encounter within the horizon of the earlier and later" (473).[19] Heidegger suspects that the Socratic philosophers struggled against the intractability of the concept of time, setting the problem aside by developing the "structures of being" (49).[20] Time was consequently treated as "one entity among others" (49). The mistake in treating time like another object in being is that, as Heidegger sees it, being itself is "inexplicitly oriented toward time." Time is reduced to Aristotle's "eternal now" as the *present* of being, including past and future nows. It is, for Aristotle, therefore understood as a steady progression, a "flowing stream of 'nows'" (474). Thus, by following Aristotle (who is following Plato), philosophy subtly but enduringly sided with the *present*. This relegates the past to a series of *former* presents and the future to a series of *future* presents. This move Heidegger finds "naïve," precisely because the very idea of being, in which the eternal now had been given preference, was already dependent on the repression of a more original sense of time, not bound to this flow of eternal nows.

Heidegger wishes to unsettle the traditional way that time is formalized and universalized. In its ordinary sense time is marked by clocks, calendars, and sundials. "World time" or "public time," as Heidegger sometimes called it, is infinite. This is the sort of time that marches on without regard to the coming-to-be and ceasing-to-be of any particular being. Time, in the ordinary and public sense, consists of an eternal sequence of nows that is intensely impersonal. Each now is replaceable by the next moment. All of the apparent irregularities perceived by individuals are leveled off and evened out by the synchronizing march of the clock's hands.[21] Plato, notes Heidegger, called time the "image of eternity" (*BT* 475). This Platonic position takes its bearings from the fact that no matter how one divides up time it is always *now*. Now is always available, without interruption, as "present-at-hand" (*vorhanden*). That which is present-at-hand, in Heidegger's lexicon, is perceptible through analysis but only when it is extracted apart from its everyday, functional usefulness, from the lived realm in which it is instrumental, or "ready-to-hand" (*zuhanden*). We cannot think "now" without thinking of a sequence of similar nows that stretch forward infinitely. But to this sequence of abstract, infinite nows the individual future of Dasein is irrelevant. This is why the ordinary sense of clock time is for Heidegger, and eventually Levinas, already an abstraction from a more primary experience of time.

For Heidegger, this means time has been treated like other abstract objects that are present-at-hand. In doing so, the Greeks defined time according to the logic of being. The irony of defining time as another entity within the horizon of "being" is the inexplicit temporal orientation of the Greek idea of being. In other words, Aristotle and company erred in their reflections on time by presuming that they conceived being *atemporally*, all the while unaware of their deep dependence on a time more original than the eternal now.

For Aristotle, and certainly for Plato, change is defined according to its eternal opposite: permanence. Of this premise Aristotle writes: "It is clear, then, that time is not change, but at the same time that it does not exist without change. So in our attempt to discover what time is we had better start with this fact and try to see what aspect of change time is. After all we do notice change and time simultaneously."[22] Time has *something* to do with the fact that all noneternal things move and change. To be in time is simply to be something other than eternal. This is certainly more sophisticated than Plato's less nuanced appeal to the eternal, but Aristotle's starting point leads to a circular argument that firmly establishes temporality as a privation of eternality. The perspective throughout Aristotle's reflections on time is consistent: he seeks to define time from the vantage of the nontemporal.

This presents a deeper puzzle: can time be thought of other than by contrast to its perceived opposite? Heidegger points out that the Greek concept of eternity is developed "without any understanding" of the "fundamental ontological function of time" (*BT* 48). Time is the privation of eternity, but the Greeks utilize the idea of eternity as though it did not already presuppose an understanding of time.

The weakness of Plato's and Aristotle's respective accounts of time is that they inadvertently sublimate the concept of time to their understanding of eternity. Heidegger's *Being and Time* does more than just identify and verify this naïve omission; it initiates a bold reconsideration of both time and being in light of this critique.

For Aristotle, time also functions to *synchronize* otherwise dissonant events. In reflecting on locomotion, he notes that two objects can be moving at very different rates, and hence be changing very differently, but remain simultaneous.[23] Time quantifies and qualifies movement and motion. Aristotle furthermore grants time an exteriority to change, given that time *is* regardless of any particular instance of motion. He also argues from locomotion, which can measure distance according to the passage of time, that time should be limited to homogeneous units. Time cannot be measured with units of space, but one time period should be measurable against another. The rules of geometry and spatial measurement are determinative; we should count time in the same way that we count dogs and horses.[24] Time is a subsidiary to being; it is meant to play by the rules and regulations common to the logic of spatial physics and metaphysics. Heidegger traces this common presumption through the ages, and although his readings of Augustine, Aquinas, and Kant may be selective, his suspicion seems to be born out by their writings. Heidegger was audacious enough to challenge this entrenched and deep-rooted tradition and boldly claims, "The question of being does not achieve its true concreteness until we have carried through the process of destroying the ontological tradition" (*BT* 49).

Heidegger rarely pauses to allow his readers to glimpse the sources behind his unique critique of Western philosophy. However, in his lecture series printed as *Metaphysical Foundations of Logic*—given shortly after the publication of *Being and Time* and attended by Levinas during the 1928–1929 academic year—Heidegger reflects on the relationship between his understanding of time and the way it is discussed in Husserl's *Logical Investigations.* Heidegger writes, "That which Husserl still calls time-consciousness, i.e., consciousness of time, is precisely time, itself, in the primordial sense."[25] Here, Heidegger credits Husserl with seeing "these phenomena for the first time, with the aid of the intentional structure." Heidegger also points to the expression of time in Husserl as the genesis of his own thoughts. But Husserl, as Heidegger reads him, limits the scope of his reflections on time to that which occurs "in the subject," to *internal* consciousness, whereas Heidegger suspects that the relationship of time divides the subject from itself. This means that for Heidegger time is not merely a feature of internal consciousness, which leaves the external world locked in the mode of Aristotle's "present." Heidegger writes that despite threatening traditional concepts of time, Husserl's innovations relate only to the "immanent," such that everything outside the subject "remains, in principle, as it was."[26]

In Heidegger's estimation, it is to Husserl's credit that he sees something peculiar about the ordinary understanding of time, even if he fails to break time free from "an efflux of the nows, just thens, and right aways."[27] Heidegger wishes to press the concept of time beyond the machinations of intentionality within the mind of a singular person. Joanna Hodge summarizes this movement against Husserl: "The critical difference is the manner in which Heidegger links the thinking of time not to the process of presentation to a single human consciousness, engaged in a series of reductions hypothesized by Husserl, but instead links the thinking of time to an understanding of the differences between the finitude of Dasein, as actually, ontically and existentielly, determinate, and what Heidegger claims to be the derived notion of time, as nonfinite and eternal."[28] Husserl therefore opens up the space of questioning, undermining the stability of traditional considerations of time. But for Heidegger, Husserl has simply not gone far enough.

Heidegger sees a number of potential benefits to be gleaned from rethinking being by way of expanding Husserl's rethinking of time. Chief among these benefits is an existential concern for authentic being-in-the-world. The failure to contemplate temporality correlates, for Heidegger, with a failure to live resolutely and authentically in the world. By thinking about time through the lens of eternity, "one *must* always *think* more time; from this one infers that time *is* infinite" (*BT* 476). In thinking about time as infinite, the original temporality of human existence is covered up. Furthermore, reflects Heidegger, we think of time through the public perspective, through the gaze of the "they" (*das Man*). The they never dies and cannot die, despite the death of any particular individual. So they is apparently not limited to temporality (*BT* 166–67). All these factors work against Dasein, subverting the ability of the individual to embrace authentic existence. To exist authentically, argues Heidegger, one must recognize in every moment one's temporality and the irreplaceable futurity of one's own death. This death, which belongs uniquely to Dasein, must be faced with anticipation and resoluteness (349–54).

Critics of *Being and Time* have pointed to clumsy exegesis on Heidegger's part in his readings of Aristotle, Augustine, Kant, Husserl, and others.[29] Heidegger interrogates the Aristotelian understanding of time with gestures toward a more ancient understanding of temporality. John Caputo points out that for Heidegger, "Plato and Aristotle are not to be read backwards, from the standpoint of modernity, but forwards, as a falling away from the early Greeks, who now assume a place of historical privilege."[30] Caputo has pointed out that Heidegger creates his own mythological version of ancient Greek thought, calling this Heidegger's "myth of monogenesis."[31] Heidegger has surely pointed to a fundamental problem in Aristotelian temporality, but he overlays his own analysis with a simplistic version of temporality in the Presocratics. Heidegger presumes that all of philosophy forgot, or at least neglected, the *same* common and original sense of time. His allegiance to this self-made mythos provides him with little incentive to see diversity in the treatment of time before or after Aristotle. It also may help Heidegger shrug off the more immediate influences on his thinking about time, including Bergson and Husserl.

In light of these criticisms, the enduring value of *Being and Time* may relate more to the questions it raises than the quality of its interrogation. For all of its brilliance and insight into the *problem* of thinking about time across the history of philosophy, *Being and Time* is short on overt reconstructions of the concept of time. Heidegger acknowledges that *Being and Time* is primarily deconstructive (*Destruktion*) and promises a sequel to provide clearer insight into the nature of time in the wake of his critique (*BT* 43–45).[32] Such a sequel never arrives. For this reason, *Being and Time* has been called "that astonishing torso" for failing to make good on its promise to redefine time.[33] Heidegger's thesis about time does indeed appear to be bolder than his ensuing work would seem interested in pursuing. He proposes that "*the central problematic of all ontology is rooted in the phenomenon of time, if rightly seen and explained,* and we must show *how* this is the case" (40). But how might we rightly *see* and *explain* time, if "being is to be conceived in terms of time" (40)? This question is taken up repeatedly by Heidegger's successors, particularly Levinas and Derrida. In fact, given the heavy and sustained treatment of time in the work of Levinas, we can justifiably wonder if he does not, in his own way, seek to fulfill Heidegger's 1927 goal of understanding time, "rightly seen and rightly explained."[34]

The chapters ahead will provide further opportunities to revisit *Being and Time* in light of Levinas's agreements and critiques, but for now, it suffices to underscore Heidegger's suspicion, shared by Levinas, that philosophy has much work to do on the topic of time. For Levinas, in his first philosophical writings, the consequences of failing to think through the implications of time and temporality are more severe than Heidegger proposes in *Being and Time.* To fail to think of time adequately, for Levinas, is to open the door to violence, oppression, and evil. This fear is already clear in the first of Levinas's many essays, continuing until it is quite radically expressed in his final works.

"Some Thoughts on the Philosophy of Hitlerism"

The beginning of Levinas's independent philosophical career coincided with the rise of Hitler and the ideology of National Socialism. Levinas and all his relationships were drastically influenced by the rise of Hitler, Nazi ideology, the Holocaust, and World War II and its aftermath. In 1934, barely a year after Hitler was named chancellor of Germany, Levinas wrote a disturbingly prescient essay on the "philosophy" that was driving Hitler's rise to power. "Some Thoughts on the Philosophy of Hitlerism" (or "Reflections on the Philosophy of Hitlerism," depending on the translation)[35] has received relatively little attention among Levinasian scholars, partly because it was written long before the hallmark elements of Levinas's unique philosophy were developed. This essay, however, has a particular importance for documenting his account of time due to its twin concerns about time and violence as well as its critique of Hitlerism's deep distortion of the concepts of time, history, and temporality.

From the outset, there is a different tone in Levinas's invocation of the concept of time than we find in Heidegger. Whereas Heidegger's concerns about time reflect an apprehension about the authenticity of Dasein, Levinas's concerns about time relate more palpably and directly to the concepts of violence and politics. Indeed, the essay is something of a philosophical social commentary, a work that attempts to explain the vulnerability of Western societies to the rise of racist ideologies by exploring the relationship between philosophy and violence and between the concept of time and totalitarianism. The philosophical link is different, in many ways, from the articulations that Levinas will develop in the years to come. Still, it is undeniable that Levinas has already seen in the legacy of the Enlightenment, which he calls "modern liberalism" (*UH* 17 / RPH 67), a moral vulnerability. And as the first pages of this essay indicate, this vulnerability relates to the philosophy of time.

"History," Levinas claims, "is the most profound limitation, the fundamental limitation. Time, which is a condition of human existence, is above all a condition that is irreparable" (*UH* 14). The past, then, appears to be unchangeable, and the future to be determined by causation or destiny. Levinas makes fascinating, sweeping statements about the phenomenon of time in his opening pages. For example, it is the "burning feeling of natural powerlessness that man experiences in the face of time" that leads to the creation of Greek mythology. The Moirai, that is, the Greek gods of fate, determined the events of a person's life in advance; even the other gods were powerless against the Moirai (14).[36] The Greek heroes struggled in vain to change the thread of life spun by the Fates from the moment of birth. Time, as past, present, and future, is tragically "irreparable."

Levinas also connects the irrevocability of time to "the whole acuteness of the idea of sin" as well as "the whole greatness of Christianity's rebellion" (*UH* 14). Writing for the progressive Catholic journal

Espirit, Levinas points out that both Christianity and Judaism, in their promise of pardon, open up for humanity a freedom with respect to time. "The Cross sets one free," he explains, "And through the Eucharist, which triumphs over time, this emancipation takes place every day" (14). At this stage, Levinas is using the word *time* to refer to the manner in which one is held fast by a past and a future that are imposed on the self. Time has already, for Levinas, gained an unassailable alterity that positions the self as helpless, and hopelessly in need of "grace." Christianity and Judaism bear the "magnificent message" of modifying time and changing the trajectory of history. Time is not closed for these religions but absurdly open. "To the Atrides writhing in the stranglehold of a past" Christianity offers liberation, an unhinging of time from its fateful and irrevocable flow. The tragic trajectory of time, bearing so much power in Greek mythology, "collapses at the feet of man like a wounded animal. And liberates him" (14).

The legacy and genius of Judeo-Christian temporality is an opening of the future, an opportunity to overturn the legacy of the past for the sake of *pardon*. The past is reopened for the sake of a future that is not subordinated to the past or even bound to its efficacy. The inevitability of sin and death are challenged and opened to a future liberated from the power of the past. This liberation from determinism and fate is of wonderful benefit to humanity, but it is intoxicating and potentially misleading. Levinas points to the "world of liberalism," which "tends to place the human spirit on a plane that is superior to reality, and so creates a gulf between man and the world" (RPH 66). Modern dualism, with its emphasis on reason and the elevation of the mental over the physical, is directly related to the freedom-from-fate offered by Judaism and Christianity. Liberalism, as a manifestation of this modern dualism, takes for granted the victory over time and forgets its gratitude for being liberated from the Fates. The result of this forgetting is an unfettered emphasis on freedom, which undermines the way that humans remain bound to bodies and cultures and particularly to *time*.

Levinas argues that the consequence of this progression is a "world rebuilt by idealist philosophy, one that is steeped in reason and subject to reason" (RPH 66). Liberation through grace is first presumed and then *replaced* by absolute autonomy; no longer is humanity "weighed down by a History in choosing [its] destiny" (66). But for all the efforts to liberate humanity from the chains that tie reason to bodies, the chains remain. Whatever the heights that reason entices us to contemplate, we are repeatedly reminded that we are riveted to our bodies. Human "essence no longer lies in freedom, but in a kind of bondage" (66). Levinas is here tracing basically Heideggerian sentiments about being. The temptation of modern philosophy, according to Heidegger, relates to an obsession with eternality and a resulting blind spot to Dasein's fleshy and temporal existence. According to Levinas, modern liberal philosophy commits this error on a grand scale.

We are nevertheless bound to our bodies, and therefore to history and to the past. So we struggle to recover the soaring heights of universal truth within the confines of our riveted existence. We struggle also to achieve for ourselves the liberation from time that religion once offered, but it haunts us from the other side of our liberal arrogance. The addiction to freedom does not abate, leaving humans riveted to bodies but reluctant to be tied down by any particular truth. The search for truth, then, becomes a game with no stakes as philosophy shifts from one theory to another with no urgency to settle. Liberal modernity advocates caution and hesitation with regard to truth; a kind of vacuum is created whereby other ways of thinking are allowed to dominate the ontologically inferior realms of history and politics. In this way Levinas wonders whether early-twentieth-century Europe has become vulnerable to the "will to power," which exploits this indolent and lethargic attitude toward historical and moral truth. A combination of these forces leaves Europeans feeling unfettered by history or responsibility and ambivalent about truth or moral authority. This sets the stage for Hitlerism.

The only ineluctable truth is the fact of our bodies, our blood, our *race*. The truth-making methodology of "modern liberalism" can have nothing but a pursuit of universals. Levinas claims that "the new type of truth cannot renounce the formal nature of truth and cease to be universal" (RPH 70). So modern liberalism finds itself divorced from history, time, revelation, meaning, and truth, but nevertheless addicted to the search for universals. Levinas is aware that modern skepticism creates a vacuum in which the truth is undecidable, "Man plays with his freedom and doesn't permanently commit himself to any truth" (*UH* 19). Nothing but resounding evidence will suffice for the modern, liberal skeptic. And the only nonscientific truths that present themselves with sufficient universal power are race and force.

At this stage in the essay, Levinas makes a subtle point about the way truth is communicated and the danger of violence. In traditional philosophy, as well as in modern thought, ideas are conveyed through a sort of evangelistic method: one makes an appeal to a universal truth and then attempts to sway one's interlocutor to share this position. Plato's dialogues represent this paradigmatically. In Plato, as Levinas reads him, "To convert or persuade is to create peers" (*UH* 20). But violence operates quite differently, demanding acquiescence and expansion, forging its own universal by threat of the sword. The exertion of violence carries with it a native form of universality. Violence forges its own truth, even if this force is quiet and subtle.[37] This means the modern liberal mind has freed itself not only from religious ties to time and its irruption in forgiveness, but also from any safeguards offered by Greek philosophy. Reason, like religion, has been destabilized by modernity and turned into a game without consequences. Now only the blatant power of force carries the capacity to convince.

Levinas will leave behind several themes that he employs in this essay, but he will retain the basic suspicion that Western thought has failed to properly brace itself against "elemental Evil" (RPH 64).[38] Among the reasons for this failure is the turning of philosophical thought into a game without consequences, suspended outside of time and history. Also prized by the modern liberalist is the disembodiment of reason from time and bodies—"Man in the liberalist world does not choose his destiny under the weight of history" (*UH* 16). This freedom from history is a distortion of the beauty of *forgiveness*. And in the hands of Friedrich Nietzsche and Hitler, as Levinas sees it, this liberation unhinges humanity from any primary obligation beyond concerns intrinsic to the present, to the self, to the nation. This primacy of freedom, already Levinas's concern in these very early essays, will become a central target for his work throughout his life.[39]

Levinas sees in the rise of Hitlerism an exploitation of the vulnerabilities developed by liberalism in Europe and an enactment of the truth-making power of violence. Hitlerism embodies the violent justifications of the will to power that Nietzsche exonerates and venerates. Levinas worries that Germany is "rediscovering and glorifying" this new pathway to self-defined truth that "simultaneously brings with it its own form of universalization" (RPH 71). Hitler utilizes force and expansion, which engender their own authority, to domesticate the transcendence of history and the diversity of cultures and persons. The trajectory of violence and expansion that characterizes Hitler's rise to power also carries with it the impression of universality. These are, however, "degenerate" forms of ideas, beget of a society that has lost its true ideal of freedom.

In the vacuous absence of premodern belief or in the security of modern scientific assurance, Hitler's followers find comforting resonance in the latent universality implied by the expansion of his territory and agenda. Racism becomes an arbitrary but tangible universal, marked by the momentum of history and heredity. "The mysterious voices of blood" and the "calls of heredity and the past" catch the liberal imagination off guard (*UH* 18). They do not succeed in the common manner of modern ideas, which are propagated by reason, proof, and calculation. Instead, racist ideology establishes its universality by its very expansion, through the momentum and exaltation of those who prevail. In such an environment, it is not the ideology of a particular group that produces its expansion but the very violence of expansion that delivers authority to the ideology. In short, divorcing philosophy from time and history leaves Europe vulnerable to Hitler.

Robert John Sheffler Manning summarizes Levinas's prescient warning: "This is Levinas's warning and his dire prophecy in 1934: the philosophy of Hitlerism, the biological conception of man, the Germanic ideal, can all be expressed in short hand in the term racism, and racism has its own particular way of spreading: violence and war and conquest."[40] The successful use of violence does not merely reinforce the validity of an ideology. The application of violence itself is full of latent ideological commitments. Violence confirms and redoubles itself, universalizing both the victor and the methodology by the validating force of expansion. Force necessarily universalizes and sanctifies its methods: war and conquest.

Levinas does not cite Hitler in his essay, but his evaluation of Hitlerism and the relationship between Hitler and force is insightful and accurate. Arnold Davidson lines up the critique of Hitlerism in Levinas's essay with Hitler's declared ideology in *Mein Kampf*:

> Each of these ideas is operative, whether explicitly or implicitly, in Hitler's *Mein Kampf*. From the claim that "a healthy, forceful spirit will be found only in a healthy and forceful body" and the corresponding emphasis on physical education and training, to the doctrine that "the folkish state . . . must set race in the center of all life. It must take care to keep it pure," Hitler arrives at the conclusion that "all great cultures of the past perished only because the originally creative race died out from blood poisoning . . . Those who want to live, let them fight, and those who do not want to fight in this world of eternal struggle do not deserve to live."[41]

Hitlerism does indeed have a distinct attitude toward the future, toward its destiny. The powerful pressure of the future and the danger of dying out "from blood poisoning" drive National Socialism to safeguard its own future by way of force and violence. The danger of death, of the future, justifies the "fight" and the death of those who "do not deserve to live." Levinas's 1934 essay should be credited with identifying the genocidal ideology that Hitler's regime would soon loose upon the world.

These conclusions might seem passé if they were not printed in the summer of 1934, while the 87-year-old German President Paul von Hindenburg still lived and Hitler waited for his opportunity to seize power. Levinas's discussion of the violence of Hitlerism and the self-authorizing power of racist ideology precedes the historical embodiment of the Third Reich and its meteoric drive for international domination whose atrocities include the murder of Levinas's own parents and brothers.[42]

This underappreciated essay bears deep significance for investigating the genesis of Levinas's thinking about time. At the very outset of his philosophical career Levinas associates the radical, and still mostly *potential,* evil of Hitlerism with a failure to think properly about *time*. Levinas is originally concerned with the philosophical relationship between time and violence. This already presses him outside of Heidegger's reflections on time and temporality, even if his indictment of Heidegger remains indirect.[43] Levinas's thesis about Hitlerism is modest and reflective, but the concern is clear: philosophy's modern thoughts about time and history have left the real world of persons and bodies vulnerable to terrible violence.

Furthermore, Levinas expends considerable energy in this essay reflecting on the relationship between the current state of philosophy and the difficulty of the concept of time. He reflects on the peculiar resolution to this impotence offered by the Jewish and Christian doctrines of divine forgiveness. Levinas will later grow suspicious of the mastery of time apparently offered through religious forgiveness, yet here, Levinas writes that pardon *exasperates* time, domesticating the sting of time's flow and leaving it panting like a wounded beast at the feet of those who have been forgiven (*UH* 14). By the end of his corpus, time will fall prey to no such submission.

Levinas offers a prophetic warning about the rise of Hitler, whose pathway to power and terror is paved by the oblivious attitude of modern philosophy toward time, bodies, and history. We also see a strong critique of Western idealism and its tendency to detach ideas from flesh, placing secondary importance on the material realm of bodies and their sighs of pain. Much will change about the way Levinas discusses philosophy and its implications for ethics, but this is a remarkably early and direct association of violence with the distortion of time.

On Escape

In his 1935 essay *On Escape* (*De l'évasion*), Levinas continues to analyze the human condition with respect to the tension created by the raw fact of being, and particularly the manner in which we are tied to our lives, our histories, and our bodies. As in his essay on Hitlerism, Levinas remains absorbed with the sense that philosophy has sidestepped a serious problem by failing to address the difficult problem of being. In freedom, the thinking person can seemingly transcend the ravages of being and the restrictions afforded by bodies and histories, but this freedom and transcendence is shocked, or nauseated, by "the brutal fact of being that assaults this freedom" (*OE* 49). However free we are internally, our actual lives are bound in innumerable ways. In response to this discord between freedom and bondage, traditional philosophy has revolted, fixating its attention on the purification of the I from all things external to it. Once cleansed of all foreign impurities, the I can then find peace with itself; it "closes on and rests upon itself" (49). But this revolt is not without consequence.

One can only dwell so long in the realm of private peace before somehow bringing the concept of the self back to bear on the transient realm of food, money, science, and *time*. In the public square, the I strives in vain to mirror the realm of internal tranquility. Why should the peace evident to the enclosed and self-reflective I not be replicated for all of being? The public realm of being is loaded with unknowns and insecurities, all of which are threats to the well-being of the self-sufficient I. This nourishes and drives an audacious dream of replicating internal stability within the brutality of transient being. The work ethic of capitalism, suggests Levinas, is driven by this very desire to secure and tame the "unknowns of time and things." The "bourgeois," as Levinas calls him, "would like to cast the white mantle of 'inner peace' over the antagonism that opposes him to the world" (*OE* 50). This makes him a worried conservative, always trying to shore up the present to stave off the uncertainties of the future. Levinas summarizes: "Yet, prosaically materialistic, he prefers the certainty of tomorrow to today's enjoyments. He demands guarantees in the present against the future, which introduces unknowns into those solved problems from which he lives. What he possesses becomes capital, carrying interest or insurance against risks, and his future, thus tamed, is integrated in this way with his past" (50).

Levinas ponders the source of this anxiety that underlies human existence. It is not so much that life presents us with suffering or the possibility of suffering but that "the ground of suffering consists of the impossibility of interrupting it, and of an acute feeling of being held fast" (*OE* 52). This threat, which the bourgeois struggles to address without regard to the absurdity of such efforts, is the instability of the present. But beyond the instability of the present is the more disturbing sense that the struggle for security itself is founded on an unreflective and naïve hope in external equilibrium within being. For Levinas, being is like a snare; the harder one tries to escape, the more absurd the struggle. There is a profound need to escape, a need that drives the nervous constructivism of the bourgeois, which works tirelessly to secure comfort for tomorrow. Pain and suffering underscore the sense that we are riveted to ourselves. Alphonso Lingis notes: "Quite early Levinas studied the immanence of pain. To be pained is to feel one's own substance, as a passive affliction, in the torment of wanting to escape oneself. For to escape pain would be to be able to transcend it towards the world, or to be able to retreat behind it and objectify it. The inability to flee or retreat, the being-mired in oneself, is the suffering of pain."[44] The emphasis on time at the outset of *On Escape* is telling. The internal, mental world of the I is timeless; it is a self-enclosed present that functions in isolated independence from the world.

This inability to escape is never clearer than in the phenomenon of suffering. For the early Levinas, as Sarah Allen has written, "it is the constant threat of physical suffering and death that opens up otherwise separated and independent human beings to each other, casting them together in relations of love and fecundity that go beyond their own needs and projects."[45] These relations, in *On Escape,* are about safeguarding oneself from the dangers of the unknown, the future. We have good reason to prefer the timeless present, which Levinas will eventually explore as "the instant." But we are always forced by pain, hunger, need back into the perilous world of temporality.

The purpose of enterprise, then, is protection and conservation of the painless present, which the bourgeois believes himself to control. The unknowns of the future, which include the possibility of pain and the certainty of death, threaten to disturb "the uncontested equilibrium of the present where he holds sway" (*OE* 50). Here one can already detect a steady reliance on Heidegger's critique of the history of philosophy. The present, the now, is secure, knowable, and tamed. The future, in particular, is frightening and threatens the stability of the seemingly stable present. The future is most frightening when it reveals its lack of predictability, when it cannot be confined into predictions available in the data of the present.

There is abundant evidence of the influence of Heidegger in this discussion of time and just as much evidence that Levinas wants to escape the environment of Heidegger's thought even as he operates within a basically Heideggerian phenomenology. *On Escape* is replete with evidence of Levinas's approval of Heidegger's critique of philosophy, alongside his implicit growing dissatisfaction with the posture in which *Being and Time* leaves us. By 1947, Levinas will say explicitly what is only implicit in 1935: "If, at the beginning, our reflections were to a large degree inspired—for their notion of ontology and the relationship that man has with being—by the philosophy of Martin Heidegger, they are driven by a profound need to leave the climate of that philosophy, and by the conviction that we could not leave it for a philosophy qualified as pre-Heideggerian" (*EE* 4).[46] *On Escape* concerns itself with the essentially Heideggerian question about the meaning of being and the absence of temporality in philosophy's consideration of the nature of being. The fundamental desire to escape the heaviness of being, which *On Escape* traces from pleasure to shame to nausea, "allows us to renew the ancient problem of being qua being" (*OE* 56). This question, claims Levinas, is the "heart of philosophy." Heidegger and Levinas here agree that traditional philosophy has forgotten to think about being in light of *time*. On this agreement, through all his strident critiques of Heidegger, Levinas will never waiver.

There is an obvious affinity between Levinas's sense of being riveted to oneself and the way in which Heidegger's Dasein relates to being, yet Levinas already seems to see that Heidegger's struggle for authenticity does not aspire to break the rivets that secure *Dasein* to *Sein* (Being). For Heidegger, one ought not to seek escape from being but should instead face it squarely, courageously, and resolutely. Heidegger may raise similar warnings about the weakness of traditional ontology, but according to Levinas, his solution is no less ontological than Aristotle's. The question that titles Levinas's 1951 essay, "Is Ontology Fundamental?" is first posed here in *On Escape*. As much as Heidegger reconfigures traditional ontology to account for temporality and individuality, Heidegger's proposed solutions constitute, for Levinas, a step backward from the forgotten virtues of Western idealism. Heidegger's Dasein must accept its thrownness (*Geworfenheit*) into being, in all the awkwardness that being-unto-death entails, and live authentically, projecting (*Entwurf*) forward into Dasein's ownmost future. The Greeks revolted from being and its brutality; Heidegger embraces the brutality of change and death, resolutely.

Levinas wants to think of being in light of Heidegger's critique and, through Heidegger, rethink what it might mean "to transcend, in thinking, the horizon of Being."[47] For Heidegger, as Levinas reads him, ontology will suffice. Yet Levinas also wishes to retain some vestiges of the idealism of traditional philosophy, which allows him to think beyond ontology. "The value of European civilization," writes Levinas, "consists incontestably in the aspirations of idealism" (*OE* 73). In practice, modern philosophy has failed to think of ideals in light of the temporality of being; this much Heidegger has established. But Heidegger becomes a stepping-stone for Levinas, who wants to reconsider the beyond being in light of being's temporal horizon.

Levinas incorporates the subject of time into his sharpest critiques of the acceptance of being. "Escape," he points out, "will not appear to us as a flight toward death or as a stepping outside of time" (*OE* 57). The phrase "stepping outside of time" refers plainly to the traditional philosophical position that advocates self-sufficiency and internal, timeless tranquility. So when Levinas promises that his understanding of escape is not a "flight toward death," what position is he denying? He appears to be attacking Heidegger for his fixation on death.[48] Heidegger's philosophy can hardly be read to advocate a rush toward death, but Levinas is interested in demonstrating that Heidegger does not aspire to think beyond being. Levinas seems to be presenting two unsatisfactory ontological alternatives, the revolt against being advocated by traditional philosophy and Heidegger's less flinching attitude toward being's brutality. Both options, as Levinas positions them, represent a flight from time.

Levinas's steady attack on Heidegger is emphasized in a stunning passage in the conclusion to *On Escape*. Here he equates the acceptance of being, which summarizes his reading of Heidegger, with *barbarianism:* "Every civilization that accepts being—with the tragic despair it contains and the crimes it justifies—merits the name 'barbarian'" (*OE* 73). Heideggerians might protest Levinas's reading on this point. Heidegger never directly advocates an "acceptance of being" *per se.* It would be more accurate to accuse him of promoting *resoluteness,* which is a kind of acceptance of being-unto-death. Heidegger does not seem to care, in fact, whether one accepts being. Being *is,* and for Heidegger, the proper response to being must reflect the temporality of being itself. So Levinas delivers here a glancing blow to Heidegger, perhaps intentionally softened for personal reasons. But beneath the semantics of acceptance is the lasting sense that Heidegger is settling for a philosophy of being, an ontology to guide all of philosophy. Ontology is, for Heidegger, *fundamental.* And Levinas here is sufficiently bold to call such approaches barbarian.

Levinas's rift with Heidegger therefore relates to the most basic facet of Heidegger's early philosophy. The invocation of barbarian ontology certainly echoes the warnings issued in Levinas's essay reflecting on Hitlerism.[49] These are subtle nuances, perhaps, but as direct as we might expect Levinas's critiques to be in 1935. Gary Mole summarizes the suspicions expressed by Levinas in *On Escape* concerning the totality inherent in ontology: "Ontological totality, Levinas would suggest, is the first step to political totalitarianism."[50] Yet, as Mole goes on to point out, Levinas is far from his mature formulations of these ideas, even if we can see already in 1935 the beginnings of a lifelong pursuit of a nonontological way to escape being.

Though the rift with Heidegger has begun in *On Escape,* the more resounding and direct critiques of Heideggerian philosophy are still to come. Noticeably absent, at this early stage, is the dominating concern Levinas soon develops for the realm of the intersubjective. In *On Escape,* the non-I creates a problem for philosophy that hides from time in individualized idealism. But the non-I here refers to all things material, not specifically the other person. This means the specifically human other is not distinguished with any force from the world outside the self in general. It is important to note that Levinas is already differing with Heidegger on the function and definition of *time* and its significance for ontology well before he frames his critique in primarily ethical terms.

On Escape opens up what Jacques Rolland calls a "space of questioning" in which to ponder "a new path" beyond the pressure of being to which we are held fast (*OE* 4).[51] Rolland believes this new trajectory represents the birth of Levinas's own distinct approach to philosophy and phenomenology.[52] We need not bend to the pressure of establishing an artificial beginning for Levinas's unique ethical philosophy. It is nevertheless important to note that in *On Escape,* while Levinas labors beneath the weight of being, he does not yet point to the other person as the new path beyond being's crushing pressure.

"The Work of Edmund Husserl"

At the age of 23, Levinas was already working fluently with the latest trends in Husserlian phenomenology. Given the tremendous influence of Heidegger on Levinas, one could easily overlook the fact that Levinas begins his philosophical career as a Husserlian and always wished to consider his work a form of phenomenology in the Husserlian tradition. More particularly, it is important to consider the way time is treated in Husserl, or more accurately, in *Levinas's* Husserl,[53] specifically his 1940 essay on Husserl's phenomenology, "The Work of Edmund Husserl" (*DEH* 47–87). This essay appears at an important juncture in Levinas's own development, and it includes several pages dealing with the innovative treatment of time in Husserl's *The Phenomenology of Internal Time-Consciousness.* From this essay, several key components in Levinas's reading of Husserl unfold that will become instrumental in his own developing understanding of time.

By 1940, Levinas was considered an international expert on Husserl. It was Levinas's award-winning 1930 dissertation, *The Theory of Intuition in Husserl's Phenomenology,* that placed him among the first handful of scholars who introduced Husserl to France, where it quickly gained a substantial foothold. Levinas wrote this book while studying in Freiburg in the 1928–1929 academic year, the same year that Husserl had retired, though he remained at the university teaching a seminar. Levinas also contributed mightily to the expansion of Husserl's influence by cotranslating *Cartesian Meditations: An Introduction to Phenomenology* from German into French.

"The Work of Edmund Husserl" provides a helpful summary of the key components of Husserl's phenomenology, but it is particularly helpful in identifying the early departure of Levinas from Husserl. Though the exposition seems rather straightforward, Cohen points out that a great deal of Levinas's own ideas slip undetected into this explication of Husserl. A number of Heideggerian elements are also present in the reading of Husserl offered in *The Theory of Intuition,* though Levinas is both aware and forthright about the use of Heidegger's insights to "sharpen the outline of Husserl's philosophy" (*TIH* lv).

The substance of Levinas's critique of Husserl in 1930 is not irrelevant to this study; Levinas is concerned even in his dissertation with the *formal* or *theoretical* nature of Husserl's approach to phenomenology. Husserlian theory, Adriaan Peperzak summarizes, "claims to identify universal, often purely formal, elements and structures which determine the life of all human consciousness and of being 'in general and as such.' "[54] Levinas suspects, and certainly his studies with Heidegger influence this suspicion, that these lofty and theoretical postures are vulnerable to distortion and misreading. Here, Levinas names a difference from Husserl that will be sustained throughout his career: Husserl trusts too carelessly in representation, objectivity, perception, evidence, and especially *theory.*[55] It is Husserl's absorption in these themes that blinds him to his own innovations regarding affectivity and sensibility, and his absorption therefore undermines a potentially landmark advance on the philosophy of time. Levinas sees in phenomenology a return to concrete existence and a shedding of the formal logic of philosophy that might obscure concrete experience. Levinas's concern is that Husserl does not sufficiently escape philosophy's proclivity to intellectualism.

Husserl understands reality to be constituted by a multitude of overlapping layers of meaning, though the relationship between these layers of meaning is often obscured or overlooked. Failing to separate or unravel these complexities leads to a skewed perspective on truth, a poor description of what is real. Husserl strives to isolate phenomena by removing assumptions; when a phenomenon is isolated in this way, it can be better known and understood. The more effectively one can erect brackets around a phenomenon, eliminating expectations and presuppositions about how an event may occur, the more accurate one's descriptions will be. This process of isolating phenomena Husserl calls *epoché,* and Levinas adopts the methodology more than is sometimes appreciated.[56]

Levinas, knowingly or otherwise, will consistently gravitate toward this Husserlian solution when analyzing complex phenomena. Levinas will differ from Husserl regarding the purpose and goal of phenomenology, but he appears to consistently trust the *epoché* to guide philosophy toward better analyses. On the question of time, then, there is a sense in which Levinas attempts to be more faithful to the efforts of Husserl than Husserl himself. Levinas questions whether Husserl, and Heidegger especially, sufficiently interrogate the concept of time.

In "The Work of Edmund Husserl," Levinas exposes the relationships between the ego, time, and freedom. For Husserl, explains Levinas, the ego is not available for typical phenomenological analysis as a graspable being with observable qualities. Instead, one must describe the ego by way of examining intentionality. The history of the ego leaves a trail, indications of the position from which actions arise. To pursue the nature of the ego is to ask questions not about what an ego *is* but to examine what it *does.* The ego's activities and history indicate the accomplishment of acts, intended activities that demonstrate the "freedom" of the ego. Intentionality, summarizes Levinas, "is nothing but the very accomplishment of freedom" (*DEH* 76).

Husserl develops his theory of "internal time consciousness" to address the fact that phenomenology and traditional philosophy have not adequately addressed the role of *time* in the freedom of the ego. Levinas points to an important consideration of the present in Husserl's concept of *Urimpression*. Husserl invokes this term to refer to the "present with its retentions and protentions"; it is the moment of gathering and intellect (*DEH* 78). This idea is more important to Levinas than it was for Husserl, who makes relatively little out of the idea of the *Urimpression*. Husserl does continue to use the basic idea of a "pure present," a kind of ground zero for conscious life. Consciousness begins in *Urimpression*, where the ego constitutes itself by gathering impressions and focusing intentionality. Levinas will later use the terms "instant" and "hypostasis" to refer to this analysis of the present. As the ego gathers and constitutes itself, it projects itself into a new present, a new instant. This active movement of the ego into a projected future, or *protention*, is an idea that becomes an important component of Heidegger's analysis of time. The new *Urimpression* receives the former moment in passivity and then gathers itself for the activity of protention.

Levinas claims that for Husserl, both retention (of the past) and protention (into the future) are constituted by intentionality. Consciousness and intentionality are grounded in the present, in the self-presence of the ego. The temporal movement into the next instant is an act of freedom, which renews and repositions the ego in a new present. The *Urimpression* is the "now-point," the source point for all freedom, all memory, and all activity.[57] And here the ego has a particularly interesting relationship to time. Levinas gives Husserl credit for taking a stand against the tendency to reduce time to "a simple content, a quality like, for example, color" (*DEH* 77). The relationship of the ego to the past and the future is a matter of the ego's freedom; the past is what is seized and carried forward in retention, and the future is what is chosen in freedom through protention. This means that time is constituted *internally*, by the ego. As Levinas puts it, "Time is engendered by the very moment of the subject's freedom" (77). Internal time is therefore the production of the ego. And this sense of time is overlooked and overshadowed by the dominant themes of external time.

Levinas will carry forward the Husserlian analysis of the instant, wielding his own phenomenological analyses to point through and past Husserl in rethinking time. Husserl places the power of time within the machinations of the transcendental ego: the ego *creates* time in the process of self-constitution. The future is that which is forged by the activity of protention, as the ego moves from the present moment to the next present. Levinas will soon deliver a strong critique of this presupposition, calling into question the assumption that the ego has the internal power to break free from the *Urimpression*.

Levinas's treatment of the concepts of "light" and "representation" are also significant for our analysis of his developing understanding of time. For Husserl, as Levinas reads him, the basis for "every intention" is representation, and representation is built on the model of *light* (*DEH* 60–61). Light belongs most obviously to the field of space, measurable according to visible distances. Derrida will call light the paradigmatic and unavoidable metaphor of phenomenology after Husserl, which depends on light more rigorously than "any other philosophy" (VM 85). So when in 1940 Levinas questions the stability of light and representation, he is challenging the stable core of Husserlian phenomenological analysis. All intentions, retentions, and protentions are for Husserl available to the light and made accessible to observation by way of vision, or at least in the mode of visual appropriation. But Levinas thinks that the mode of representation is itself an advanced and secondary form of intellection. He explains, "Self-consciousness, better than being a simple acknowledgement of intellection, *is* intellection, and consequently is light and freedom. It is carried out in the inner consciousness of time" (*DEH* 77). Levinas suspects that something is happening *before* self-consciousness, before light and representation are activated.

Levinas thinks Husserl is aware of this *before*, of this passivity that is deeper than the typical epistemological tension between passivity and activity. Such a passivity takes on a "positive mark," pointing to "nonattentive consciousness" and "implicit thought" (*DEH* 76).

Husserl returns to the deep freedom of internal time, where meaning is formed by way of self-conscious intellection. Husserl has liberated time, therefore, from the external demands of clocks and calendars and has returned it to a less formal manifestation in the way an impression is formed by the ego. But Husserl's time is still too formal for Levinas, in that it is unable to gesture toward a "domain" beyond the one in which "the ego manifests its freedom" (*DEH* 76).

The internal time of Husserl is, for Levinas, too closely bound to the *present*, an insight that is central to Heidegger's break with Husserl. Light is possessive, grasping a phenomenon according to how it comes into vision. But the reliance on light reduces phenomenology to the evaluation of the illuminated and denies epistemological access to anything that fails to become illuminated. Tilottama Rajan summarizes Levinas's concerns: "Light, intention and representation form a matrix that allows Levinas to see phenomenology as blind to the unthought."[58] Phenomenology is, therefore, confined to a field that Levinas considers an incomplete cross section of human intellection. Worse yet, it can be read to suppose that represented insights are the only insights available for consideration. So Levinas claims that there at least *may* be something *older* than light, something that arises in a mode unfamiliar to a discourse that can only speak in the mode of self-presence and objectivity. And this something is the passivity that precedes even the passivity of receiving light. But Levinas has only here begun to glimpse where this search for a better understanding of the *Urimpression* will take him.

In the 1940 essay, and perhaps already in more subtle ways in *The Theory of Intuition*, Levinas stakes out a new direction from Husserl's phenomenology. He is already critical of assumptions that trouble the phenomenological method of Husserl. It is, in part, Heidegger's critique of Husserl that leads Levinas off in new directions. But while he departs from Husserl with Heidegger, his trajectory will lead him back against Heidegger, and in the later stages of his thought, Levinas returns to a more sympathetic conversation with Husserl.

John Drabinski offers an intriguing perspective on this complex relationship between Levinas, Husserl, and Heidegger. Drabinski points out that the standard interpretation of Levinas's departure from Husserl hinges on his discontent with the "intellectualism" of

Husserl's phenomenology, a critique that can be derived from *Being and Time*.[59] This typical reading of Levinas's departure from Husserl has some credibility; Husserl routinely uses "the most un-Levinasian terms in describing the aim of phenomenology."[60] In Husserl, philosophy is rooted in transcendental subjectivity, which is itself moored in "presence to self." Levinas will work against both ideas in his early and late writings, joining Heidegger in the suspicion that philosophy must be much less certain about the availability of a transparent view of oneself. For Heidegger, after all, being-in-the-world means being ahead of oneself. By all appearances, then, Levinas's 1940 "critique of Husserl is wholly Heideggerian" even though by 1940 it was clear that Levinas was moving against Heidegger in other essays.[61]

Drabinski sees holes in this typical summary of the relationship between Levinas and his mentors. In particular, Drabinski suspects that what attracts Levinas to Heidegger are the new options Heidegger offers for the expansion of Husserl's thought. Levinas had already written in 1930 that the originality and power of *Being and Time*, "even though it is in many respects different from Husserlian phenomenology, is to some extent only its continuation" (*TIH* lvi). Drabinski thinks that what is "revolutionary about Heidegger's work, for Levinas, is Heidegger's *husserlianism*."[62] In other words, although Heidegger helps Levinas think past Husserl, Levinas's path does not look much like Heidegger's. This is particularly clear on the concept of time, which Levinas moves to define in ways that are specifically opposed to time in Heidegger. Drabinski summarizes: "I would argue that Levinas's concern with Heidegger is, in fact, not very Heideggerian. There is little or no positive talk of Being or historicity, of facticity or resoluteness. What Levinas calls "being," for example, in [*Existence and Existents*], bears little resemblance to the meaning of Being in general.... Historicity plays little role in Levinas, except in the sense of the nonnarrative history of the I, which places oneself in the immemorial."[63]

Levinas becomes fascinated with the Husserlian analysis of affectivity, even if Husserl himself envelops affectivity in transcendental subjectivity. As Drabinski suggests, the pursuit of a better articulation of Husserlian affectivity inaugurates a project to which Levinas will faithfully return throughout his career: "Though Levinas's work will shift in terminology and even in the implications of his findings, affectivity—whether sensibility, proximity, materiality—remains the point of departure. Although this turn to affectivity certainly bears the marks of Heidegger's understanding of *Befindlichkeit* in *Being and Time,* the language of Levinas's analysis is distinctly Husserlian."[64]

Levinas will concern himself less with Husserl in the 1950s, as his differences with Heidegger and Jean-Paul Sartre become more central concerns. It may not be coincidental that in this decade, when he seems least interested in Husserl, he is markedly less interested in the concept of time. But Levinas continues to consider himself a disciple of Husserl and thinks of his work as phenomenology in the tradition that Husserl inaugurates. In 1977, Levinas says that "in spite of everything, I think what I do is phenomenology, even if it is not according to the rules laid down by Husserl, even if the entire Husserlian methodology is not observed."[65] We will also discover that Levinas's definitive turn toward the concept of time and diachrony in the 1960s coincides with, and is perhaps driven by, a renewed interest in the sense of passivity latent in Husserl's understanding of time and the *Urimpression.* Both his early and late conversations with Husserl confirm that Levinas is primarily interested in extending Husserl's work. As the Husserlian scholar Christian Lotz summarizes, "Husserl and Levinas are in far closer agreement than the literature in the field would seem to indicate."[66] And in 1983, Levinas reflects, "It is undoubtedly Husserl who is at the origin of my writings" (*OS* xvi).

A chasm separates Levinas's early essays and reflections from his work in the late 1940s. Levinas's public scholarship grounded to a halt when in 1939 he was drafted by the French military. Then, in 1940, he was captured by the Nazis and shipped to a labor camp for military prisoners. Only after his 1945 release would Levinas learn of the murder of his family members at the hands of the Nazi Secret Service. He had spent a very long five years in Stalag XIB at Fallingbostel, a city between Hannover and Bremen in Germany.[67] Forced to work in the cold forests by day, Levinas devoted his remaining energy to reading and even writing. His first original book, *Existence and Existents* was published shortly after his release. To that work, and its implications for his growing interest in time, I now turn.

Two

The Freedom and Horror of the Instant

Time, far from constituting the tragic,
shall perhaps be able to deliver us from it.
—Emmanuel Levinas, *Existence and Existents*

Introduction

We can only guess at the significance of this reality: Levinas's first book-length contribution to philosophy, *Existence and Existents* (1947), was largely written and conceived while he was in captivity.[1] We should not be surprised, therefore, at the raw and haunting images that Levinas now combines with familiar themes from his earlier essays. Levinas incorporates here evocative metaphors such as anonymous being, insomnia, the raw doctrine of the *il y a* ("there is"), and a host of other graphic concepts. Salomon Malka notes that captivity was "decisive" for Levinas. The years in the *stalag* exposed him to "the most simple things, the ordeal of loss and of liberty, the sensation of time, deliquescence, misery, absolute passivity, fragility, precariousness—everything that continually tormented his work."[2] Levinas did not write of the death of his family, except in the formal dedication of *Otherwise than Being*, his second and final major work. Levinas rarely referred to his time in the camp, but Malka's assessment is spot-on: Levinas never strays far from the gritty themes that dominated his thoughts in Fallingbostel. The postwar publications are full of gravity, approaching darker questions and more serious themes. In 1963, Levinas would write that his philosophical and personal biography "is dominated by the presentiment and the memory of the Nazi horror" (*DF* 291).

Levinas obviously continued to reflect philosophically while he was a prisoner of war. When his camp was liberated in April of 1945, Levinas carried out with him a considerable amount of notes and musings. Many of these have been published under the title *Carnets de captivitéet autres inédits,* which provides a trove of insights into Levinas's philosophical transition during his captivity and in the years immediately afterward. These were critical developmental years for Levinas; he emerges from captivity quite prepared to step definitively onto a philosophical pathway that is uniquely his own.

Particularly noteworthy with regard to Levinas's understanding of time are the final sections of *Existence and Existents* in which Levinas develops a tightly argued, innovative philosophy of the instant. In his translator's introduction to this work, Alphonso Lingis writes, "Levinas's work contains not only wholly new analyses of the forms of time—of the present, the past, the future—but also a new concept of the work of time" (*EE* xxii). After an extended discussion of Levinas's account of the instant, it is important to further explore some salient issues that arise from these initial investigations in two of his essays that appeared in 1948, "The Transcendence of Words" and "Reality and Its Shadow."

Levinas, as we have seen, was philosophically nourished in his undergraduate and graduate studies by the French and German phenomenological tradition. The innovations of Bergson, Husserl, and Heidegger opened up a number of opportunities to reconsider the philosophical meaning and importance of time. Their innovations, which fascinated Levinas, can be considered a philosophical breakthrough in a milieu that had been dominated for more than a century by Immanuel Kant and Kantian philosophers.[3]

The German and French appropriations of phenomenology were slow to take up the question of the other.[4] This appears to have been an unintentional consequence of moving past Kant, for whom the problem of ethical intersubjectivity was a driving and central concern.[5] Phenomenology, after all, attempts to eliminate presuppositions and to think with clarity and objectivity. People, as phenomena, are notoriously complicated, and morality has historically been dependent upon a whole series of larger presuppositions; ethics has depended on larger philosophical systems. Consequently, one finds little in the way of sustained ethical theory in Bergson, Husserl, or Heidegger. Perhaps these thinkers continued to presuppose the moral structures of Kant's philosophy even as they moved past his epistemology and metaphysics.

Whatever the reason, Samuel Moyn points out that in Levinas's educational background the sources for "a theory of ethical intersubjectivity" were "few and fragmentary."[6] In both French and German phenomenology, a concern for ethics had simply failed to arise from the more scientific and modernist concerns of early phenomenology.[7] We should not be surprised, therefore, to find that it takes some time for Levinas to arrive at a position that can enthrone "ethics as first philosophy."[8] The turn to intersubjectivity is substantial and unique. Levinas is clearly in debt to his philosophical mentors and his tradition, but Moyn proposes that Levinas develops the deep ethical concerns of his philosophy in spite of these traditions. Moyn also suggests that even Levinas's parents were not helpful in this regard, raising him in a "Russian Jewish bourgeois culture" that lacked some of the rigorous moral concerns common to other strands of Judaism.[9]

Aside from the obvious exposure to trauma, fear, death, captivity, and forced labor, there may be some significance to the sheer separation in which Levinas developed philosophically while in captivity. Cut off from the academic world, Levinas's thought evolved on its own, in painful and cold isolation. He was able to do some reading, including Hegel, Marcel Proust, and Jean-Jacques Rousseau, but he was obviously unable to keep pace with the developments in phenomenology and existential philosophy.[10] We can justifiably wonder if this isolation may have enhanced the originality of the two works published in 1947, *Existence and Existents* and *Time and the Other*.[11] These two works announce, demonstrate, and determine a sustained focus on the meaning and importance of time for Levinas.

In isolation, Levinas continued to reflect on time, particularly Bergson's innovations on time.[12] He also developed themes that looked comparable to those being developed by Sartre during these same years, a similarity Levinas will later contest as it becomes clear to him that Sartre fails to break with the Hegelian dialectic. In fact, Levinas will work rigorously to distance himself from Sartre (see *CPP* 1–13). Levinas sets to work immediately after his release, publishing *Existence and Existents* before he has acquainted himself with the current trends in philosophy. He invokes the stalag not "as a guarantee of profundity nor as a claim to indulgence, but as an explanation for the absence of any consideration of those philosophical works published, with so much impact, between 1940 and 1945" (*EE* xxvii).

We need not determine which of the transitions apparent in Levinas's philosophy after the war are a product of his wartime experiences. For this study, it suffices to say that Levinas will return immediately and attentively to the topic of time. Shortly after his release, Levinas noted that "everyday life is played out at the crossroads between life and nothingness."[13] As he writes in *Existence and Existents*, "It is in times of misery and privation that the shadow of an ulterior finality which darkens the world is cast behind the object of desire. When one has to eat, drink and warm oneself in order not to die, when nourishments become fuel, as in certain kinds of hard labor, the world also seems to be at an end, turned upside down and absurd, needing to be renewed. Time becomes unhinged" (*EE* 36–37). The early 1940s were utterly sobering for Levinas, and in a manner that appears to press him to write about time.

Il y a and Insomnia

By 1947, as evidenced in both of his publications that year, Levinas had developed a mature philosophy of the instant, a key concept adapted from Bergson in particular. For Levinas, however, the instant takes on a unique set of characteristics. Through this concept, and its impact on the philosophy of time, Levinas attempts to introduce to phenomenological analysis the foreign concepts of transcendence and radical hope. He will continue to work throughout his life as a self-avowed phenomenologist, though the moves he makes in *Existence and Existents* already, in the minds of some thinkers, situate him outside the phenomenological tradition. This sentiment is scathingly articulated by Dermot Moran, who moves chronologically through Levinas's corpus and attempts to demonstrate how each phase in Levinas's writings positions him farther and farther away from "the very essence of Husserl's conception of philosophy as rational self-responsibility."[14]

Despite this critique, Levinas, phenomenologist or otherwise, uses phenomenological tools to isolate the topics he wishes to analyze. Part of what makes Levinas a phenomenologist, at least at this stage in his career, is his methodological approach to philosophical questions. In *Existence and Existents,* he seems to utilize something like Husserl's *epoché* in his selection of analogies and examples. Levinas investigates experiences that reveal the instability of knowledge masked by everyday phenomena. Light, for instance, masks the insecurities and instabilities of darkness, and it deceives us into thinking that all knowledge comes by way of vision and light. The theme of darkness, therefore, makes its way into the analyses he offers here.[15] For Levinas, insomnia, in particular, provokes images of a swirling darkness without any of the comfortable illuminations that make objects knowable and discernible during the daytime. Without sleep, caught in darkness, we are forced to face that which we normally neglect. This is the phenomenological method at work; Levinas wants to bracket light and discover what remains.

Levinas expends significant energy in *Existence and Existents* performing phenomenological analyses on the relationship between "the I and the world" (*EE* 27). This relationship may at first appear straightforward, as though at every given moment one is already positioned in immediate relationship with the world of sensibility, but Levinas suspects that something important about the I is overlooked in this assumption. Levinas looks for ways to perform creative bracketing to bring the difference between the internal self and external world into clearer relief. We normally experience food, drink, shelter, and the natural world prereflectively. Our being-in-the-world is intertwined with other beings. The world, as Levinas defines it, "is what we inhabit, where we take walks, lunch and dine, visit, go to school, argue, carry out experiments and investigations, write and read books" (36). The choice, cognizant or otherwise, to be in the world of these objects and experiences is a movement toward the desired, and desire is understood as knowing and aiming at what one wants.[16] Apart from desire for the external world, the self could rest content on itself, as Levinas indicates in *On Escape*.[17] But internal existence functions like a frozen instant, where the objects of the world become like a winter landscape frozen in place. We are driven from the solitude of our private instant by desire for the other, for the world, for the temporary satiation of hunger and thirst.[18] Being in the world means being attached to the things of the world (*OE* 27). This movement outward is instinctive and is the very "meaning of living" (*EE* 36). We do not eat, drink, or play in order that we may live; these actions *are* living. Levinas goes on to maintain this notion of "living from" later in *Totality and Infinity* (*TI* 110–15).

Because these everyday experiences constitute being in the world, or living, we seldom become aware of the distinctive way the self relates to the world. So Levinas analyzes circumstances that press the relationship between the self and the world into sharp focus. This exercise is intended to demonstrate instability in the way the I experiences the world. Levinas will also use this method to question the way in which philosophy has instructed us to think about the internal relationship to oneself. According to Levinas, the "things" of the world can be overshadowed by "times of misery and privation," which jolt and destabilize the relationship between the self and the world it desires. In the face of starvation, food becomes fuel alone, ingested "in order not to die" (*EE* 36–37). Privation makes me aware not only of my destabilized relationship to the outside world of materials, but also aware of the normally seamless relationship between intentionality and sensation. I no longer choose my sensations intentionally; instead, they begin to choose me. This division between intention and experience breaks the "unity of the I" (*OE* 49).

In these moments the external world, which I presumed was mine, turns back against me. The world that once existed for my sake now exists against me.[19] What was once a "joyous appetite" becomes a desperate hunger (*EE* 27). When intentionality and "what is given" are bound seamlessly, we presume that living is a matter of "desire, a movement to take hold of something, to appropriate something for oneself" (38). Levinas has already named "time" as the phenomenon whereby the weight of being is alleviated.

In the pressure of hard labor or the privation of food or rest, time no longer redeems the present from the past. The world is no longer what I desire and seek, but that which falls back on me in my affectivity. There is a certain *activity* inherent in my own capacity to absorb and arrange the information delivered by my senses. Yet these everyday freedoms can also be seen as labors, as elemental efforts to construct and arrange a world of experiences that I *ex*-perience. I move outside of myself and encounter that which is, by all appearances, *not* within me. It feels very much like I experience the world directly, in immediacy, without the hidden mediation of my self-relation. In immediacy, I unreflectively identify myself with my experiences; I am what I do, feel, and experience. In privation, this correspondence is utterly threatened. I still *am,* but what am I?

In a similar way, Levinas finds in the phenomena of sleep and insomnia an unsettling of both the self-relation and the relation between the subject and the world. After a lengthy discussion of sleep and wakefulness, Levinas turns to a theme that will periodically inhabit his reflections throughout his career: insomnia (cf. *EE* 61–64; *TO* 48–49; *TI* 258–59). In the night of restless vigilance and in the absence of light, one experiences the world in reverse. In the daytime, one chooses to attend to one thing or another. One has little awareness of the raw fact of being because one is absorbed in this or that vision, activity, pursuit, or reflection. Importantly, the daytime is marked by *intentional* vigilance, a watchfulness over the world that is intended by the thinking self. The invocation of intentionality, alongside the importance of sight and vision, are obvious correlations with Husserl. But insomnia reverses this more obvious epistemology of sight.

By definition, insomnia defies the desire to sleep, to escape consciousness. Insomnia is the maddening desire for an escape that cannot be found. Rather than a time I intentionally endure or an experience I move through in duration, insomnia strips me of all these powers, and leaves me *timeless* and captive to the night. Intentional consciousness typically acts as a shelter and buffer to hide and obscure the raw "there is" (*il y a*) that resides beneath the bright and noisy beings that consume our attention. In insomnia, all things lie unrevealed, and one glimpses the idea of being without beings, existence without existents. Held against the night by insomnia, I am caught struggling with and against the remnants of experience that whirl around in my mind. A random thought carries me along for a while but demonstrates its transience by dissipating as soon as I realize I am thinking it. What Levinas means by insomnia is not this fight against the night but what is left when I finally concede that I am powerless against the night that closes in around me. Whatever I wish it to be, the night simply *is*. Whatever I wish to be, I simply *am*.[20] My intentionally is futile. It serves me well until I see it for what it is: a mechanism that protects me from the horror of my relationship to existence. The night is here, or better yet, now, with no regard for me whatsoever. In this moment, finally, brackets have been erected to isolate a most important reality in Levinas's work: affectivity.[21] The instant lacks "rhythm" and is unable to "break into being" (*EE* 62) Insomnia locks me in a moment, pinning me to an unavoidable affectivity that leaves me stripped of defenses. Enchained to the night, and to the raw fact of being, I can only wait. The irrelevance of ordinary time to my most basic experience is revealed, along with my reification of clock time.

We discover here, in this unusual and partly hypothetical exercise, that we have overlooked something fundamental about time, presuming we control time's motion. The blatant turn against Heidegger is now underway for Levinas; for Heidegger time is irrevocable movement and ecstatis, part of the essential structure of being and through which Dasein moves ecstatically, aware or unaware. In Heidegger, there is always movement, for better or for worse; Dasein "always goes beyond the *hic et nunc* [here and now]" (*EE* 104).

It is intriguing that Levinas believes the anxiety indicated by Heidegger is inadequate to express the horror of the *il y a*. Levinas's sense of anxiety is a "horror of the night," a dread that he considers much deeper than "Heideggerian anxiety" (*EE* 58). The anxiety inherent in Heidegger's attitude toward death remains, above all, unique and personal to Dasein—Dasein's time is its own, just as death is Dasein's "ownmost" moment, a moment that is utterly and irrevocably unique to it.[22] Levinas is proposing the phrase *il y a* because of its utter anonymity.

For Levinas, Heidegger's being-toward-death is already too abstract, not sufficiently concrete or embodied. Tina Chanter summarizes:

> In his early work, Levinas criticizes Heidegger for not acknowledging the specificity of the instant in its materiality, and his critique rests on reworking the ontological difference by providing it with the concretion of hypostasis. Levinas would thereby draw back the abstraction achieved by Heideggerian ecstasis to the immediacy of a concrete present. . . . By dramatizing the hypostatization of the subject, Levinas seeks to elaborate the dynamic in which the subject becomes an existent. It is no accident that in doing so Levinas is able to provide an account of the concrete materiality of the subject that Heidegger could never accommodate.[23]

Heidegger's Dasein is not really *in* the instant at all; Dasein's identity is out in front of itself.[24] Death is Dasein's most obviously unique moment, providing individuation, becoming, and remission, even if the finality of this remission makes Dasein anxious. Levinas opposes this with the *il y a,* an existence without evasion. Levinas's "there is" is an existence without remission, without hope of remission. The frozen present of the night is devoid of the ecstasies, incapable of *jouissance,* stripped of the going-forward that seems to inhabit intentional life. Here in the instant we know the "presence of the present," and the irony and tragedy of our so-called "freedom" is unveiled (*EE* 78). We are free, but we are masters of a domain that is empty and void.

There is no "taking pleasure" because there is no *taking* of anything. I am instead taken. Nothing goes forward in insomnia; it is the night that closes in. There is a feeling of raw eternity, of the immeasurableness of a night that now measures me. The futurity of Heideggerian anxiety at least provides some release from the pressure of the "now," pressing back against the anonymity of being by focusing on Dasein's ownmost, unique possibility.[25] Levinas's anxiety is the terror of the frozen and unmoving tundra, stripped of all nourishment and movement. The *il y a* is the terrible loneliness of immortality.

The powerful concept of *il y a* inhabits a number of Levinas's writings, playing a pivotal role in *Time and the Other* and *Totality and Infinity,* as well as some of his confessional writings.[26] The *il y a* is often considered one of Levinas's first unique and innovative philosophical moves. This idea enables Levinas to express a reversal of intentionality that is at the heart of his break with both Husserl and Heidegger. By challenging the epistemology of light, Levinas is attempting to identify something deeper, something *prior* to vision and intentional awareness. He is digging beneath the structures of knowledge that are common to epistemology and Western philosophy in general. Our awareness of the terror of the instant, of the *il y a*, is masked by this light in which knowing is usually manifested. The *il y a* makes us mindful that all representations rest uneasily on structures that fade away in the night, and this fading reveals a reversal of intentionality. The experience toward which Levinas is gesturing is the antecedent of all events, the *before* that is not a recoverable or prior moment.[27]

In *Existence and Existents,* insomnia is Levinas's choice metaphor to point toward this experience of the pure "there is." In the night of insomnia, I am stripped of all visual distractions, "sobered up" from their intoxication. Though Levinas does not say as much, insomnia performs an *epoché,* bracketing out the elements that make being personal and meaningful. Insomnia is wakefulness without shields and evasions. The insomniac stares into the blackness and becomes *subject* to a presence of being itself, which is unlike any other phenomenon. Levinas writes, "Wakefulness is anonymous. It is not that there is *my* vigilance in the night; in insomnia it is the night itself that watches. It watches. In this anonymous nightwatch where I am completely exposed to being, all the thoughts which occupy my insomnia are suspended on *nothing.* They have no support. I am, one might say, the object rather than the subject of an anonymous thought" (*EE* 63). Open eyes imply a watching, a vigilance, but the wide eyes of insomnia have no subject, no aim, no thing to occupy the gaze. Insomnia offers a phenomenon for analysis by way of negation: an instant which is not redeemed by the next instant but is instead locked helplessly in the infinitude of frozen time. The core of Levinas's early understanding of time is this philosophy of the instant.

Levinas and the Instant

The instant, as we will momentarily investigate in the work of Bergson, is often defined by negation, as an abstraction from the ideas of duration and change. Philosophy has "throughout its history" understood the instant as an abstraction, a derivative of time, as a "cross-section of duration" (*EE* 72). The isolation of the instant is not always readily apparent to us, particularly because of the loud hands of the universal clock. Like the light that obscures the insecurity of my self-relation, I am shielded from the *il y a* of time by the "universal tick-tock where impassive moments succeeded one another."[28] Modernity, with its emphasis on uniform and universal rationality, reinforces this sense. Levinas wonders, in a note from 1946, if the modern world has not turned the human being into "a mechanism among other mechanisms, small clocks reproducing the beating of astronomical time." Modernity led us to believe that the concepts that operate in science "were part of the world itself. The concept of uniform and inhuman time—Saturn devouring his children—dominated the universe."[29]

Here Levinas invokes the ancient myth of Saturn (Cronos), who devoured his children to prevent the flux, fecundity, and transience of temporal reality. Cronos preferred the tranquility of eternal bliss to the imperfect movements of time and temporality. The Cronos myth expresses a nostalgic longing for permanence, which becomes transposed onto modern science. For the modern scientific imagination, time is universal and unchanging, the measured and rhythmic march of the eternal now. It is the wisdom of Bergson, Husserl, and eventually Heidegger, to see that the reliance on the universal time of science masks a more primitive and original sense of time beneath the steadily churning hands of a universal clock.

In the night of insomnia, and in the horror of the *il y a,* I realize with anxiety that all of the experiences that colored my life, which *seemed* to be external and independent, were actually features of my own internalized existence. I am left without *time*, in an instant that I cannot redeem by moving time forward. These experiences are mine for analysis in the instant which is my present, my now. For Levinas, the present of the subject bears no resemblance to the eternal now preferred by Cronos and Plato, it is shared with no one; it synchronizes nothing. Husserl and Heidegger already suspected that time is more originally configured within the experience of the subject, but for these thinkers—and Sartre—the extraction of individualized time from the external world is a liberation.[30] For Levinas, quite the opposite, the present, the instant, is a captivity.

Levinas does not challenge the internalized freedom inherent in the instant, at least not at this stage in his career. For Levinas, we are indeed free to sort and arrange the data we have retained by memory within the privacy and the isolation of the moment. But we are also trapped there, rearranging the furniture of our minds. So as identified in the previous discussion of insomnia, during the light of day, this isolation is unthreatening. I can always look in another direction, get another fix of information to reinforce the false impression that my experience of the world is immediate. I can watch the hands of the clock, which reinforce the sense of time's motion or I can see the sun cross the sky, apparently organizing the creatures beneath its rays according to its bright gaze. But in insomnia, I am trapped with the remnants of these thoughts. The hour hand of the clock, so helpful in fostering my delusions of immediacy during the daytime, works *against* me during insomnia. The clock was supposed to pass these hours while I slept. Yet I remain trapped, with only *myself* to experience. This experience is unusual and atypical, perhaps not properly named *an experience* at all.[31] If there is experience here, it is the experience of needing to be rescued, redeemed.

Bergson and *Durée*

Levinas, in these deliberations, is driving deep the wedge between Saturn's "universal and inhuman" time and what he deems a more original sense of time. What is left, in the moment where universal time demonstrates itself to be an abstraction, is the instant. These questions about the isolation of the instant bear significant debt to the work of the French philosopher Bergson.[32] In different ways, both Levinas and Heidegger follow the lead of Bergson, who in the first decades of the twentieth century had steadily challenged the prevailing understanding of time as the compression of infinite instants. Levinas's admiration for Bergson had its genesis in his university years at Strasbourg (DEL 13). Despite his later fascinations with Husserl and Heidegger, Levinas retained high admiration for Bergson's work. In 1946, as he was preparing *Existence and Existents* for publication, Levinas wrote, "Five years ago, January 4th of 1941, at the age of 82, one of the greatest philosophical geniuses of all time, Henri Bergson, passed away in Paris. The titles he acquired throughout his career—Professor at the College de France, Member of l'Institut and of l'Academie Francaise, Member of numerous foreign scholarly clubs, Nobel Laureate—paled before his actual glory. While alive, he joined the company of first-rate minds. At the eternal banquet of Plato, Descartes, Spinoza and Kant, he sat down as a peer."[33]

If there is a tone of philosophical insurrection in Levinas's oeuvre, it may be traceable to Bergson, for whom Plato and Aristotle were fair game, particularly on the concepts of time and eternity. Bergson questioned the abstraction of the Platonic ideals and the way Platonists have too often settled for "snapshots" of real, temporal existence, which is what Bergson calls "Becoming" (*CE* 317). And change, or becoming, "is far more radical than we are at first inclined to suppose" (1). For Bergson, a misguided understanding of the eternal, which functions awkwardly in the world of matter, is revealed in the gaps that appear between the permanent ideals: "In right, there ought to be nothing but immutable Ideas, immutably fitted to each other. In fact, matter comes to add to them its void, and thereby lets loose the universal becoming. It is an elusive nothing, that creeps between the Ideas and creates endless agitation, eternal disquiet, like a suspicion insinuated between two loving hearts. Degrade the immutable Ideas: you obtain, by that alone, the perpetual flux of things" (317). The flux of things is the *real,* for Bergson.[34] The passage of time, he points out, is typically marked by artificially piecing together a collection of eternal instants. Led by a scientific desire to understand time, which is reinforced by the atemporality of Plato's *eidos* (Forms, Ideas), philosophy has settled for defining time according to a collection of *spatial* instants.

Imagine a photographer who takes pictures of the moon as it crawls across the night sky. Each picture is a frozen instant, a *present* ossified in the image. The spatial relationships can be measured as the moon moves by various degrees from each horizon. If the photographer were to string the pictures together on the wall, they would have the sense of time, the appearance of duration. But for Bergson, this is an artificial domestication of independent moments.[35] The set of photographs does not measure time, but *space*, and the change of space between each photograph leaves an artificial impression of time. In any given instant, the future and past are uncertain; clouds could occlude the moon, a meteorite could mar its surface, or some calamity could befall the photographer.

For Bergson, the more genuine way to think about time is as *durée*, typically translated "duration."[36] *Durée* is the sense of time's motion that cannot be measured because it is constantly vulnerable for interruption. This sense of the passage through time precedes the capacity for measure and is rendered abstract under the time-measuring tools of science. *Durée* may be rendered measurable in hindsight, but it is a function of experience that has to be wrested away from the timepieces of physicists. The instant conceived intellectually is already a degradation of the experience of time as *durée*. The arbitrary nature of an instant is revealed by the absurdity of ripping a single note out of a melody; duration is the experience of the whole melody as it unfolds.[37] The instant, which seems to win a victory over transience and change, is actually stuck in a now that is blind to the prior and the ensuing moments. The photographic instant, for Bergson, is utterly isolated from any of its prior or ensuing moments.

Space, suggests Bergson, is *static*, or stationary. At any given evaluation, the moon can be described according to its spatial position relative to the photographer and relative to other objects in space. Time, on the other hand, is mobile, making it a moving target for measurement and evaluation. As soon as one attempts to measure an instant, the moment has passed. Time does not stand still for evaluation. One cannot take a picture of time's movement, of *durée*. Duration is, therefore, not available for the themes and theories native to the sciences. Science must settle for an abstraction of time, which for Bergson is just the measurement of space.

Levinas retains Bergson's sense of the isolation of the instant and its isolation from duration and the flow of time, but he challenges Bergson's dismissal of the photographic instant from phenomenological significance (*EE* 21–22). Bergson thinks about the instant in terms of his overall understanding of time, *durée*. The instant is, for Bergson, a deficient expression of time, whereas Levinas proposes to think of time from the perspective of the instant: "We do not seek in an instant anything else other than the very dialectic of time" (73). Bergson points beyond the intellectual evaluation of an instant to something more intuitive and less mathematical.

For Bergson, we are to understand the nature of time by way of intuition, which he contrasts to the *intellect* (*CE* 267). Human evolution, he reasons, has rather randomly disposed humans to be a combination between intuition and intelligence. And according to Bergson, we have leaned far too heavily on our intellect and have underdeveloped our intuitive natures: "Intuition goes in the very direction of life, intellect goes in the inverse direction, and thus finds itself naturally in accordance with the movement of matter. A complete and perfect humanity would be that in which these two forms of conscious activity should attain their full development. And, between this [perfect] humanity and ours, we may conceive any number of possible stages, corresponding to all degrees imaginable of intelligence and of intuition.... In the humanity of which we are a part, intuition is, in fact, almost completely sacrificed to intellect" (267).

Traditional reflections on time have, therefore, blindly lined up a series of instants and pretended that time is the relationship between an infinite number of photographic nows. Instead, suggests Bergson, time as *durée* points to the way an instant freely gives birth to the next instant. Duration is the extension-in-time of an object or person, and it is unique to the subject. By continuing to type words on my keyboard, I conceive of a duration, a movement through time. Duration appears to science and mathematics like a measurable collection of instants, but only retrospectively, after my fingers stop moving and the event has become what it was. In the same way, the steady track of the moon across the sky can be evaluated by virtue of its spatial relationships, but it cannot be measurably evaluated by way of its duration. Time is the very mobility of the moon's movements, not the movements themselves. This makes time as duration *scientifically*, though not "spiritually," ineffable and indescribable.

By appealing to intuition, Bergson suggests that duration can be a useful way to talk about the inner life of individual objects and persons. The moon's inner life is unknowable, but a person can reflect on his or her unique experience of the moon's movement through time. This explains the widely different senses of time's passage that accompany human experience. To the enthusiastic and devoted student, a philosophy lecture can pass by in a flash. To the unengaged, bored student, the same *spatial* experience can have a long and dreadful duration. This diversity of experiences is due to the ecstatic nature of time. Time will not hold still long enough to become an object of study, and it must therefore be relegated to the realm of the intuition, or it will remain unintelligible.

Levinas had the utmost respect for Bergson's innovations concerning time. Levinas compares Bergson's attack on universal time to the "divine act of Jupiter" (Zeus), who dared to challenge Saturn's (Cronos's) appetite: "He was the first to raise a hand against the cold time of science before which all philosophers had bowed. In this universal tick-tock where impassive moments succeeded one another, like points in space where man disappears in a cloud of seconds, Bergson opposed the immediate data of consciousness, the becoming of our lives, the concrete *durée* of our lives."[38] Bergson is aware of the oddity of moving from an intellectual understanding of time to the more intuitive sense of time. He insists that one does not arrive at intuition by moving through intellect but by investigating that which is *missed* by intellectual evaluations.[39] In this way, Bergson has clearly paved the way for Heidegger's critique of the traditional thinking about time, though Heidegger is less generous than Levinas regarding his debt to Bergson. For example, Heidegger pays a backhanded tribute to Bergson by repeatedly stating that he is upending the history of thinking about time "including Bergson."[40] Though Levinas and Heidegger may share an equal debt to Bergson for opening up a reconsideration of time, it is Levinas that is the more careful and grateful reader of Bergson.

The Powerless Freedom of the Instant

Heidegger essentially agrees with Bergson that time is ec-static, that it is not a collection of states, and that philosophy has forgotten the slippery nature of time. Like Bergson, Heidegger suggests that we have been conditioned throughout the history of philosophy to think of time through the lens of spatial relations. They agree that the Aristotelian preference for eternal time helps occlude the more primitive sense of time. For Heidegger, as well as Bergson, this universalizing tendency pushes us away from recognizing the uniqueness of individualized existence and relegates us to an evaluation of the scientific sense of infinite, mechanized, abstract time. The critique of Bergson in *Being and Time* has the effect of heightening the originality of Heidegger's own treatment of time.[41] It is, in fact, on Bergson's suggestion that continental philosophy began to rethink the concept of time *apart* from the concept of space.[42]

Levinas points to these similarities in Bergson and Heidegger on their understandings of the instant. For both Heidegger and Bergson, the instant is radically individualized; it is utterly isolated from other instants, and it is capable of moving itself forward into the next instant. For Bergson this is *durée,* and for Heidegger it is care (*Sorge*).[43] In both cases, the instant is granted the internal power of procreation. But Levinas wonders if these thinkers have too quickly established the relationship between freedom and the isolated instant. Why should we presume that one instant gives birth to the next? Why grant the instant regenerative power, the virility of its own beginning (see *TO* 67)?

And here we see that Levinas is attempting to think more precisely about the concept of the instant than either Heidegger or Bergson attempted. He does so in part with the help of Husserl, who helped him see some of the limitations in Bergson's attempt to think about human consciousness. As early as 1929, Levinas was aware of the dominance of spatial imagery in Bergson, despite his emphasis on time. In a footnote to an essay on Husserl, Levinas suggests that we must "move beyond the alternative before which Bergson placed us: either consciousness must be studied like space, grasped by the intellect in well-defined concepts, or it must not be studied by the intellect" (*DEH* 182n20). Levinas does not wish to abandon intellection and pursue Bergsonian intuition, which he believes Bergson too quickly turns over to the realm of the spiritual.[44] In light of the full trajectory of instants, Bergson presumes that time is the mystery of their dynamism. But the instant remains, for Bergson, an abstraction useful only to science, since reality is "composed of the concrete élan of duration, ever turned to, and ever biting into, the future" (*EE* 72).

In an interesting correlation with Levinas's work in *Existence and Existents,* Bergson uses the instability of the photographic instant to make a strong argument for the "free will" of humans.[45] This move is strikingly similar to the one Levinas is making here in *Existence and Existents,* where he explains that the instant is the moment of "mastery" and "freedom," even if this freedom ends up being isolated and fleeting. The evasiveness of this freedom is not produced by time's passage but by the lack of anything or anyone over which to exercise this so-called freedom. One has mastery over the instant, but no dominion. This is an ironic, even paradoxical liberation, like an inmate freely rearranging a prison cell. The instant and its mastery are "an evanescence" (*EE* 73). Elisabeth Louise Thomas summarizes Levinas's treatment of the freedom of the instant: "Mastery is an interval, terminus or stance which harbours another event and does not describe the advent of a free being. The subject of hypostasis discovers it is not free but alone responsible for the consummation of infinity in each instant. This is a Promethean fatality—it cannot be evaded despite a certain power being invested in the subject of hypostasis. In the hypostasis of an instant in which master is manifested the *il y a* returns, the 'I' is bound to existence."[46]

Heidegger pauses longer than Bergson on the unique individuality of the instant, which is a cause for concern-filled focus on Dasein's ownmost possibilities. But the instant is again, in Heidegger, best understood according to its unique, ecstatic futurity. In *Existence and Existents,* Levinas is not fundamentally opposed to the possibility that an instant must be understood according to its futurity. In fact, he applauds Heidegger for challenging the ancient presumption that the relationship between time and existence is "evident and simple" (*EE* 72). Heidegger, sounding rather like Bergson, perceived the circular reasoning in philosophy's reflections on time (72). Levinas credits Heidegger for going beyond Bergson, even as he suspects that Heidegger has not moved far enough.[47] Indeed, Heidegger has allowed us to rethink the idea of an instant with renewed appreciation for its gravity for ontology. For him, who uses the terminology "moment of vision," the instant has a kind of pregnant and internal possibility that belongs to itself.[48] This unique possibility is part of its thrownness, of which Dasein becomes authentically aware in the present moment, characterized by "vision." This moment of vision analyzes thrownness according to projection. Thrownness and projection are evaluated together as two aspects of Dasein's authentic being-in-the-world. The moment, properly and authentically embraced, is accompanied by resoluteness, which "looks at those Situations which are possible in one's potentiality-for-Being-a-whole as disclosed in our anticipation of death" (*BT* 396).

In *Existence and Existents,* Levinas implicitly challenges the relationship between these two features of the instant. One is *thrown* into an instant, but need we assume without question that an instant gnaws and bites its way into the future (*EE* 72)? For Heidegger, as Levinas has perceived his work, the instant grants a perspective, or mood, in which Dasein can embrace its thrownness and project itself forward into the dizzying future. For Heidegger, this "moment" is laden with internal power and possibility.[49] Levinas reconfigures time according to the mystery of the solitary subject, much as Heidegger recommends, but even in this accord, Levinas is more reluctant than Heidegger to rush to any particular understanding of the past or the future. Levinas questions this power of the instant to dislodge itself forward into duration. In effect, he questions something Heidegger never considered—can the self, Dasein, make time *go*? Can *I* effect duration? Or must the instant, and therefore the isolated subject, be rescued from pure isolation?

After exploring the helplessness of insomnia and the affectivity with which one faces raw existence, Levinas proposes a reconsideration of time apart from the presupposition that it moves because the subject *makes it move*. He is suggesting that what is discovered in the experience of the *il y a* is something that dwells forever beneath our everyday experiences. In reasoning thus, Levinas joins himself up, if only temporarily, with a few voices from philosophical history who were unafraid to suggest that an instant needs external assistance to be reborn. "Unlike the theories of Bergson and of Heidegger, here [the instant] is devoid of the power to be beyond itself" (*EE* 74). The inability of an instant to "preserve itself in existence," for Levinas, gives rise to the doctrine of "continuous creation,"[50] which is taken up by René Descartes and Nicolas Malebranche: "The theory of continuous creation in Descartes and Malebranche refers, on the phenomenal level, to the incapacity of an instant to join itself up with the following instant. Unlike the theories of Bergson and Heidegger, here it is devoid of the power to be beyond itself. An instant is in this specific sense without dynamism" (74).

Heidegger and Bergson, whatever their differences elsewhere, still equip the subject with the fundamental power of motion, a potentiality that is internal to the subject. Bergson thinks the doctrine of continual creation is just a distortion required by mathematical minds that cannot imagine how one instant could move to another. Yet Levinas finds intriguing insights in Melebranche and Descartes regarding the incapacity of creatures to create another moment in which to live. For Malebranche, this amounts to a kind of occasionalism; he believes that particular bodies cannot be efficient causes and therefore all movement is a form of divine ordination.[51] Levinas writes of Malebranche,

"But what is profound in Malebranche's views is that instead of situating the true dependence of creation on the Creator in its origin and in its liability to be reduced to nothingness by a new decree of the Creator, Malebranche places it in its inability to preserve itself in existence, in its need to resort to divine efficacy at each instant" (*EE* 74). Levinas will not settle for Malebranche's straightforward theological resolution to the pressure of existence. But Levinas realizes that he is hardly the first philosopher to ponder the power of self-preservation. Levinas presents a stark contrast between Malebranche and the self-origination of Heidegger's moment of vision—Malebranche is challenging the very capacity of Dasein to project. He upends the power of a *durée* to endure.

Malebranche's statements are extensions of similar reflections by Descartes, who points out a difference between the fact that an object has local motion and the very principle force of motion itself. We can speak easily enough about local causes that give rise to a particular movement, but for Descartes, this says nothing about the overarching production of motion in the world.[52] Things move, of course, but this does not provide any necessary or direct understanding of motion itself. Descartes would not reflect on "the force or action that produces [motion]" in his official publications.[53] He did, however, develop a theological explanation for this "production" in his written correspondence with Henry More. In a 1649 letter to More, Descartes points to God as the force of motion, and he explains his reluctance to publish this theory for fear of depicting God as the "*anima mundi* united to the matter of the world."[54]

Thus, both Descartes and Malebranche provide us with a phenomenological understanding of the instant that underscores the "incapacity of an instant to join itself up with the following instant" (*EE* 73). By revisiting the ideas of Descartes and Malebranche, Levinas is in this manner challenging the capacity of traditional phenomenology to treat the concept of time. In doing so, he is faithfully bearing forward the torch of Bergson, who wished to divest science of the delusion that it could measure and quantify time in the same way it had objectified space. But Levinas is also turning against Bergson, who looked past the instant and to *durée* in order to establish free will and the freedom of the subject.[55]

Phenomenology, at least in the sense advocated by Husserl, remains committed to a certain sense of scientific analysis. Both Malebranche and Descartes excuse the question of motion from the scope of scientific inquiry by introducing a transcendent force, God, to explain the movement of one instant into the next. But unlike Bergson, Descartes, and Malebranche, Levinas wishes to retain a place for time within phenomenology. In order to do so, however, Levinas will need to introduce a foreign concept into phenomenological inquiry: transcendence. But how can an instant, on the phenomenal level, bear witness to the transcendent? It is not just Descartes and Bergson who have fallen silent at the face of this mystery; philosophy and phenomenology have routinely hesitated to introduce the intractable concept of transcendence. And for Levinas, the type of transcendence offered by the notion of ecstatic time in Heidegger will not suffice to explain the principle of motion that allows Dasein to move.

For Levinas the instant is a kind of prison, a position of isolation that does not contain *time* as its property. Bergson and Heidegger challenged the universalizing and abstracting function of world time, but they nevertheless imparted a sense of time to the individual and the instant. It is worth pausing to note that Levinas has at least identified an unnamed presumption in the Bergsonian and Heideggerian critique of the traditional treatment of time: on what ground are we to suppose that each instant bears the property of time? And has not Heidegger, for all his celebrated critique of philosophy's tacit subordination of time to being, repeated Aristotle's error by letting time be a feature of the moment of vision? Levinas is raising a question about the hidden assumptions of Heideggerian time, and the question is warranted by Heidegger's investigation itself. In this vein, Levinas suggests that time arises as a gift, as the possibility of pardon, as reason for *hope*.

Levinas's language has already begun pointing at subjectivity; the instant is correlative to the self, rooted and lodged in a timeless eternity until redeemed from the subject's imprisonment by the movement of time. For the Platonic and especially Neoplatonic tradition, as for the mythology of "Saturn devouring his children," time stands in negative opposition to the beauty of impassible eternity. Temporality is cast as the enemy of perfection, and it therefore comes to kill rather than give life. But caught in the isolation of an instant, what the individual needs most is *time:* "Time, far from constituting the tragic, shall perhaps be able to deliver us from it" (*EE* 78).

Hope

Levinas introduces the phrase "economic time" to describe the future that is apparent and latent in every given instant. This is the sense of time that fits our life in the world, our transactions and interactions within existence. Then Levinas embraces and extends the Husserlian, Heideggerian, and Bergsonian suspicions that philosophy has obfuscated the concrete, individualized relation to time. And Levinas, using Descartes and Malebranche for support, proposes that time is *external* to the self, to the instant, to the moment. The *present,* as Levinas begins to call the instant, is irreparable. This is an intriguing suggestion, and in some ways, it demonstrates how far Levinas now situates himself from traditional philosophy. If philosophy has only ever managed to think time through the present, as Heidegger suspects, then Levinas's proposal is almost the exact opposite. Time is the one thing that the instant, and therefore the subject, *lacks.*

The instant, for Levinas, carries the gravity of its past and the limitations of its future, unable to alter or repair what has come to pass. The present may contain optimism, as a person who surveys pathways and yearns toward the most favorable of static alternatives. But hope scandalizes the present by defying the "gravity of the instant in which it occurs" (*EE* 91). As Lingis suggests in his introduction, hope relates not to an inherent potentiality of a given instant but to what a moment cannot contain or engender: the "possibility of beginning anew" (xxiv). The future is not the fulfillment of possibilities but the overturning of present possibilities in favor of the coming of a future that the instant cannot know as possible. This crafts a kind of eschatological redemption of the instant. As Lingis summarizes, for "Levinas, the lure of the future is essentially the lure of pardon" (xxiv). This pushes Levinas up against the limits of philosophy, let alone phenomenology, restricting philosophy's scope to the tightening confines of an instant. To know anything, to sense anything, requires that Levinas appeal to radical transcendence, and the form of transcendence Levinas envisions will attempt to outstrip the versions that philosophy had previously utilized.

The need to transcend the instant is configured by Levinas as radical *hope*. Hope settles for neither a static interpretation of the future nor a fixed past. The moxie of hope is based in the fact that it directs itself *from* the instant toward both the past and the future. To have hope is to settle for neither an irreparable past nor any anticipatable future. And because time itself arrives as a gift from beyond the instant, it defies the logical closure of the moment. Hope therefore liberates the present from the fixed channels of economic time, but in doing so, it unravels what seemed to be a fixed past. This is contrasted by Levinas with "compensating time," which is represented by the drying of tears and, at best, a "forgetfulness" that numbs the pain of the past (*EE* 92). Economic time can compensate for damages, like a day of leisure that "does not sanctify the week, but compensates for it" (92). Hope, suggests Levinas, transcends compensation: "But this compensating time is not enough for hope. For it is not enough that tears be wiped away or death avenged; no tear is to be lost, no death be without a resurrection. Hope then is not satisfied with a time composed of separate instants given to an ego that traverses them so as to gather in the following instant, as impersonal as the first one, the wages of its pain. The true object of hope is the Messiah, or salvation" (93).

Hope refuses to resign the future to an extension of the past contained in the weight of the instant. To hope in this way is to take leave of rationalism, which can only rearrange the facts of the instant. Levinas points, as if to confirm his rebellion against rationalism, to the idea of a resurrection to exemplify the way an instant is given new life through what it can *hope for* but never *achieve*.[56] What better model for the insufficiency of the instant than a complete pardon from the momentous weight of past injustices? Not coincidentally, these images are received in affectivity. Levinas's clever inversion of the freedom of the Bergsonian instant requires that salvation come from otherwise than within the confines of the subject, locked in its present. As Didier Franck writes, "If the whole of the subject is found here, in the present instant, then resurrection can come only from an other person who can pardon me of the evil that I am. This is why the irruption of the other person signifies the rupture of the ego, for to be otherwise than oneself is to be pardoned of Being and to no longer be definitive existence. Freedom thus consists 'in having one's being *pardoned*.'"[57] Franck emphasizes his interpretation that "Being is evil" and claims that this idea "presupposed the whole length of Levinas's thought."[58] This critique, however, voiced here by Franck and ardently elsewhere by Phillip Blond, is questionable in light of Levinas's overall discussions of being.[59] But Franck's dour assessment of Levinasian being allows Franck to emphasize the sheer need for a pardon that comes from outside the subject. One does not grant oneself pardon, nor resurrect one's own death. Like time, pardon can only be received as a *gift*. As Levinas winds down *Existence and Existents,* he begins to escalate the intersubjective nature of these reflections. Time is not a gift from an anonymous being, not a recovery of some internal potentiality, nor a discovery made by the self in the world of matter. Time is a gift from the other; it is the other who gives me time, and in giving time, gives life (*EE* 96).

In this sense, hope is never internally fulfilled; one hopes not in what can come to pass but in what cannot come to pass. Hope ushers the subject to the brink of salvation, but that salvation must come from beyond the subject, beyond the instant, beyond history. As Jeffrey Bloechl puts it, "In hope, says Levinas: I am brought just to the threshold of transcendence—there and no further."[60] And the figure of transcendence, even here in this early work, is configured eschatologically, as the Messiah. Levinas invokes the concept of the Messiah here to indicate a radical exteriority of salvation that can be hoped for but never attained, an exteriority that cannot even be properly named according to the vocabulary available to the subject in the instant. Hope in this register is not hope in a possible resolution to pain. Hope here is not about reparation or an "*afterwards* in economic time" (*EE* 93). To hope is to hope in the repair of the irreparable, the redemption of a present that is heavy with its past and weighed down by its burdens and suffering (93–94).

In order to articulate the function of hope and its relationship to time, Levinas turns to a theme that he will expand in *Totality and Infinity:* the caress (cf. *TI* 34–35, 258–59). In the moment of suffering, Levinas points out a caress is not compensation for pain but something categorically different. The consoler does not promise an end to the suffering, "does not announce any compensation, and in its very contact, is not concerned with what is to come *afterwards* in economic time" (*EE* 93). Rather, the caress addresses and liberates the "very instant of physical pain, which is no longer condemned to itself" (93). The caress brings "fresh air," liberation from the insularity of the instant. In this way, the other person, approaching in a caress, opens up the movement of time. Hope irrupts at the moment of caress. Thus for Levinas, "to hope then is to hope for the reparation of the irreparable" (93). Hope is a manifestation of a broken economy, an escape from the inescapability of the present. Levinas describes hope as: "the impossible exigency for salvation which must concern the very instant of pain, and not only compensate for it. Does not the essence of time consist in responding to the exigency for salvation? Does not the analysis of economic time, exterior to the subject, cover over the essential structure of time by which the present is not only indemnified, but resurrected? Is not the future above all a resurrection of the present?" (94). Time is therefore bound to hope, which is emblemized as the broken economy of a caress.

This philosophical understanding of hope and the instant provides the foundation for Levinas's understanding of time. Richard Cohen connects Levinas's understanding of the instant to "the subject's primordial 'enjoyment' (*jouissance*), which is again both an independence, a happiness, and a dependence, a burden."[61] Even if the instant is where Levinas's unique considerations about time have their genesis, we can expect even this original independence to be revoked in his later book *Otherwise than Being*. The affectivity of the instant, already stark in *Existence and Existents,* will be radicalized in Levinas's final writings, transformed into a passivity too ancient for even the raw affectivity of the instant.

Already in *Existence and Existents,* Levinas is prepared to boldly associate time with the other person: "Is not sociality... time itself? The dialectic of time is the very dialectic of the relationship with the other" (*EE* 96). The alterity of the next instant "comes to me only from the other." Levinas's philosophy builds from this explicit break from Heidegger and Bergson, and from the history of philosophy that these thinkers have already critiqued. We find the concept of time, in *Existence and Existents,* to be utterly critical for understanding the disposition of the subject with respect to the world, to the other, and to salvation. The pathway into life leads through the gift of time, the gift of pardon, and the gift of freedom, all of which reach the subject in the utter affectivity of an instant.

Light, Sound, and Words

Before moving on to *Time and the Other,* Levinas's second 1947 publication, it is helpful to follow these themes into some of the less celebrated articles Levinas wrote in this same period. First, then, we examine the nature of the instant as it relates to the tension between light and sound. The article "The Transcendence of Words: On Michel Leiris's *Biffures*" (1948) includes an interesting, if passing, reflection on the character Robinson Crusoe from Daniel Defoe. The essay is a review of an autobiographical work by the French author Michel Leiris.

Having explored in *Existence and Existents* the way light relates to knowledge in Western philosophy, Levinas makes one of his first forays into understanding sound and language. Light, by which Levinas primarily means the illumination that enables *sight,* gives the impression of "possession at a distance" (*EE* 38). For the world to be knowable, it must be illuminated, and since Plato, this preference for light has correlated with a distrust of the other senses. We know things by their illumination. In Heidegger, the unveiling of truth is a kind of movement into the light. As Levinas explains, "Care in Heidegger, which is not founded on perception, nonetheless comprises an illumination which makes of it comprehension and thought" (40). Optics and physics aside, light is philosophically significant because it serves to provide access to phenomena through their appearance in the world of visual perception. So for Levinas, "light makes objects in the world, that is, makes them belong to us" (40).

Traditional philosophy has made heavy use of the comparison between illumination and knowing. The eyes create a perceivable horizon, unlike the ears and the other senses. Sight allows us to relate to events without being caught up in them; light allows voyeurism, an ability to withdraw unscathed from an event while nevertheless retaining the event as knowledge. This sort of knowing is a "way of being on the hither side of being" (*EE* 42). In *Existence and Existents,* Levinas questions the manner in which light covers up and obscures a more primitive sense of being than our eyes can perceive or our sight-centered epistemologies can detect; in "The Transcendence of Words," Levinas moves a step further to discuss the phenomenon of *sound.*

Vision, claims Levinas, gives to the subject a world that is "completely here, and self-sufficient" (*OS* 147). Sound, on the other hand, does not offer itself as a relation to the whole of being. Instead, sound "offers itself to intuition." Sight has no need for critique, no need to second-guess its impressions. Through sight, the world is taken into the possession of the seer. Sound, however, "can be given" (147). Regardless of the accuracy of Levinas's phenomenological bifurcation between human senses, he is offering here an illustration of the temporal difference between the world that is taken up in an instant, sight, and the world that comes from beyond. Sight, as he defines it, is contained in a "here" that mirrors the heaviness of the frozen moment. But sound hails from beyond this enclosure; it comes from the other. The ears carry messages that disturb the eyes and disrupt the beautiful and pacific vision toward which the eyes choose to gaze. Sound is less easily screened, more likely to unsettle and surprise; it is "all repercussion, outburst, scandal" (147). In *Existence and Existents,* Levinas gives a similar designation to describe the visionless night of insomnia: "We could say that the night is the very experience of the *there is,* if the term experience were not inapplicable to a situation which involves the total exclusion of light" (*EE* 52). Without sight, one cannot measure the approach of the world in advance; hearing is a paradigm of affectivity.

The tranquility of the picture, painting, and photograph are not quite satisfactory; sight fails to satisfy "a need to enter into relation with someone, despite and above the achievement and the peace of the beautiful" (*OS* 147). The reverberations of nature's sounds will not suffice because we desire *critique,* because they are not verbal, not symbols, not words. "Pure sound is the word" (148), precisely because it comes from beyond that which is given by light. One need not direct a literal or proverbial gaze toward the other in order to receive the gift of this symbol; it comes to the subject regardless of posture or vigilance. Throughout his career, Levinas will retain this suspicion that light and sight contain a steady preference for "being" and presence.

Enter Robinson Crusoe, whose fictional shipwreck and island isolation provides a perfect analogy for Levinas's 1947 deliberations about time, the instant, sight, and sound. Crusoe is impoverished on his island not for lack of vision but for lack of spoken words. He surrounds himself with ties to civilization "through his use of utensils, his morality, and his calendar" (*OS* 148). But these ties do not suffice. Levinas does not mention that Crusoe struggles mightily to train a parrot to speak, teaching it the sympathetic phrase, "poor Robin Crusoe."[62] The parrot, obviously, can reproduce the sounds of words but not truly speak. His words are noises of nature, which disappoint Crusoe's desire for the "scandal" of sound (147). The words of the parrot resonate with being and are a part of the instant in which Crusoe is riveted. They are not from beyond, and they are not "outburst;" the bird chirps "poor Robin Crusoe," but its noises are the "ineffable sadness of echoes" (148).

Levinas notices that on this island Crusoe endures a circular existence in which the only words he hears are echoes of his own. He freezes, locked in a world that is noisy but silent, beautiful but empty. Despite the echoes of social life and civilization that Crusoe nurtures in his years of isolation, "he experiences in meeting Man Friday the greatest event of his insular life—in which a man who speaks replaces the ineffable sadness of echoes" (*OS* 148). The written word, Levinas already claims in 1948, transforms and disfigures verbal communication into "frozen words" that are mere "documents and vestiges" of speech. In *Otherwise than Being,* this very theme will blossom into a robust philosophy of the diachronic relationship between saying and the said. But here, we already find Levinas expressing with new clarity: "The subject who speaks does not place the world in relation to himself, nor place himself purely and simply at the heart of his own spectacle, as does the artist, but in relation to the Other. This privilege of the Other ceases being incomprehensible once we admit that the primary fact of existence is neither the *in itself,* nor the *for itself,* but the *'for the Other.'*... By the proffered word, the subject that posits himself exposes himself and, in a way, prays" (149).

To correlate the prolonged discussion of the instant in *Existence and Existents* and the reflections on sound in "The Transcendence of Words," it is the yet unintelligible words of Friday that finally give *time* to Crusoe. For unlike his bird, or the duplications of society he constructed on his island home, Friday could *speak* to him. Friday's first words are alien, foreign, dangerous, and even frightening. Friday must learn to speak English, of course, and unlearn the habits of cannibalism and barbarism. Here Levinas misses the chance to use Defoe's story to further underscore the radical nature of transcendence because Friday's words represent not the repercussions of Crusoe's memories of society and language, but something radically *other.* Crusoe endeavors to colonize Friday, to domesticate him to the language of familiarity.

Levinas is particularly interested in how Friday introduces a sociality to Crusoe's world. But he is the other, and as such, he transforms what Crusoe now does on the island into human existence. The other transcends the frozen present of Crusoe's existence and brings it forward. Crusoe is freed by Friday to be *for* Friday, regardless of how we may feel about Defoe's colonialism. In *Time and the Other,* Levinas will return to Crusoe, again briefly, and this time point to a sense of solitude even deeper than the "factual isolation" of the fictional castaway (*TO* 43).

For Crusoe, the beauty of solitary existence is transformed into terror at the absence of words, which is for Levinas an absence of transcendence. In an intriguing connection to Bergson, Levinas claims that sound appeals to intuition, that aspect of humanity that Bergson considers overwhelmed by intellect. Vision, in this essay, is the work of the intellect. But instead of turning to the realm of the spiritual, as Bergson seems to advocate in *Creative Evolution,* Levinas turns to another manner of sensibility: sound.

Image, Art, and Idolatry

Even here, as Levinas makes use of the novel *Robinson Crusoe* to underscore the unnoticed primacy of sound, we need to pause to track an important transition in Levinas's thought that plays out on the question of *art.* Levinas celebrates language in "The Transcendence of Words," but this appreciation does not extend to sung or spoken forms of art. Though these are full of sound, which he praises as appealing to "intuition," they fail to do more than bewitch: "all the arts, even the sonorous ones, create silence" (*OS* 147). In this resolute rejection of art and representation, Levinas sustains a discourse that begins in his critique of Husserl and eventually expands to hold a central place in Levinas's body of work. Representation is always secondary, considered, re-presented. And as such, it seduces us into believing in the relationship between the represented object and its referent. Over Levinas's career, the invective against representation will become increasingly associated with time in his later years. Representation, after all, *aims* to make things present. The very modality of art, as Levinas sees it, functions by way of domesticating the transcendence of that which is represented.

Levinas displays a thoughtful, if thoroughly scathing engagement of the value of art in his 1948 essay "Reality and Its Shadow." This essay is notable for its subtle encounter with Sartre and its further development of the phenomenology of the instant. But more importantly for this study, the essay helps confirm the central thesis of this chapter: Levinas has begun to develop intersubjective, ethical philosophy with the tools offered by his original understanding of time.

His harsh analysis of art hinges on the way art covers up, distorts, and conceals the phenomenon of *time*. Art forces things into the present.

Levinas begins this essay by questioning some presuppositions about the function of art and the way art is evaluated. He appears to have Sartre in his crosshairs as he challenges the power and transcendence of artistic work. Levinas demonstrates concern with the elevation of the image in the realm of art and art criticism. His concern in "Reality and Its Shadow," at any rate, is to undermine the priority Sartre gives to images and imagination. What Sartre attempts in *The Imaginary* (1940) is a "phenomenology of the image."[63] This meticulous discussion of perception, impression, images, and imagination provides a foundation for Sartre's later existentialist work on the human condition.[64]

Sartre claims that images and imaginary objects are incapable of teaching us anything. What we experience when we gaze at an image is a conglomeration of our own past impressions.[65] It is important, therefore, to avoid the temptation of thinking we have arrived at some new knowledge by gazing at an image. One sees in a painting what one intends to see. The impression is steered by internal *choice*, which for Sartre will lead to an unprecedented expression of irrevocable ontological freedom. Nothing and no one can wrench us free of our imaginations. Yet upon contemplation, we discover the hidden complexities of the relationship between the artwork and the object it attempts to represent. In contemplating artwork, what one discovers is *oneself*. Images critique the observer by revealing not what a person intends to see but that which one cannot anticipate and for which one cannot fully prepare. Sartre is less interested in the image than what *reflection* on the image can teach us.

However, what is unique about an encounter with other people is that unlike other objects, people *look back at me*. This experience of being seen makes me aware of myself in a way no other object can, for I can never see myself as an "other." For Sartre, this means that when I reflect on another person, I have already taken in the vision of that person as object and possession.[66] I can only in a secondary sense regard that person as person, but, even in this, the other remains a feature of my experience, a part of my self-awareness. What appears before me is a series of visual images, and the other person becomes a particularly complex group of images.[67] For Sartre, the other person is merely the strongest challenge for the imagination, the most entertaining and engrossing of myths. Seeing is possession, and we have the terrifying freedom of doing as we please with our possessions.[68]

Sartre positions each subject within a private consciousness, surrounded by a host of images *taken* for the purpose of reflection.[69] We gather, arrange, and interpret the objects around us, even if our imagination always fails in the attempt to reproduce the reality of what is imagined. But artwork provides a strange instance for this process of gathering and possessing. Artwork is neither *object* nor *person* but a hybrid of these. And one can do nothing with artwork without imagination. It would be folly, of course, to embrace a photograph of my daughter and suppose that I had hugged her. It would be folly as well to treat a photograph of my daughter as though it were a mere object, a collection of unrelated pixels. Art calls on our imagination. The consciousness of the artwork (Sartre calls it *image consciousness*) is a product of the artist's attempt to provide an image of his or her internal imagination. The artist has always *failed*, of course, in the attempt to capture the object in a representation. But, as Jeffrey Bloechl summarizes, "This very failure and the effect it has on the spectator constitutes what is for Sartre the lasting value of art."[70] Images are vital to human existence because they require imagination, which is the fundamental property of human consciousness. To be human is to imagine, and to contemplate images is to become oneself.

Sartre's thesis alarms Levinas. First, Sartre has essentially agreed with Levinas's suggestions in *Existence and Existents* and "The Transcendence of Words" that sight constitutes a possessive grasping. But Sartre has ascribed high value and authority to what Levinas deems a fallen, artificial manifestation of the world. For Levinas, art is a disengagement from the world, a suspension of reality that freezes time into an instant for the sake of observation. What Sartre (and presumably those Greeks with a taste for aesthetics) overlooks is a false kinship between the eternity of a statue and the Platonic eternity of ideals. Auguste Rodin's *The Thinker* is frozen and unchanging and not because it has been sustained by museum curators; the frozen moment of *The Thinker* is eternal in a sense similar to the eternity of any instant. As Levinas puts it, "A statue realizes the paradox of an instant that endures without a future" (*CPP* 9). But *real* instants, the sort discussed in *Existents and Existents,* wait for the salvation that comes through "time and the other" (*EE* 96). *The Thinker* is a special kind of tragedy; his posture of concern and contemplation was intended to depict Dante's own reflections at the sight of hell below. *The Thinker,* therefore, sits forever with his eyes locked on hell; he will never rise, walk, or act upon the scene he observes. There is a future implied by this contemplation, but it is a future "forever to come" (*CPP* 9). Levinas contends that the most important thing about an instant is its *evanescence,* its redemption and salvation in time. A statue pretends to halt reality, it leaves the impression of a frozen present, in all the horror of the instant expressed in *Existence and Existents.* A statue strips the real of its most important feature: time. In the process, the absurd hope in the renewal of time is obscured by the glorification of the now that is cast in marble.

For Levinas, the contemplation of artistic images, so pivotal for Sartre, presses us further from reality rather than toward reality. For an image does not represent immobilized reality, given over to our careful analysis. Representations are a fall from reality in that they foster the impression that we control the reality that we see. Art performs a crafty seduction, exploiting our weakness for mastery and perspective.

In a sense, an artistic image gives the impression of the mastery of time. We feel, when viewing *The Thinker,* that we have gone beyond his pensive moment, that we are disengaged from his pose. And we are accustomed to thinking that this disengagement allows us to "go *beyond,* toward the region of Platonic ideas and toward the eternal which towers above the world" (*CPP* 2).[71] But what is missing in *The Thinker* is more important than what is present. Robbed of the gift of futurity, stuck in a "meanwhile," *The Thinker* becomes something "never finished, still enduring—something inhuman and monstrous" (11). The fate of the statue mirrors the fate of the subject, who is imprisoned by the "instant." My present is just as devoid of futurity; I cannot move until I am *given* time. *The Thinker* reproduces the terrible affectivity and impotence of the present.

Art cannot, for Levinas, guide us beyond being; it settles for trivial and spurious muddling within the economy of being. And this makes art a great indulgence, a frivolous venture into the world of fate and hopelessness. Has a more scathing indictment of artistic enjoyment been delivered than Levinas's stunning lines in "Reality and Its Shadow"? He writes: "There is something wicked and egoist and cowardly in artistic enjoyment. There are times when one can be ashamed of it, as of feasting during a plague" (*CPP* 12).

Levinas, therefore, follows Plato in exiling the poet, at least from the core of philosophy. "For Levinas, as for Plato," writes Bloechl, "philosophy is our guardian against the treachery of form and images. It recalls us from the pedestal or stage."[72] The interplay between art and imagination, so important to Sartre's depiction of individual freedom, is for Levinas an evasion of the other, a stopping-short of reality. By halting in the examination of a timeless moment, "art lets go of the prey for the shadow" (*CPP* 12). Sartre prizes the contemplation of a time-stripped image, an image wrapped into the private world of the ego. This, for Levinas, is akin to idolatry; it mistakes a trace for its maker. When monotheism decries idolatry, it safeguards God from being trapped in the possessive, grasping "present" of the visible. Art, granted any weight whatsoever beyond frivolity, is idolatry. Art is pagan. This inherent paganism need not spell art's doom, but it certainly undermines the high importance of images as advocated by the young Sartre.

At stake here is the nature of freedom and the relationship between time and freedom. For Sartre, the instant is inherently *free*, full of self-mastery. The internalized moment of imagination, which suspends the world in order to reflect on it, is the most irrevocable and sturdy of human freedoms. For Levinas, this repeats an ancient philosophical error. Although he agrees that an instant is marked by a mastery of the existent over raw existence, he will not concede that this freedom is blissful or even entirely liberated. Instead, the instant is weighed down by the existence it holds in the present. The instant also carries a "melancholy over the eternal course of things" indicated by "the fact that the present contains a knot which its fading out will not untie" (*EE* 77). Art celebrates the autonomy of our epistemology. We are indeed frightfully free to know and see the world in the isolation of autonomy. But this freedom is an evasion. Levinas claims that art is by nature necessarily disengaged, that art celebrates an embarrassingly luxurious vacation from reality, and that the freedom art secures evades the "world of initiative and responsibility" (*CPP* 12).

Jill Robbins has pointed to another layer to Levinas's treatment of art that warrants mention at this point. There is a clear objective on Levinas's part to disparage the relationship between artistic expression and the ethical relation, and this critique of the value of art is mostly sustained throughout his career. Another story is told, however, when one looks at Levinas's own work and its routine use of artistry, poetic expression, not to mention his repeated references to works of art to aid his arguments.[73] Robbins demonstrates that there is a deeper tension at work in Levinas's mostly negative comments on art. This problem is exacerbated by Levinas's style of argument; he uses a particularly poetic hand to disparage poetry.[74] Robbins calls this element in his work a "tension," though she does not reflect at length on the way this instability is addressed in Levinas's final works. Gabriel Riera reflects more directly on new developments in *Otherwise than Being*, particularly the distinction between the saying and the said, which require a much more nuanced analysis of poetic and prophetic language.[75]

Levinas's critique of art, especially in "Reality and Its Shadow," is best seen as a byproduct of an agenda that now takes clear precedence over any appreciation for aesthetics. Responsibility will eventually become the most important term in Levinas's philosophy. We can note at this point, however, that for Levinas the evasion of responsibility is already a feature of evading *time*. Levinas's critique of art calls for the renewed importance of art criticism, which should keep us mindful of the triviality of art and the everlasting Platonic ideals that are timeless in a way that no image can render. Unlike art, which seems to induce indolence and irresponsibility, he believes that art criticism exercises "the muscles of the mind" (*CPP* 13). One can justifiably question whether Levinas has effectively engaged and dismissed all forms of artistic expression, but regardless of one's opinion of Levinas's derogatory critique of images, "Reality and Its Shadow" is abundantly clear about what separates the real from the artificial: time. To discuss the true exegesis of art, Levinas claims, would require exceeding the discussion of art and addressing the logic that drives exegesis. To do this, he concludes, "one would have to introduce the perspective of the relation with the other without which being could not be told in its reality, that is, in its time" (13).

Though Levinas is clearly taking issue with Sartre in this essay, there is a detectable reaction to Heidegger present here as well. For Heidegger, Dasein's most important sense appears to be vision. He uses the words *vision* and *look* to refer to the way Dasein analyzes those "Situations which are possible in one's potentiality-for-Being-a-whole as disclosed in our anticipation of death" (*BT* 396). Authenticity is rooted in "clear vision" for Dasein's possibilities.[76] In building a case against vision and images, Levinas is undermining the ability of Dasein to escape its enchainment to itself.

The central thrust of this essay, however, undermines Sartre and Heidegger by positioning the exercises of sight and imagination as a feature of the present. For Sartre, imagination is, after all, the paradigm of self-presence. For Heidegger, being-toward-death relies on a moment of vision. But by miring imagination in the present, Sartre has sealed it off and given it over to the whims of fate. For Levinas to press forward his central thesis, the other person must present more than just a particularly challenging *image* for me to appropriate. So it is by tearing into Sartre's sense of the imaginary that Levinas wishes to show that the encounter with the other bears a powerful, central feature that is absent in art and imagination. This feature is the twofold relation of the subject, to time and to the other.

THREE

From Darkness to the Other

> The aim of these lectures is to show that time is not the achievement of an isolated and lone subject, but that it is the very relationship of the subject with the Other.
> —Emmanuel Levinas, *Time and the Other*

In a note to himself, probably written in 1942 in the Nazi *stalag,* Levinas outlined some of the philosophical work he had in front of him. The top three items on this note were "1. Being and Nothingness, 2. Time, 3. Rosenzweig."[1] While Levinas desired to engage Sartre's work prior to publishing *Existence and Existents* (*EE* xxvii), it was not until 1946 that he read Sartre's magnum opus. Traces of this reading appear in the four-part lecture series published as *Time and the Other;* however, in this text, his other priorities clearly take precedence as Levinas furthers his philosophy of time and the influence of Rosenzweig on his thought can already be seen.

Existence and Existents concluded with the invocation of a profound *need* for intersubjectivity; in *Time and the Other,* intersubjectivity becomes the principle interest. Few would contest that this transition into the realm of ethics and intersubjectivity is a permanent one for Levinas. The difference between Levinas's *ideas* found in these two 1947 works may be subtle, but the difference in structure between them demonstrates an escalating interest in the way philosophy addresses lived, intersubjective relations, particularly observed by his reference to the sick, the widow, and the orphan. Additionally noteworthy themes in *Time and the Other* include Levinas's bold claim to depart from Parmenidean thought, his critique of a number of key Heideggerian concepts, and the role Rosenzweig's thought plays in this text.

In *Time and the Other,* Levinas provides his first extended treatment of the intersection between time and ethics. By claiming that "time is not the achievement of an isolated and lone subject, but that it is the very relationship of the subject with the other" (*TO* 39), Levinas is declaring something that will be expressed again more radically in the latter stages of his career. In *Existence and Existents,* Levinas moves through a discussion of time, or the lack of time in the instant, and concludes that something alterior to the self is necessary to rescue the ego from its captivity in the instant. Levinas concludes that we are powerless in the instant, that "Time and the other are necessary for the liberation from it" (*EE* 104). *Existence and Existents* seeks to demonstrate that philosophy has missed something important, something obscured by intentionality and activity. The virility of my daily, visual grasp of the world can make the other trivial, as someone that I only experience things alongside, a fellow master of the universe of data and sensation. But this is "the gravest sin," Levinas claims (101). In the *Time and the Other* lectures, Levinas turns his attention with specificity to that which we *need* in the powerlessness of our present: the other person. The significance of the other person here emerges and reveals itself to be far more important than any other phenomenon external to the isolated self.

This transition also sharpens Levinas's turn away from Heidegger and appears to indicate an early reliance on Rosenzweig. However, with Rosenzweig we must initially operate by inference. Rosenzweig is "too often present to be cited" in *Totality and Infinity* (*TI* 28), and he appears in Levinas's philosophical *à faire* list in the early 1940s. Clearly his influence on Levinas's work begins long before 1961. Indeed, an insightful exegesis of *Time and the Other* by translator and editor Richard Cohen reveals an early but pivotal dependency on Rosenzweig, particularly for the sake of thinking beyond Heidegger.

To Break with Parmenides

First, we must note the daring scope of Levinas's argument in *Time and the Other,* which is rather breathtaking. He outlines his thesis by promising to think of the dialectic of time in a way that is "in any case not Hegelian" and "if this can be dared, break with Parmenides" (*TO* 42).[2] The promised break with Parmenides is bold indeed, and Levinas is aware of his audacity. For Levinas, this is a pledge to abandon one of the most steady and reliable tools in Western philosophy. He claims that since Parmenides, philosophers have confidently examined a part of being, or one instance of being, through the lens of the whole of being. Philosophy has leaned, in one form or another, on "the unity of being proclaimed by Parmenides" (85). All pluralism, for Parmenides, must merge into an ultimate unity.[3] Basic, rational principles of some stripe must unify the many into one. In this sense, as Levinas will take pains to point out, Hegel is the paradigmatic fulfillment of the Parmenidean vision.

According to Levinas, philosophy has borne forward, at least since Parmenides, a deep urgency to explain plurality according to common principles. The assumption is that with sufficient perspective, all ideas, persons, diversity, and difference can be perceived according to a common relationship to the totality of being. The key to philosophy in this vein is the development of a vast perspective, a vantage from which all ideas, persons, diversity, and difference can be perceived according to some shared truth. Thus, Levinas opens *Time and the Other* by challenging Parmenides and his descendant Hegel. This promised philosophical revolution will require many decades of development, restatement, and reconfiguration. Parmenides is an intriguing choice of foes, for he denies any *past* or *future* for being; being just *is*.[4]

Levinas chooses to "break with Parmenides" on precisely the question of solitude and time. According to Levinas, a key misstep for Western philosophy after Parmenides has been the simplification of time, the reduction of the future to a feature of the present. To fail to see time as fundamentally alterior to the self is to reduce the whole of the phenomenal world to the metaphysics of presence, to the logic of the instant. Levinas has launched an attack on philosophy's presumption that time is a collection of universally shared instants progressing eternally in all directions.

Levinas, for his part, is suggesting a kind of plurality that "does not merge into unity" (*TO* 42). Cohen notes the gravity and audacity of this move: "The structure of separation puts Levinas in a difficult position relative to philosophy and its history, which has perhaps defined itself from its inception as intellectual vision of the one, the whole, the comprehensive. Levinas attempts to both reject Parmenides (the One) and remain a philosopher" (45n9).[5] This attack on philosophy since Parmenides is reminiscent of similar language used repeatedly by the Jewish philosopher and theologian Franz Rosenzweig in his influential book *The Star of Redemption,* where he frequently uses the phrase "from Parmenides to Hegel," and whose influence will be considered later. *Time and the Other* considers these same two thinkers to be the paradigmatic voices of unity, fusion, and monism (42). The first lecture sets the course in this direction, through philosophy and past the philosophical tradition that begins with Parmenides and reaches its apex in Hegel.

Levinas's strategy in this bold endeavor begins once again in his deliberations on the instant. He returns to this theme with a renewed focus on the intensity and gravity of the solitude of the instant. To demonstrate the weight and effort of being, Levinas revisits the term *hypostasis,* first used in *Existence and Existents* to refer to the way a subject relates to him or herself. In the history of philosophy, the term *hypostasis* was used by Aristotle and the Greeks to refer to that which "underlies other things and serves as a support."[6] For Levinas, hypostasis indicates the effort-laden process whereby a subject remains distinct from anonymous being.[7] In taking a position within and against the impersonal *il y a* ("there is"), a subject hypostatizes, resists and suspends the anonymity of being. Hypostasis makes the impersonal personal, founding and establishing the existent as a subject, but not without effort. Levinas uses the examples of fatigue and indolence to demonstrate the weariness that results not from any particular task but from being "burdened with itself" (*EE* 12–25). Being oneself is laborious. Jeffrey Bloechl calls this effort of being "pure event," the "effort *of* the present, and not effort in it or from it."[8]

Levinas will make sparse use of hypostasis after his 1947 publications. Indeed, he will later develop a philosophy of passivity that threatens the independence of the labors of hypostasis. But at this stage in his career, Levinas is determined to confirm the solitary nature of being that is inherent in the process whereby a subject "contracts its existing" (*TO* 43). Hypostasis presses the weight of existence on the subject, but the bare fact of existence does not come to the rescue. The idea of being is not *full* as suggested by Parmenides; the mere awareness of the *il y a* is anonymous, raw, and impersonal.[9] Hypostasis withdraws the subject into itself. There is nothing in this instant to bind the hypostatic self to the grand ideas of philosophy; there is only existence, severed, awkward, and affected.

An existential distance arises when the instant, perceptible once again through insomnia and the *il y a*, demonstrates its frailty and impotence.[10] If one finds in oneself an emblem of truth, a connection to the highest truths of being, as Parmenides would indicate, then being is *not* anonymous; it is known, familiar, and *familial*. But this is not the case, even if it appears to be the case by the light of day. One finds instead, especially in experiences of impotence and insomnia, that this connection is illusory. Levinas claims, "Idealist philosophy on the whole has been a way of grounding being on something that does not have being" (*TO* 46). The relationship to being is anonymous; I have no *meaningful* connection to being-itself. This does not mean, however, that I am disconnected from being. Far from it, in fact, I am riveted to being irrevocably. It bears on me, mightily, like the heavy darkness of the night weighs on the insomniac:

> Let us approach this situation from another slant. Let us take insomnia. This time it is not a matter of an imagined experience. Insomnia is constituted by the consciousness that it will never finish—that is, that there is no longer any way of withdrawing from the vigilance to which one is held. Vigilance without end. From the moment one is riveted there, one loses all notions of a starting or a finishing point. The present is welded to the past, is entirely the heritage of that past: it renews nothing. . . . Here time begins nowhere, nothing moves away or shades off. (46)

Yet in taking up again the themes of anonymous being, insomnia, and the *il y a*, Levinas has a more explicit target for his reflections: Heidegger. His critique of Heidegger, already expressed and sketched in *Existence and Existents*, will be pulled alongside his general critique of the ontological tradition from Parmenides to Hegel. The strategy

Levinas uses here will become familiar in the decades ahead: Levinas endeavors to show that Heidegger, for all of his efforts in this direction, fails to extract himself from this long tradition.

Heidegger and Nothingness

Levinas reveals in *Time and the Other* that he is a sophisticated and close reader of Heidegger, or at least of portions of *Being and Time*.[11] Levinas appropriates Heidegger's critique of Western thinking about time, but he also rigorously opposes Heidegger's work. Specifically, Levinas is concerned about the way Heidegger attempts to reconfigure time according to the primary concern of Dasein, which is chiefly authentic being-toward-death.

Anxiety, which Levinas attempts to identify from the slant of his analyses of insomnia and the *il y a,* refers to our reaction to an impersonal nothingness that is more empty than Heidegger's nothingness. For Heidegger, claims Levinas, nothingness "does not keep still" (*TO* 49). Levinas denies any motion of movement for the subject, and in so doing stakes out a position that opposes Heidegger's still active version of nothingness. The "there is" has been robbed of any personal characteristic; the ego, laboriously established despite the anonymity of being, relates to existence as a stranger. This is not so with Heidegger, claims Levinas.

Levinas critiques Heidegger over his understanding of *nothingness* (*Nichtigkeit*), which is a critical component of both *Being and Time* and "What Is Metaphysics?" (1929), both of which are invoked by quotation in these lectures. Nothingness is a complex phenomenon that receives different, if complementary, treatments in these two publications. In *Being and Time,* nothingness has a functional capacity, undermining the way Dasein retreats in anxiety from its authentic being-toward-death. For Heidegger, anxiety is a result of nothingness, or "nullity," which quietly erodes the effectiveness of Dasein's evasions.

In "What Is Metaphysics?" Heidegger attempts to provide a definition of nothingness itself.[12] Nothingness is not a void or an absence for Heidegger, who recoils from nihilism.[13] Instead, nothingness is the unsettling of inauthenticity, the undertow that destabilizes the many ways Dasein conceals its ownmost possibilities and Being-toward-death. *Nullity*—*Nichtigkeit* is often translated *nullity,* presumably to avoid the intonations of nihilism—functions for Heidegger as the source of anxiety.[14] Heideggerian anxiety, like Levinasian anxiety, is frequently masked by the busyness of everyday life.[15] Anxiety comes and goes, revealing the *Nichtigkeit* that is a component of Dasein's existence.[16] *Nichtigkeit* is active both in Dasein's thrownness and in Dasein's authentic being-toward-death.[17] Though the term *Nichtigkeit* is a noun, Heidegger needed in "What Is Metaphysics" to demonstrate that it is chiefly verbal in its function. Nothingness is not a "thing" among "things." It is an active component of Dasein's existence, authentic and inauthentic. Nothingness nothings; *Nichtigkeit* nihilates.[18] And the fact that nothingness actively nihilates enables Dasein to break from the they, to be liberated from "comparison with an ideal which has been set up" (*BT* 331). Nothing is not, for Heidegger, as it is for Levinas, the privation of a "something" (231).[19]

In short, Heidegger's *Nichtigkeit* is the instability inherent in temporal existence. Dasein proceeds from, through, and in *Nichtigkeit.* Nothingness destabilizes any connection with ideals or idealism. It is because Dasein is thrown *into* nothing (*Nichts*) that Dasein is released from the *somethings* of everyday existence that would constrain, define, identify, and restrict Dasein's becoming. *Nichtigkeit* is not the unpleasantness of anxiety but that which is the permanent cause of this anxiety. Levinas detects the *active* capacity of nothingness in Heidegger and singles out his peculiar phrase "nothingness nothings" to demonstrate the motion and meaning that in fact undergirds Heideggerian anxiety. Dasein has a narrative, a story that leads Dasein between inauthentic and authentic being-toward-death. *Nichtigkeit* plays a crucial role in this drama, functioning as the instability in both Dasein's thrownness and project.[20] Nothing is *busy* in Heidegger, and this busyness does not strip the subject of its identity but functions to assist Dasein in the authenticity of embracing itself as the "basis of its being" (*BT* 330). Heidegger's Dasein rests on a property of motion, a something that enables Dasein to activate the care whereby Dasein moves toward authenticity. Or as Levinas claims: "One can also find this turning of nothingness into existing in Heidegger. Heidegger's nothingness still has a sort of activity and being" (*TO* 49). Heidegger's understanding of time begins here, in the *Nichtigkeit* that reveals the temporality of Dasein and the instability of ideals and static being.

This is hardly the empty night of solitude, or the *il y a*. In fact, the very term *il y a* strips away any sense of a subject from the "there is." Dasein is, after all, literally translated "there-being." No stable connection binds Levinas's subject to a future, even death. The significance of Levinas's reference to the instant as immortal is now clear: Heidegger's "moment of vision" is a moment where Dasein authentically faces its own mortality. Levinas's "instant" is the discovery of an affectivity and an impotence beneath the virility of being-toward-death. Levinas's accusations that Heidegger expresses a fundamental virility are supported by the very activity of *Nichtigkeit* in Heidegger's lexicon (*TO* 70).

For Heidegger, the business of nothingness relates directly to the establishment of Dasein as the basis for its own being. Nothingness is therefore not anonymous or impersonal in Heidegger but the condition whereby being *becomes* authentically personal. Levinas is not afraid to use the term *annihilation* to refer to the effect of existence on the existent. The experience of the *il y a* bars the self from its typical mode of becoming, where sights and sounds are accumulated through the virility of grasping eyes and hands. In the *il y a*, I am *sans-soi* (without self) (*TO* 49). The *il y a* robs the subject of *any* evasion of being, any escape into nothingness. Insomnia is the incapacity to turn away from being; there is darkness on all sides.

In contrast to Heideggerian nothingness, and Sartrean nothingness for that matter, Levinas proposes "a notion of being without nothingness, which leaves no hole and permits no escape" (*TO* 50). The *il y a* is simply inescapable; it is utterly indifferent to the subject. The darkness of the instant is everything to me, but I am nothing to the night. It weighs heavy on me, pressing me down in the immortal instant of pure affectivity, but it doesn't know me. In *Existence and Existents,* Levinas spoke of this reversal as akin to being *prey* to the creatures of the night, the hunter turned into the hunted. Heidegger's nothingness affirms Dasein as a condition of care,[21] whereas Levinas's *il y a* is the utter absence of care, the lifting of any veil of affirmation or resonance with "being itself." The *il y a* leaves me exposed, running for my life but unable to move. Time is not what *I* move, ecstatically or otherwise; time is what comes to my rescue, what moves me.

The correlations with Heidegger here cast an eye back toward *Existence and Existents*. In a sense, Levinas uses *Time and the Other* to level the technical themes developed in *Existence and Existents* directly at Heidegger. The effect is a heightened sense that Levinas was even more intently concerned with addressing Heidegger in *Existence and Existence* than he then revealed. The brutal analysis of the *il y a* serves to underscore, most radically, the very desperate need of the ego for something or someone to redeem it from the anonymity of "irremissible being" (*TO* 50).[22] For Levinas, time can grind to a halt. For Heidegger, "time will not let itself be halted," though "halting time is something that we want" (*BT* 478). If time could stop, Dasein would have no need for anxiety and no nothingness would open up beneath Dasein's feet. It is natural for Dasein to long for this, but it is unhelpful in a world where time drags us all toward an inevitable death. What Heidegger fails to see, in Levinas's estimation, is that time is not merely a feature of Dasein's relationship to its ownmost future. Time itself, for Levinas, is a gift from the other person.

Miteinandersein

True to form, in *Time and the Other* Levinas continues to affirm aspects of Heidegger's analysis alongside his critiques. Levinas is impressed with Heidegger precisely because he has identified the failure of traditional philosophy to think about the existent without recourse to the grand, universal idea of existence (*TO* 44). The lasting value of Heidegger's *Being and Time*, for Levinas, is his critique of this tendency.

Heidegger redirects the attention of philosophy to *death*, the only indubitable event in being, and its solitary nature. Philosophy begins, for Heidegger, with reflections on the way Dasein faces the world into which it is thrown. This world obviously contains other persons, so Dasein must work out its being-in-the-world alongside other subjects mired in the same "ontological-existential" predicament.[23] This way of thinking of Dasein as being-with-others (*Miteinandersein*), according to Levinas, essentially silences the importance of the other. The other, after all, has little to do with the existential crisis Dasein must face in light of its death.[24] Heidegger is not silent on the place and role of the other; in fact, he is quite clear about the status of the other. Dasein is thrown into a world of plurality, and struggles to be authentic alongside others who are facing their own temporality with varying degrees of authenticity.[25]

Heidegger pins the anxiety of time primarily on the recognition of mortality and the inevitability of one's own death. Being-toward-death, inasmuch as death is never something Dasein can *be*, ironically makes possible "Dasein's *ownmost* possibility" (*BT* 307). Death, and the anxiety it arouses by revealing the *Nichtigkeit,* wrenches Dasein away from the they, the lostness of the "everydayness of the theyself." Death establishes the individuality of Dasein by presenting a possibility that cannot be escaped or evaded. This is the root of Heideggerian evasion; to evade authenticity, for Heidegger, is to let the trivial overwhelm the paramount. The other person, whom Dasein inevitably struggles alongside, can be counted on to tempt Dasein with trivial chatter and frivolous occupations. Heidegger states this idea with clarity: "The ownmost possibility [of death] is *non-relational*" (308). In addition to this, Heidegger is intensely concerned about the crowd, which tends toward inauthenticity and attempts to busy Dasein with activities that outstrip the paramount activities of anticipation and projection.[26]

Heidegger investigates the seemingly banal activity of "chatter" (*Gerede*), which exemplifies for him a senseless forestalling of authentic being-toward-death.[27] Other persons pacify and tranquillize the awakening of Dasein's vision. Vigilant living is therefore a being-toward-death that refuses to fritter away life in distraction and inauthenticity. Heidegger claims that when "Dasein is absorbed in the world of its concern—that is, at the same time, in its Being-with towards Others—it is not itself" (*BT* 163). It seems reasonable, then, to point to Heidegger's ambiguous attitude toward other persons as a factor in the notable paucity of ethical reflections in his corpus. This is a common, if hotly debated, critique of Heidegger's work, which is sharpened by his unrepentant participation in the Nazi party.[28]

For Levinas, and here he seems to break decisively with Heidegger, the situation of the self is anxious precisely because self-identity is not established in a circle back to the self. The anxiety of being-in-the-world cannot, for Levinas, be reduced to a matter of internal resolution. For Heidegger, anxiety arises and is addressed with little reference to the others with whom Dasein is *mit* ("with"). This can be illustrated by the drama of Dasein's awakening and reorientation in authentic care, all of which is expressed as the internal and ownmost experience of Dasein (*BT* 308).

For Heidegger, Dasein is made anxious by death, which reveals the *Nichtigkeit* hidden behind everyday existence. But neither dread nor *Nichtigkeit* is necessarily found in the face-to-face relation with the other person. Anxiety and nothingness are features of Dasein's self-examination and introspection. Heidegger makes little mention of the *death* of the other or any anxiety or dread that might result in the phenomenon of the other's death, which assumes principal importance for Levinas in *Time and the Other*.

In *Time and the Other*, Levinas unveils one manifestation of his critique of Heidegger by fixating on the preposition *mit* and its insufficiency for describing the relationship with the other. Levinas detects a subtle but serious issue with Heidegger's terminology. Heidegger's choice of the word *mitsein* ("being-with") to refer to the realm of intersubjectivity betrays a persistent regression of Heidegger's work into a synchronous understanding of time (*TO* 40–41).

This relationship with the other, described by the preposition *mit* configures the self and the other "side by side," gathered "around something, around a common term and, more precisely, for Heidegger, around the truth" (*TO* 41). This hardly seems deniable; Heidegger may have declared Dasein's unique individuality superior to any grand metaphysical concept, but *each* Dasein remains beholden to the truth revealed by authentic being-toward-death. Pluralism merges again into unity; my death may be my own and it may be like no other person's death, but it is the truth of death that defines us all. Heidegger may momentarily turn the tables on Parmenides by reintroducing temporality into the analysis of being. Heidegger will maintain a *distinction* between being and beings, between the existent and existence—but in the end, he will not allow the distinction to stand. Referring to Heidegger, Levinas writes, "Existing is always grasped in the existent, and for the existent that is a human being the Heideggerian term *Jemeinigkeit* [mineness] precisely expresses the fact that existing is always possessed by someone. I do not think Heidegger can admit an existing without existence, which to him would seem absurd" (45).

So because Heidegger will not maintain the distinction between being and beings, Levinas claims that he thereby allows the dialectic to collapse into a monism, albeit a relativistic and poorly defined monism. Levinas uses this accusation to open *Time and the Other*, alleging that, like Hegel, Heidegger traverses dialectic contradictions only to collapse them (*TO* 41–42). Parmenides is again confirmed, or so it would seem.

For Heidegger, the "private fact of one's existence" is superior to Dasein's posture *mit* the other (*TO* 40–41). In a direct challenge to the central thesis of *Being and Time*, Levinas concludes *Time and the Other* by claiming: "Solitude is the absence of time" (57). Heidegger has wrenched Dasein out of the "eternal now" of Western philosophy only to leave it dangling in the winds of timelessness. And by failing to rivet Dasein to an alternate understanding of time, Heidegger has inadvertently abandoned time altogether. *Mit*, if nothing else, is the language of simultaneity and synchrony. This common dance around "truth" *presumes* a philosophy of time, a shared present, an ontological structure of existence to which all people are held.

By reverting to a subtle ontology of presence, Levinas believes, Heidegger has undermined his own critique of Western philosophy. Instead of carving out a unique temporality that holds Dasein in distinction from being and beings, he has left Dasein existing alongside other beings in simultaneity, the hallmark of presence. Heidegger has created the opportunity to rethink time apart from a collapse to presence and synthesis, but by failing to reconfigure time, he has allowed temporality to default to the traditional universality. All human beings are bound to the same sense of time and its relationship to death.

Franz Rosenzweig

Rosenzweig's influence makes itself known in *Time and the Other,* though in a less obvious fashion than Heidegger or Parmenides. Richard Cohen points out that Levinas uses language that is "doubtlessly borrowed from Franz Rosenzweig" (*TO* 45n9), and he makes this case convincingly in several of his own publications.[29] The apparent use of Rosenzweig's terminology in *Time and the Other* provides an opportunity to investigate the indications that Rosenzweig and Judaism factor significantly into Levinas's understanding of time. These features will be more evident and warrant deeper exploration in later decades, but it is vital to see how these influences are already playing a steady role in Levinas's description of time.

Levinas does not mention Rosenzweig in print until 1961, aside from a passing reference in one of his confessional Jewish writings (*DF* 109) and a 1959 paper on Rosenzweig that was not published until 1963.[30] However, Cohen points to textual evidence that Rosenzweig is already influencing Levinas's thoughts in *Time and the Other.* This evidence, alongside the 1959 paper and his 1942 list of things *à faire* published in *Carnets de captivité,* leaves little doubt that Rosenzweig is already an active force in Levinas's thinking. Rosenzweig's work is complex and cannot be addressed in its breadth in this study, but I will use this section to introduce *Levinas's* Rosenzweig.[31]

We have some incentive for sifting through the early influences on Levinas's understanding of time, particularly in regard to the Jewish influences. The later phases of his career are marked by a sustained use of religious imagery to describe both time and the other. For instance, in the 1980s and 1990s, Levinas uses the term "holiness" to describe the utter transcendence of the other person (*GDT* 223, 265n6). Even more noticeably, the term *separation,* which Cohen believes Levinas borrowed from Rosenzweig in *Time and the Other,* is eventually correlated with "holiness."[32] Moreover, Levinas eventually names his discovery of Rosenzweig one of the most important stages in his developing understanding of time.[33] Yet the mysterious reference to Rosenzweig in the introduction to *Totality and Infinity* (cf. *TI* 28) helps us little in our attempt to unravel the impact of *The Star of Redemption* on the young Levinas.

In 1921, Rosenzweig published *The Star of Redemption* (*Der Stern der Erlösung*), his principal text. Rosenzweig devotes several pages of this book to the description of the Jewish people and the way their law, language, and eternal future differentiates Jews from those in the lands they inhabit (*SR* 298–305). The concepts of time and holiness are intertwined in this section, as they are throughout *The Star*. He explains that the Jewish people are set apart eschatologically; they are a people of a future that is discontinuous with the present. Rosenzweig is unapologetic in his application of Judaism to the key problems of nineteenth- and twentieth-century philosophy. This text had a direct and perceptible influence on Levinas's understanding of time, and the later movements in Levinas's understanding of time demonstrate an escalating dependency on Rosenzweig.

The Star of Redemption is a complicated book, weaving together philosophy and theology in a remarkably original challenge to philosophical trends of the nineteenth and early twentieth century. In sweeping fashion, Rosenzweig takes aim at all of Western philosophy, particularly as it reaches paradigmatic expression in the work of Hegel. *The Star of Redemption* can be considered an existentialist document, for Rosenzweig expresses consistent concern for the obfuscation of everyday human existence by philosophical ideals. Rosenzweig was intensely familiar with Hegel's idealism, but grew increasingly suspicious of his optimistic and sweeping theory of history. Hegel unifies the world, divine spirit, humanity, nature, and individual lives in his all-encompassing understanding of history. But Rosenzweig wonders if Hegel has not too quickly reconciled "heaven and earth" (*SR* 7). With a gesture to Kierkegaard, Rosenzweig challenges the wisdom of Hegel's drive for "integration." He is particularly concerned about the smooth integration of revelation and creation, and the scientific method that blesses the resolution of seemingly disparate phenomena.

Time after time, claims Rosenzweig, philosophy has let a vision for the "All" triumph over the plight of the individual.[34]

Rosenzweig sets out to rethink God, Humanity ("Man"), and the World with a renewed eye for potentially irresolvable separation between these concepts. Rather than letting an obsession for interconnection drive his analysis, Rosenzweig rethinks these themes with an eye for separation and difference that does not resolve into the grand syntheses of idealistic philosophy. He challenges the tendency of Western philosophy, since Parmenides, to think of "the totality of being" (*SR* 12). Rosenzweig is aware that denying the scope of totality means challenging philosophy's reliance on reason. He suggests that reason will naturally be at home in the world but that it gets ahead of itself when it tries to encompass things beyond the world. As Cohen summarizes, "Reason denies every chasm, abyss, and hiatus, denies them the very moment it must admit them, in the very instant it bears witness to them."[35] How, wonders Rosenzweig, can reason attempt to speak of what is beyond the walls of the world? Yet from "Parmenides to Hegel" philosophy has tirelessly followed an ancient extrapolation from the world to "outside the world" (22). This is a "godless" endeavor; it is intrinsically atheistic and reductionist. Hegelian idealism minimizes the disparate nature of God, humanity, and the world, and discretely declares itself lord of all three. Judaism and Christianity, according to Rosenzweig, anticipate the redemption of the world in their pervasive emphasis on love for God and neighbor. The perfect love advocated by both religions is a poor fit for the natural reality of the world of competition and struggle. The internal logic of the world simply cannot give rise to the kind of love advocated in Judaism and Christianity. The "star of redemption" is a geometrical depiction of the relationships between creation, revelation, and redemption. The center of the star burns hot with the mystery of revelation through the internalized religion of Judaism. Christianity moves evangelistically outward, like the spokes of the star.

At the "core of the star" of redemption is a very different sense of *time*. Rosenzweig writes, "'Blessed art Thou . . . who hast planted eternal life in our midst.' The fire burns at the core of the star. The rays go forth only from this fire; and flow unresisted to the outside.

The fire of the core must burn incessantly. Its flame must eternally feed upon itself. It requires no fuel from without. Time has no power over it and must roll past. It must produce its own time and reproduce itself forever" (*SR* 298). Rosenzweig develops an opposition between pagan time and the time of redemption. We can easily detect a tendency in Levinas's early works to express concerns about the points where philosophy seems to cross over into "paganism," which is a systematic concern for Rosenzweig. Levinas occasionally singles out philosophical positions as pagan, a label that for Levinas has much to do with understandings of time. For Levinas, a reference to paganism often indicates a philosophical resignation to *fate,* to the consignment of time to fatality.[36]

For Rosenzweig, paganism consigns humanity to an economy, to a "heathen world," in which humans are doomed to temporal laws of cause and effect, the time of before and after (*SR* 345–46). Paganism is incapable of truth, which is thoroughly separated from the everyday, economic time that engenders "the hostility of nations as well as the cruelty of gender, the jealousy of class as well as the barrier of age" (346). Rosenzweig attempts to attune his reader to the sense of divine time, messianic time, that bears no logical relationship to the timepieces that mark pagan temporality. He presses for an understanding of time that is otherwise than world time, not the "twelfth stroke of the world clock" but "eternity as present at every hour" (306).

For Levinas, paganism has to do with a resignation to existence; the best we can do in paganism is to examine being for its best features and celebrate these facets of being. Paganism, in resigning itself to existence and the experiences of being, is incapable of any thought that is not self-generated. The pagan cannot encounter the other or see beyond (or before) the violence generated by the forces of being. Levinas demonstrated this concern already in "Some Meditations on the Philosophy of Hitlerism," where he is concerned with the "forces of fatality" and their inevitable recourse to power and racism (*UH* 18). Fate rules and justifies itself by expansion, which is the way of nature. For Levinas, from the very beginning, paganism represents a permanent temptation for philosophy, the lure to think on the "hither side" of existence, on the "hither side of time" (*CPP* 3). The terms *pagan* and *paganism* are recurring themes in Rosenzweig, referring to a philosophical temptation strikingly similar to the one configured by Levinas.

In *The Star of Redemption*, Rosenzweig uses the term *pagan* to decry just such a blending between the realms that he strives to keep separate. The "pagan world," writes Rosenzweig, is the realm of an unfortunate "temporality" in which it is "impossible to love one's neighbor as oneself" (*SR* 346). In *pagan* time, we are trapped between "before" and "after" and never fully present to one another. In eternity, we are able to love fully because time has been defeated. Referring to the Christian doctrine of salvation, Rosenzweig points to the eschatological future when the absence of pagan time clears away the path to "brotherliness" (346). Levinas, for his part, will avoid the notion of an eternity in which all persons are present to one another, but he will assume to a large degree Rosenzweig's negative sense of the temporal operations of the world. Rosenzweig will be particularly helpful when it comes to resisting the impulse of traditional philosophy to reduce messianic time into a logical relationship with world time.

Levinas, who may differ more from Rosenzweig than Levinasian scholars have let on,[37] almost certainly utilizes Rosenzweig to think about a transcendence that "breaks with that which it transcends."[38] More profoundly, Levinas will come to use Rosenzweig to formulate his understanding of messianic history, as well as the anarchic past. In *Time and the Other*, the concept of separation, which was influenced by Rosenzweig, serves to further divide the subject from any meaningful relationship to existence. The unity of the subject can be sustained, if laboriously, through Levinas's understanding of hypostasis. But the subject finds him or herself radically separated from being. This is what scholars in this field refer to as the "ontological difference"; it is the divide between the existent and existence itself, the gulf between the ego and being. And by utilizing Rosenzweig, Levinas finds a philosophical tool for thinking about separation that is not again subject to reduction to sameness.

Death Reconsidered: The Widening Rift with Heidegger

Such stark separation between an existent and existence is opposed to Heidegger's analysis, which would constitute such separation as an evasion of one's authentic and unique relationship to being. But this pits Levinas not just against Heidegger but also against the Western philosophical tradition in general. Levinas is proclaiming a solitude and separation that borders on solipsism: "Solitude is not tragic because it is the privation of the other, but because it is shut up within the captivity of its identity, because it is matter. To shatter the enchainment of matter is to shatter the finality of hypostasis. It is to be in time. Solitude is an absence of time" (*TO* 57). This separation allows Levinas to think about two different sorts of relations. There is the relation between a subject and existence, and there is a relationship between a subject and the other person. In the first relation, the subject has mastery of existence; the subject has attained a kind of novelty and victory over anonymous existence. But this relation is also bondage to oneself, enchainment within the limits of private hypostasis. One may *seem* to transcend this bondage and be liberated from the private solitude of one's existence. This is what vision and light seem to give us, a world that breaks open our solitude. But for Levinas, sight and light only reduce the exterior world to the internal logic and language of self. The eyes seize and sort and reduce the visible into a feature of prior knowledge; vision is grasping. Likewise, every attempt to project oneself in hope of grasping novelty and possibilities is doomed to fail. Wherever I go, there I am, reconstructing the world that I find with the familiar language and images that come along with me.

Levinas realizes that in order to truly break with Parmenides, he must do better than Heidegger; he must identify and safeguard a plurality that does not sink again into unity. For "in Heidegger there is a distinction, not a separation" (*TO* 45). The distinction allows philosophy to consider the relation with the other person in a manner not bound by the logic of being, not secondary to a theoretical depiction of being-itself. The distinction slips away as Heidegger focuses only on Dasein's relationship with existence; Dasein defines *Sein,* and little needs to be said of the other person who plays a trivial role in the relationship between Dasein and its being-toward-death. Levinas suggests that Heidegger loosened the bond of Parmenidean totality by denying the right of Western idealism to impose on Dasein's way of being-in-the-world. But he pulls at the seam Heidegger has opened in the relationship between existence and existents. Alongside Rosenzweig, who used the term *separation* to denote a similar irresolvable difference, Levinas proposes that philosophy reconsider its roots in thinking of the many vis-à-vis the one.

Not far into the *Time and the Other* lectures, it becomes clear that Levinas intends to strike at the very root of Heideggerian ontology. Levinas deems that Heidegger has failed to consider existence apart from existents, and in so doing he has also failed in his monumental tasks to reconsider both death and time in light of his own existential and phenomenological critique of Western philosophy. So Levinas challenges Heidegger's assessment of the phenomenon of death. By questioning his portrayal of death, Levinas is positioning himself as a strident and vocal critic of the constructive moves of *Being and Time.* This is not to imply that Levinas has abandoned the project initiated by Heidegger; *Being and Time,* as Levinas will always underscore, has delivered a brilliant and influential critique of Western philosophy. Levinas will continue to embrace that critique, for the most part, while steadily and scathingly pointing to problems in the constructive ontology that Heidegger offers as a replacement for the negligence of traditional philosophy.

We could detect the beginnings of a "break" with Heidegger's work in *On Escape* and perhaps in "Some Meditations on the Philosophy of Hitlerism." But here, in *Time and the Other,* as well as in *Existence and Existents,* Levinas is no longer subtle. He is willing to say directly: Heidegger was right about a serious negligence in traditional philosophy, but he was dangerously wrong about death and time. *Time and the Other* is far from shy about these challenges, even as it still lacks the comprehensive ethical concerns that Levinas will develop in the 1950s and reach mature expression in *Totality and Infinity.*

For Heidegger, as Levinas reads him, the challenge for Dasein is to come face-to-face with the "thrown" nature of being-in-the-world and then to face the unique potentiality for being that belongs exclusively to Dasein. The "mood" characterized by Heidegger is one of resolution, strength, and decision. Resolution must resist the temptation to resign oneself to indifference or abandon oneself to "one's thrownness" (*BT* 396). Because death belongs uniquely to Dasein, Levinas points out that death becomes "an event of freedom" for Heidegger: "Being toward death, in Heidegger's authentic existence, is a supreme lucidity and hence a supreme virility. It is Dasein's assumption of the uttermost possibility of existence, which precisely makes possible all other possibilities, and consequently makes possible the very feat of grasping a possibility—that is, it makes possible activity and freedom. Death in Heidegger is an event of freedom" (*TO* 70). Levinas opposes this development by pointing out that death, and more principally *suffering*, is notable not for the freedom it establishes but for the opposite: passivity. Death is remarkable not because it is my own, but because death remains something "absolutely unknowable" (71). Death is not something that I resolutely embrace but that moment where I am passively seized by the unknowable. Death takes place outside of the light of knowing; it represents the limit of idealism, which relies on the epistemology of vision. The irony of Heidegger's treatment is the serious philosophical capital given to something that is as mysterious as death. For Heidegger, the uniqueness and singularity of my own death indicates my fundamental freedom to relate to my thrownness in authenticity or inauthenticity.

Levinas contests this position by pointing at *suffering* as the "limit of the possible" (*TO* 70). In relating to death, the subject relates not to something that is uniquely its own, but something that "is in a relationship with what does not come from itself." Death cannot take place beneath the light of knowing; it resides beyond the grasping world of vision and sight. Death announces itself as a kind of impossibility, for death can never be present to the subject. Levinas points to Epicurus, who quipped of death: "It does not then concern either the living or the dead, since for the former it is not, and the latter are no more."[39] Death is a feature of the future that lies beyond any grasp of the present, beyond the hypostatic present over which the self has mastery. Instead, death marks the end of "the subject's virility and heroism" (*TO* 72). Levinas believes that he has made good on his promise to think about plurality that does not merge into unity; the self and death can never be *one*. He can therefore state: "Right away this means that existence is pluralist. Here the plural is not a multiplicity of existents; it appears in existing itself" (75).

By locating time *outside* of the self, and its mastery of the world of the present, Levinas can consider death anew. Heidegger wraps temporality into his definition of Dasein's relationship to being. By introducing an irresolvable separation between the world of the subject and the world of the other, Levinas has relocated death outside the power of the subject. In every instant the self is *not dead*, and if dead is *not a self*. In other words, there is no actual experience of death, of being dead; one cannot have a relation with this event. The relation with death is an impossibility because death eliminates the subject from the relation. Death "announces a subject over which the subject is not master, an event in relation to which the subject is no longer subject" (*TO* 70).

We can say much more, claims Levinas, about the phenomenon of suffering, which provides an announcement and anticipation of death. Unlike death, suffering is an event that can be grasped in the dynamics of the instant, in the privacy of hypostasis. Suffering is abundantly *present*, manifested as an inescapable bondage to pain that is very much my own. In pain, writes Levinas, the subject finds "itself enchained overwhelmed, and in some way passive" (*TO* 71). In physical pain, a subject is radically *alone*. Suffering is an inversion of the virility with which Heidegger's Dasein approaches death. In suffering, one faces the immanence and proximity of death, but this facing is passive rather than heroic. As death approaches, this passivity increases and reduces the self to a state of incapacity, to "infancy" (72). Death is defined by the absence and elimination of futurity, an "instant" without pregnancy or expectancy, whereas suffering portends and mimics what death promises: the very end of virility and hope.

Yet as long as there is a moment and instant left, the characteristics of the hypostasis remain. Hypostasis gathers the past together and looks hopefully toward its renewal in a new moment, toward the future. As the self remains master of the instant, so in the instant there reappears hope. Levinas points to Macbeth's death, which became all but certain when he discovered that Macduff was "of no woman born" (*TO* 73).[40] At first, Macbeth is resigned to the imminence of his death, and he tells Macduff, "I'll not fight with thee." But in the next instant, hope is reborn, for as unavoidable as death may seem, every instant is pregnant with the hope of another instant. The Heideggerian hero, Levinas implies, finds in death the freedom that belongs to no other. The hero, for Levinas, "is the one who seizes chances . . . who always glimpses a last chance" (73). So Macbeth takes up his sword and fights; "Yet will I try the last. Before my body I throw my warlike shield. Lay on, Macduff, and damn'd be him who first cries, 'Hold, enough!' "[41] To breathe again is to hope again. As the Latin proverb proclaims *Dum spiro, spero:* while I breathe, I hope.[42] Heroes seize new responsibility before death announces its final immanence by returning the self "to this state of irresponsibility, to be the infantile shaking of sobbing" (72).

Against Heidegger, death is not the coming of an event that belongs to the self or to the ownmost possibility of self. Quite the opposite, "the approach of death indicates that we are in a relation with something that is absolutely other, something bearing alterity not as a provisional determination . . . , but as something whose very existence is made of alterity" (*TO* 74). Death is a mystery because it lies beyond sight, beyond any horizon of the self. But suffering, and the death it portends, can make one aware of the world that is exterior to the self. Like death, the relationship with anything truly exterior to the self must be a relationship other than the grasping and possessive epistemology that is quite natural to existence and to the world of light.

Suffering and death, like insomnia, are phenomena with the unique capacity to open a window past the walls of interiority that surrounds the self. Normally, things are only provisionally alterior to the eyes; distances can be closed and vision can be improved. But no such maneuver closes the distance to death; nothing pierces the night of insomnia and offers security against the *il y a*. These experiences bracket out the glaring light of being and provide a glimpse of the limits of being. That limit makes us aware of an encounter that can only occur in passivity. All degrees of activity, virility, and projection grind toward a halt in suffering and insomnia. Death confirms this radical passivity; "Death is thus never assumed, it comes" (*TO* 73).

Levinas has therefore questioned the very centerpiece of Heidegger's ontological framework. But he has also embraced Heidegger's overall critique of the history of thinking about time. Heidegger refuses the subordination of time to the dominant logic of being; Levinas concurs. But Heidegger, as Levinas reads him, resituates time as a component of Dasein's capacity. Levinas therefore returns to the concept of death not as a moment faced in strength and heroism, but as the undoing of the subject's virility. Levinas is suggesting that the unraveling of virility we experience in suffering, insomnia, and death is symptomatic of a fundamental passivity that is normally overwhelmed by activity.

In *Time and the Other*, Levinas calls death "other" and outlines the various ways that death defies any encapsulation into the internal world of the same (*TO* 69–79). At this stage in his philosophical career, Levinas has constructed a kind of metaphysical schematic that resists the impulse to consider otherness as a temporary or artificial boundary to knowledge. He is pressing a philosophical proposal with far-reaching implications. Levinas suggests a radical division between the self and the world beyond the mastery of the self, a separation that does not dissolve by gaining greater insight or broader perspective. By isolating the phenomena of insomnia, suffering, and death, he has attempted to show that something startling lies beneath the basically Platonic fixation on the world of light. Levinas will spend the rest of his philosophical career attempting to articulate this *something* that lies elusively below, beneath, behind, beyond, or above the light of being.

Levinas examines episodes of passivity to suggest that the world over which the self has mastery is limited to a solitary, present instant. This instant is loaded heavy with the weight of being, with the pressure of its past, and with the latency of its future. But it cannot budge of its own accord, however laden it may be with hope. The subject has the power to exist, to face the raw and anonymous abyss of existence and *be*. But this existence is insular and enchained, "free" only to divvy and sort its own experiences according to reminiscence and internalized images. The present offers an artificial mastery of everything under its control and grasp. This present takes the form of an instant, which has the timeless tragedy of a statue. For Levinas, this timelessness constitutes real tragedy. The Greeks have long emphasizes the tragedy of *time*, which rots, decays, and destroys. But for Levinas, time is precisely what redeems us from the terror of an eternal present. Time comes to the subject passively, from the outside, as a gift and a new birth. Time redeems the present from its self-captivity.

Fortunately, the relationship between the self and anonymous existence is not its only relation. The second and utterly separate relation of the subject is to the world outside the self, to the other. For Levinas, this relation with the other is not with something that can be assimilated with the other things the subject can *know:* "The relationship with the other is a relationship with a Mystery" (*TO* 75). The alterity of the other is a relationship with something that exceeds or transcends all typical ways of knowing. And Levinas is well aware that he is suggesting a dialectical situation that "is not phenomenological to the end" (78). The move past Heidegger and Husserl is a move out of the field of phenomenology, or at least an unconventional experiment at the borders of phenomenology. Levinas is, in fact, arguing that the normal means of phenomenological analysis are inadequate for thinking about true alterity, and insomnia, suffering, and death are helpful examples of this inadequacy.

The Suffering Other

Levinas therefore utilizes typically Heideggerian images, though he turns them around on Heidegger. Death *does* play a pivotal role in a subject's everyday existence. The discussion of death in *Time and the Other* is merely another example, an *epoché*, where certain phenomena come into greater relief. Rather than establish the very ground rules for authentic existence, death merely demonstrates what Levinas has already said about the nature of an instant: like insomnia and suffering, death eliminates the distractions and complications that nourish our fantasy that we are fundamentally *active* rather than passive in our existence in the world. Death, for Levinas, is passivity. But death is far from the only example of radical passivity.

In the final pages of *Time and the Other,* Levinas begins his transition to a philosophy of otherness that is abundantly practical: "I have just described a dialectical situation. I am now going to show a concrete situation where this dialectic is accomplished" (*TO* 78). The concrete situation that accomplishes this pure alterity is the face-to-face encounter with the other.

Levinas then proposes that "time itself refers to this situation of the face-to-face with the Other" (*TO* 79), that the alterity he has identified between each instant is concretized in the subject's encounter with a face. The dialectic that before described "the whole abyss . . . that separates the present and death" is the same abyss that separates the self from the other. The relationship with the future, and therefore salvation from the present, is "accomplished in the face-to-face with the Other" (79). Time is therefore not a fundamental feature of the subject, nor the all-governing march of world-time, but the very relation between human beings. Time redeems the present with a new moment, which arrives from the other person in supreme novelty. Time is a feature of the intersubjective, the mysterious abyss that separates my present from the time of the other.

Levinas is aware that the relationship to another human often appears as significantly less than mysterious.[43] The other arises in the world of the self as an object alongside other objects, given to the fields of vision and light. At first glance, the other is just a complicated feature of the world offered up as objects to knowledge. But Levinas has taken great care to demonstrate that these deliberations occur within an unbroken solitude that he calls "the instant." At best, we lay eyes on objects that are difficult to assimilate with reminiscence of past experiences. But in this world, we rarely *see* anything new. Death, suffering, and insomnia are interesting precisely because they strip away the decency of the visible, everyday world and expose the subject to unsettling phenomena that are covered up by the brilliance of vision. This "decent" world of light and objects bears forward the illusion of symmetry, reducing the other person to a component of the "present" of the subject. Diminished to a feature of the instant, only that which is graspable about the other is of principal importance. The mystery of the other person is trivial and inconsequential. Difference is minimized by the world of light and decency. Here difference is akin to spatial distance; mystery merely beckons for closer perspective and deeper analysis. Levinas writes, "The exteriority of the other is not simply due to the space that separates what remains identical through the concept, nor is it due to any difference the concept would manifest through spatial exteriority. The relationship with alterity is neither spatial nor conceptual" (*TO* 84).

By binding the relationship with the other to the alterity of time, Levinas has located the other person *permanently* beyond the field of assimilation. In the mastery of one's relationship to existence, the subject can craft and nourish a myth of symmetry between beings and objects. But in the face-to-face relationship the subject encounters something that destabilizes and unsettles the myth of symmetry. The other person, who confronts me with a face, shames my efforts to domesticate the mystery of his or her alterity to a feature of my present.

For Levinas, this is already clear in the faces of those who suffer, specifically the sick, widowed, orphaned, and poor.[44] The faces of the destitute and the poor provide particularly important windows into the pure alterity of the other. For Levinas, this emphasis will escalate as his career progresses. The suffering of the widow and orphan permanently and completely escapes the epistemological tools of traditional philosophy, which guide the subject toward the quantification and evaluation of such suffering according to reason. Levinas sees in the other an alterity that never breaks down into symmetry, and therefore, he sees a multiplicity to being that resists the Parmenidean pressure for ultimate unity.

Levinas culminates *Time and the Other* with a discussion of caress, which was first invoked in *Existence and Existents* and will play a pivotal role in the later sections of *Totality and Infinity.* He also presents a preliminary discussion on *eros*, voluptuousity, the feminine, and paternity. The themes of eros and the discussion of the feminine relate quite directly to Levinas's discussion of time, underscoring the way the other withdraws from the grasp of the subject even in the erotic and enfleshed encounter between persons. Like the caress, these themes will also play more significant, and problematic, roles in *Totality and Infinity.*

At this point, it is vital that we see the intimacy between the concepts of time and the alterity of the other. Levinas uses *Time and the Other* to first establish the irreconcilable alterity that we can detect in our reflections on death. He then turns this notion of alterity toward the everyday encounter with the other person and shows that not only death but also the other person dwells in a futurity that will never be recovered into the subject's present. For Levinas, the philosophy of time can establish an exteriority that is not reduced to interiority. The alterity of time makes it possible for Levinas to break with Heidegger, not to mention Parmenides, and to position the self in asymmetrical relationship with everything outside of the instant over which the self has mastery.

The problem, then, with Heidegger's treatment of the other person is not necessarily the lack of ethics, for which he has often been attacked. The problem with *Being and Time* is that for all his effort to create a unique temporality for Dasein, Heidegger nonetheless continues to presume a basic symmetry in the relationships between persons. Each person, for Heidegger, exists alongside (*mit*) other persons, one person *with* another. Dasein and its others struggle for authenticity alongside one another. This produces an "association of side by side, around something, around a common term and, more precisely, for Heidegger, around truth. It is not the face-to-face relationship, where each contributes everything, except the private fact of one's existence" (*TO* 41) Levinas suggests a commonplace, everyday exposure to the alterity of the other in the primitive sense of time that is manifested in the face of the other. This kind of alterity is not just a freedom that exists alongside the freedom of the subject. Levinas links Heidegger's common dance around truth back to Plato, whose Eleatic notion of being crafts a republic that imitates a timeless world of light and ideas. In Heidegger, Dasein achieves its authentic existence in the world *Miteinandersein*. The timeless beauty of Plato's *eidos* becomes the focus for philosophy. The other person, therefore, stands *beside* me as we gaze together toward the truth. "It is the collectivity that says 'we,' that, turned toward the intelligible sun, toward the truth, feels the other at its side and not in front of itself" (93). Heidegger, for all his innovative critiques of Western philosophy, has replicated Plato's mistaken relegation of the other to trivial, or secondary, in the pursuit of higher truth.

Levinas, for his part, embeds the mysteries of everyday existence in the mystery of time. The alterity of time is glimpsed in the alterity of death and the passivity of suffering, but we come to know time primarily in the face-to-face encounter with the other. So the great mystery of philosophy is not to be explored *alongside* the other person, but in, through, and especially *for* the other person (*TO* 93–94). All these themes, toward which the later parts of *Time and the Other* are clearly building, will receive significant development and extension in *Totality and Infinity*.

Levinas has used these brief lectures to establish a theory of time that considers temporality utterly exterior to the self and its powers. Time has historically been understood as the bane of human existence; for Levinas, it is instead to be understood as salvation. And Levinas has furthermore bound the alterity of time to the alterity of the other person, declaring that both resist any inclusion or reduction into a Parmenidean unity. He has, perhaps, bitten off quite a bit to chew, and it will take an entire career for Levinas to hone, develop, and defend these themes.

Signs of Spatiality

Before moving forward into Levinas's work in the 1950s, we should note that he has already displayed an acute awareness of the fact that there is much work to do on the concept of time. He doubles back on himself in several places, aware of the difficulty of speaking of alterity without simply *expanding* the reach of the grasping subject. Levinas recognizes the vulnerability in his use of spatial language to describe alterity, but he seems at a loss for alternatives. *Totality and Infinity* will continue to manifest this ambiguity. As we will see, the transition from *Totality and Infinity* to *Otherwise than Being* has much to do with Levinas's rethinking of this spatial imagery.

In *Time and the Other,* as well as in *Existence and Existents,* Levinas looks for signs that traditional epistemological frameworks are inadequate to approach the kind of separation that he here envisions primarily as *interior* and *exterior* worlds. Traditional philosophy has presumed that the interior and exterior worlds were similar; Levinas proposes that they may be irreconcilably different. This proposition opens up a chasm between the self and the other, a spatial separation Levinas will soon describe with the metaphors of infinity and distance.

Levinas seems aware of the allure of spatial imagery and its tendency to nullify the dynamics of time and reduce all things to the present. In an important passage that anticipates the move from *Totality and Infinity* to *Otherwise than Being,* Levinas writes, "We recognize the other as resembling us, but exterior to us.... The other's entire being is constituted by its exteriority, or rather its alterity, for exteriority is a property of space and leads the subject back to itself through light" (*TO* 75–76; cf. 82–83). A fascinating ambivalence has begun to reveal itself here in Levinas's selection of metaphors. In some regards, the word *exteriority* is a perfect fit for the way Levinas wishes to situate the other person with respect to the subject. The other is *outside* the subject, utterly exterior to the world of appropriation and projection that I refer to as "me." The outside is a realm completely different from the internal realm. I have mastery over the internal, which Levinas will soon begin calling the world of the "same." But the opposite is true of that which is exterior to my hypostatic foundation.

The language of interiority and exteriority is so pivotal for Levinas that he will set up *Totality and Infinity* using these central metaphors. The very subtitle of *Totality and Infinity* is *an Essay on Exteriority,* and these metaphors will later contribute to some of the problems with *Totality and Infinity.* Furthermore, we will find that the spatial imagery of the early works and *Totality and Infinity* will be radically challenged by the later stages of his thought. It is important to track

Levinas's heavy use of the term *exteriority* in *Time and the Other*. But Levinas himself anticipates the weakness of this spatial metaphor. Near the end of his lectures he states: "The exteriority of the other is not simply due to the space that separates what remains identical through the concept, nor is it due to any difference the concept would manifest through spatial exteriority. The relationship with alterity is neither spatial nor conceptual" (*TO* 84).

Detecting this ambivalence in *Time and the Other* underscores how the transitions made in the later stages of Levinas's thought are in large part a recovery of some of his early and stark claims about time and alterity. This is not to suggest that *Otherwise than Being* replicates these earlier writings, but to point out that Levinas is already aware of the shakiness of spatial imagery. The use of spatial imagery, in this passage, worries Levinas precisely because at the spatial level the other merely takes his or her place as a feature of the solipsistic world of the same. True alterity is not accessible through vision, which Levinas believes to be intrinsically reductive and possessive. Alterity must be found *otherwise* than by sight and by way of distance; here, as in Levinas's later works, the distance is better framed as a function of *time*.

Metaphors and Reality

Levinas commentators often neglect to discuss the fine line between the existential experiences described by Levinas and the theoretical ideas toward which he is gesturing. It is highly likely that Levinas intentionally obscures this line or that he hopes that it does not exist at all. He joins Heidegger in suspecting Husserl of settling too often for theory over a philosophy that concerns itself with everyday life.[45] The opposition to theoreticism can be seen as a steady concern of Levinas's across his career; he does not wish to provide a philosophy that has a noticeable divide between the theoretical and the practical.

When he describes, in *Existence and Existence* and *Time and the Other*, experiences like insomnia and the *il y a*, Levinas is indeed attempting to point *through* unusual experiences at something that lies covered underneath. This is a strategy common to Husserl and Heidegger as well. For Heidegger, and Levinas, we are typically unaware of the nullity that unsettles our being-in-the-world, but we become aware of it through anxiety, which arises unpredictably and unevenly in life. But for Levinas, philosophy has fallen prey to thinking about knowledge and sensibility according to the way it seems typically to function. The type of anxiety that unsettles Levinas's understanding of light and sensibility is different than Dasein's anxiety in Heideggerian work. Nonetheless, both thinkers are using uncommon, unusual experiences to point to something relatively obscured by everyday life. This is the heritage of Husserl and a method taken up by many other philosophers including Jean-Paul Sartre, Maurice Merleau-Ponty, and Jacques Derrida. The strategy is to assume that when lived experience defies the rules of Western epistemology, even if only in flashes or moments of instability, this slippage is meaningful. For Levinas, the whole house of cards is threatened by these moments.

Insomnia, for Levinas, is not about the medical condition, a psychological event, or even the actual experience of insomnia. One is not as trapped by insomnia as his example suggests. In my worst night of insomnia, I can still move, turn on lights, sing and dance, or otherwise wrest control back from the night. Insomnia represses the ability to do this, perhaps to extreme degrees, but it does not preclude them entirely. But this does not invalidate Levinas's reflections on insomnia, precisely because he is not really talking only about actual sleeplessness. This becomes obvious in his reflections on death and suicide. Levinas calls death an impossibility (*TO* 50–51), which is plainly false in the everyday world of death and dying. What is impossible about death is the inability of a subject to evade the weight of being. For example, even in "keeping the power to die," Shakespeare's Juliette fails to rebel against the *il y a*. Suicide is ironic because one ceases to be in order to rebel against "being itself," which is already indifferent to Juliette's existence.

The point, for Levinas, is that despite the onslaught of destabilizing examples, somehow the subject's unity still *holds*. This is surprising, perhaps, because Levinas has worked mightily to rattle even the internal relationship of the subject to itself. But in the night of insomnia, all I have left is that *I am*. Levinas's reduction to the shivering, solitary self in the face of the *il y a* provides an existentially charged version of Descartes's reduction. There is nothing to be known with any stability in the *il y a*; everything beyond solipsism is a leap.

Despite using examples from everyday life, however, Levinas is indeed retaining a theoretical edge to his reflections. The experiences that he describes only attempt to approximate something that lies beneath nausea, deprivation, hunger, labor, insomnia, and the *il y a*. This is another feature of Levinas's work that is constantly under construction; he is forever working against the calcification of theory that attempts to encrust everything one attempts to say about existence. For this very reason, one can constantly read Levinas against himself. Significantly, at this stage of his career Levinas turns to the intense practicality of the face-to-face encounter and the intersubjective relationship. Here Levinas finds more than just a set of phenomena that belie an abiding theoretical truth about existence. In the face-to-face, we find that the other always unsettles any fall into theory.

Levinas still presents this face-to-face relation in thematized ways, as in the relation of the subject to the child and to the feminine (*TO* 80–94). The fall into thematization on these matters undermines not just Levinas's reflections in *Time and the Other*, but especially his reflection in *Totality and Infinity*. Levinas is most susceptible to these hurtful thematizations, as we will continue to point out, when he drifts away from the tools he has developed in his reflections on time. But what is most clear at this point is that in *Time and the Other* Levinas has made a turn away from general phenomena, which is weakened by its theoretical nature, and toward the more vivid and intensely practical face-to-face encounter. This transition is far from complete, and may not even be complete in his latest writings. But Levinas has indicated a direction for his philosophical itinerary: away from themes and theories and toward the intersubjective.

Four

The Recession of Time

Creation is the fact that intelligibility precedes me.
—Emmanuel Levinas, "Freedom and Command"

The 1950s present a fascinating turn in the development of Levinas's unique philosophy of time. Returning to Jacques Derrida's analogy that compares Levinas's thought to waves crashing higher on the shore, we might say that the tide recedes for nearly a decade in terms of Levinas's development of his concept of time. He declares, as the *thesis* of his 1947 lectures series, that time is "the very relationship of the subject with the other person" (*TO* 39). But Levinas then seems to mysteriously suspend his deliberations on time for years on end. The contrast is surprising. In the dozens of papers and presentations that Levinas makes in the 1950s, time is rarely mentioned and even less often invoked with any philosophical force, making way for other concerns.

This is not to suggest that Levinas discontinues his reflections on time entirely or that he reverses his earlier positions on the alterity of time and the other. Indeed, the extensive use of time in *Totality and Infinity* (1961) indicates that Levinas continued to develop his unique understanding of time. Nevertheless, the 1950s present an intriguing complication to the development of Levinas's philosophy of time.

Keeping *Totality and Infinity* in mind, Levinas's development of time in the 1950s can also be observed in a few paradigmatic essays including "Is Ontology Fundamental?" (1951), "Freedom and Command" (1953), "The Ego and the Totality" (1954), "Philosophy and the Idea of Infinity" (1957), and "Lévy-Bruhl and Contemporary

Philosophy" (1957). However, the lack of explicit work on the concept of time may contribute to some of the blind spots in *Totality and Infinity*. As we will see later on, by neglecting the concept of time, Levinas slips into spatial and ontological imagery, leading to the troubling analogy of the "dwelling" and its problematic expression of the feminine. Levinas attempts to addresses these problems in *Otherwise than Being*, in large part by returning to some of his early reflections on time, if only to radically eclipse them.

"Is Ontology Fundamental?"

Though Levinas will interact with a host of different concepts and thinkers in this decade, his writings most often and most expansively address the insufficiency of Heideggerian ontology. If anything, it seems that Levinas has made his disagreements with Heidegger on the concept of time clear in *Time and the Other* as well as *Existence and Existents*, and so he now directs his attention toward the nature of ontology, and particularly what he deems to be Heideggerian ontology. Levinas publishes the important essay "Is Ontology Fundamental?" (*BPW* 1–10), in which he attempts to demonstrate the perils of positioning ontology as the foundation for philosophy.[1] This exercise is essentially a questioning of Heidegger, or at least of early Heidegger, and it is noteworthy for its explicit appeal to *responsibility*. Whatever ontology relates to, and whatever we may say about our comprehension of being, ontology simply fails to sufficiently consider the relation with the other. "The relation to the other person is therefore not ontology," writes Levinas (7). Particularly noteworthy in this essay is the appearance of the face as a spatial and visual phenomena, which begins a steady ascent to the forefront of Levinas's philosophy, along with the initial signs of Levinas's shift away from his earlier fixation on time.

Throughout his career, Levinas overlays his philosophical discourse with countless metaphors and analogies. Perhaps more than other philosophers, Levinas is aware of the inability of language to articulate the phenomena toward which he gestures. This tendency also allows us to track the notable shift in his vocabulary away from temporal imagery and into spatial imagery during this period. The differentiation from Heidegger that Levinas stresses in "Is Ontology Fundamental?" is important in that it lays the groundwork for several sections of *Totality and Infinity,* but the essays from the 1950s also demonstrate that Levinas nearly abandons the use of temporal metaphors in favor of spatial ones. So focused on his critique of Heidegger's ontology, Levinas even seems to have forgotten the caution he voices in *Time and the Other,* where he worries about the use of the term *exteriority* because of its spatial overtones.

In "Is Ontology Fundamental?" Levinas repeatedly uses the spatial language of openings, horizons, and planes. The self and the other are positioned in separate, nonoverlapping *fields.* Levinas retains the internal freedom he loaded into the concept of the instant in 1947, but here he speaks of it as a *where* instead of a *when:* "The encounter with the other consists in the fact that despite the extent of my domination and his slavery, I do not posses him. He does not enter entirely into the opening of being where I already stand, as in the field of my freedom. It is not starting from being in general that he comes to meet me" (*BPW* 9). The instant has been transformed into a kind of isolated *place* that the subject organizes and arranges in freedom but never actually *shares* with the other. The "separation," which Levinas borrows from Franz Rosenzweig and applies to time in *Time and the Other,* is now almost thoroughly converted to a kind of spatial isolation (*TO* 45). And although the spatial analogies of horizons, fields, and planes function well enough to differentiate Levinas from Heidegger, they also function to obscure the unique understanding of time that previously supported Levinas's discussion of alterity.

Levinas particularly utilizes "Is Ontology Fundamental?" to sustain an extended conversation with *Being and Time,* as Heidegger refers to the text as "an investigation in fundamental ontology" (*BT* 238). Levinas questions the primacy of ontology for philosophy, specifically by claiming that ontology obscures the most important of all phenomena, the encounter with the other person. Heidegger uses ontology, the science of being, to attack the kind of philosophical intellectualism that allowed traditional philosophy to distance itself from everyday existence. He insists that reflections about being must originate at the place where such reflections matter, in the existential situation into which Dasein finds itself thrown.[2] For Heidegger, ontology is fundamental because Dasein's primary concerns are ontological.[3] Heidegger appears to have effectively refocused the concerns of philosophy on the individual, and away from the universal. Levinas never ceases to regard this as an important philosophical advance, though in "Is Ontology Fundamental?" he questions the success of Heidegger's bold rebellion against Western philosophy.

Levinas does not appear to have a problem with Heidegger's rejection of idealism, especially as it seems to engender intellectualism. "Ontology," Levinas points out, "is the essence of every relation with being and even of every relation in being" (*BPW* 4). These relations, to being itself and among beings, are riveted to a common understanding of being and its machinations. Anything that is incomprehensible according to ontology is dismissed, overlooked, or considered *inferior* to that which is comprehensible within the logic and reason of being. If, as Heidegger claims in *Being and Time,* ontology is fundamental, then that which fails to register ontologically also fails fundamentally.[4] The connection between being and comprehension is the essence of ontology for Levinas. Ontology presumes that everything in being is ontologically comprehensible. Everything rests on "the ontological relation," Levinas claims, even "Being-with-the-Other" (*Miteinandersein*) (*BPW* 6). By subordinating every relation within being to Dasein's relation to being, Heidegger is placing ontology in a fundamental position.

By allowing ontology to remain fundamental, claims Levinas, Heidegger has unwittingly reverted to the Western tradition of thinking about *beings* in light of a universal understanding of *being.* Heidegger has allowed Dasein's individual situation to be the source of his ontology, but this ontological structure is then presumed beyond the scope of Dasein's individuality.[5] The reconstruction of ontology, which moves through Dasein's situation, repositions universal claims about being out in front of the other person.

In order to think about the Dasein's relationship to existence, Heidegger must therefore impose a broad comprehension of being through which the situation of Dasein can be interpreted and understood. This may look different than Plato's universal *eidos*, but it is itself a kind of ideal and remains grounded in the tradition of thinking about the individual in light of the universal. On these grounds, Levinas challenges the fundamental nature of ontology, claiming that there is something more fundamental than the structures of being. Ontology misses the critical feature of the relation with the other, which cannot be configured by way of ontology because the other defies comprehension (*BPW* 7).

Levinas claims that the relation with the other, outside of the comprehension of ontology, is best understood in the mode of *religion*, though he invokes this term without "pronouncing the word *God* or the word *sacred*."[6] The encounter with the other person is incomprehensible because the other escapes "comprehension and possession" through language. Levinas deems it a critical flaw of ontology that it must always recoil from the incomprehensible; this is not so with religion. Levinas's invocation of religion is an attempt to break free of traditional epistemologies. If something incomprehensible can be important philosophically, then ontology is incapable of even approximating such a field of philosophical inquiry. In contrast, religion, and Levinas certainly has in mind Western monotheistic religions, celebrates the incomprehensibility and transcendence of God (*BPW* 7). For Levinas, the transcendence indicated in liturgy is not ontological, for worship does not begin with the presumption of comprehension.

Religion, as Levinas treats it here, humbly admits its limitations to *this* side of being and, furthermore, is prepared to admit that the relation with the other is dissimilar to the relation to "things" (*BPW* 8). Ontology, which is concerned with the comprehension of being, readily takes its bearings from the most comprehensible phenomena. Heidegger may have challenged the way we look at the things in our world, and he may have reconfigured the way we think about their being—between that which is ready-to-hand and that which is present-at-hand, for instance—but he has not ceased to define them according to their *being*, and therefore, according to their comprehensibility with respect to being as it is generally conceived. The basic structure of being, whether it is considered with respect to Dasein's being-in-the-world or with respect to the way Dasein takes up the tools and things of being, remains bound to comprehension. Yet both language and the other person who speaks defy comprehension and therefore exceed the ontological relation.

Heidegger's ontology, Levinas claims, is ruthlessly reductive. For Dasein, the other person is afforded the "ruse" of particularity and independence, but this is for Levinas a hunter's trap.[7] The supposed freedom afforded to Dasein for its individuation and authentic being-toward-death is quickly reinscribed within the field of reason and comprehension (*BPW* 8). Heidegger's ploy, as Levinas reads him here, is to press human beings into "the categories adapted uniquely for things" (8). Once we are supposedly liberated by the sheer individuality of Dasein's authentic being-in-the-world, Heidegger then reinscribes this liberation within the broad comprehensibility of being. This is fundamental ontology; all people and things are beholden to a central understanding of being.

Levinas demonstrates that fundamental ontology gives philosophy back to idealism and wrenches philosophy away from the particularity of the face-to-face relation. Poised to reconsider the relationship between the subject and the other person, Heidegger instead returns our attention to meditations on the nature of being as it manifests in Dasein's internalized self-relation. He positions the other person as merely alongside others. The relation with the other person, the very site of philosophy's limit, is again subordinated to the grand science of being.

Is Levinas Right about Heidegger's *Mit*?

Levinas combats this move in Heidegger by enlisting a surprisingly spatial set of images to demonstrate the other person's exteriority to the realm of being and its demands for comprehension. The other person, he claims, is not bound to my power or my property, nor is the other person beholden to any common third term that might allow the other to be encountered "upon a familiar foundation" (*BPW* 7).

Levinas's analysis appears to relate to part 4 of *Being and Time*, which pauses to discuss the unique case of the "Dasein of Others"

(*BT* 153–63). At first glance, Heidegger's treatment of other people seems to resist Levinas's summary. Other people are not like other things that Dasein encounters in the world, Heidegger tells us; they are neither ready-to-hand nor present-at-hand (154).

Heidegger offers a helpful example: when I walk along a beach and see a boat that is anchored at the shore, I encounter something as being. Heidegger calls the being of a "thing" being-in-itself (*BT* 154). The boat may be utterly strange to me, but as it exists ready-to-hand, it is encountered without separation from the one "who undertakes voyages with it." The boat is still "indicative of Others." I encounter the boat as Dasein; moving ecstatically into a world that is available to others, even as it is already also "mine." The things of the world are objects that bear the peculiar characteristic of being bound to the Dasein of others. Heidegger admits that he omitted these concerns from the first sections of *Being and Time*. For the sake of simplicity, he there described Dasein's encounters with the world only in terms of ready-to-hand or being-at-hand. But here, he begins "characterizing the encountering of Others" (154). When I see a boat tied to the shore, I happen upon an object that has a being-in-the-world that is bound to the Dasein of another. And this makes other persons a unique case in Dasein's analysis of its environment.

Heidegger establishes that people *appear* as part of the environment, even if they do not appear as mere things within it: "Others are encountered environmentally" (*BPW* 155). Furthermore, Heidegger does not wish to universalize Dasein's being-in-the world, or even Dasein's Dasein (being-there). Other persons cannot strip Dasein of its particular orientation; but neither can I, in my being-there, presume a great deal about the Dasein of others. Here Heidegger seems particularly well situated to gesture toward the alterity of the other person, toward the incomprehensibility of the being-there of others. But as Levinas suspects, Heidegger is willing to go much further and extend the implications of Dasein's analysis of the world to the others who share the world and lay competing claims on the things of the world: "Thus in characterizing the encounter with *Others*, one is again still oriented by that Dasein which is in each case one's own" (154).

Heidegger worries that we might do too much to help the people who are "Daseins with us [*die Mitdaseinden*]," and therefore rob other Daseins of the necessity of their individualized care (*BT* 156). One must not disburden the other completely of his or her concerns. Citing examples like "welfare work," the provisions of "food and clothing, and the nursing of the sick body," Heidegger's concern is that by "leaping in," Dasein might do the other a disservice by completely disburdening him or her (158). He fears this might leave the other person worse off, dominated and dependent, even if the other is not aware of this dependence. So he suggests that we "leap ahead" of the other in order to find ways to facilitate the other person in becoming authentic (159). For all the complications of Heidegger's lexicon, he seems to be advocating a kind of teach-a-person-to-fish ethic. The being-there of others is a part of Dasein's environment and, therefore, part of Dasein's concern. Others are not to be ignored; their appearance in the world requires solicitude. Claims that Heidegger has no ethical structure in *Being and Time* are misguided; he has plenty to say about how Dasein should be with others. The presumption, in the way Heidegger advocates that Dasein offers help, is that other persons share the common ontological problematic. They need help becoming authentic, and should be addressed as if this were the case. Heidegger presumed that because I share a common, fundamental ontology with the other, *his or her problems are fundamentally similar to mine.*

Levinas's accusation is that Heidegger has allowed the logic of Dasein to dominate the way the self encounters the other. The needs, cares, sufferings, and joys of the other are configured by Heidegger through the analysis of being that is Dasein's own. Heidegger is not without an awareness of this problem. He reminds us that when we encounter others we will be tempted to concoct all sorts of explanations of the "Being-present-at-hand of Others" (*BT* 155). The *vorhanden* of the other person is therefore outside the scope of comprehension for Dasein's analysis. But this does not stop Heidegger from suggesting that other Daseins are thrown into being in the same way and that they should therefore be aided with this assumption in mind. Levinas appears to offer a careful reading of Heidegger, which he first began in *Time and the Other* and revisits here. The presumptions that Heidegger makes about the being-there of others are precisely what Levinas has in mind when he uses the phrase "fundamental ontology." There is an ontological foundation upon which Dasein and its neighbor encounter one another, the foundation of Heideggerian being. Levinas appears to have this right. For all of Heidegger's critiques of the default mode of "presence" in Western philosophy, by employing the preposition *mit* (with), Heidegger appears to create a common moment in which Daseins face down the same anxiety, the same nullity, the same need for authentic facing of one's own death.

Levinas proposes that we know far less about the other person than Heidegger's configuration will allow. In "Is Ontology Fundamental?" he consistently uses "horizon" and "field" to denote the insular world of the self. The "face" becomes a "breach," a fissure in this otherwise seamless "horizon" (*BPW* 10).[8] The face signifies the insufficiency of ontology as a starting point for philosophy. Ontology can think no further than "mere reflection on the self or on existence" (*BPW* 10). By beginning with Dasein and reflecting on existence, Heidegger offers only "the tale of a personal adventure, of a private soul, which returns incessantly to itself, even when it seems to flee itself" (10).[9] This strikes Levinas as a game of power, where the relations between persons are encapsulated into the struggle of being.

Levinas suggests that ontology works fine for an analysis of the field over which I have power, which is really only the field of representation that appears paradigmatically in art. But ontology is dangerously misleading when applied (as fundamental) to the encounter with the face of the other. Because the other person remains irreducible to the comprehensive sweep of reason, "the relation with the other is not therefore ontology" (*BPW* 7). To address the other is to speak in powerlessness, reverence, and toward a transcendent realm that escapes comprehension and representation. For Levinas, repeating an idea he invoked in "The Transcendence of Words," speech to the other is best understood as prayer" (*OS* 149). Ontology simply cannot abide prayer, whose address necessarily exceeds the scope of being. To speak to the other, as to pray, is to neglect the need to have full comprehension of the other in terms of "universal being" and to address the other as "the particular being that he is" (*BPW* 7).

Horizons and Openings

"Is Ontology Fundamental?" has significance for its clarity and shift in terminology. There is a striking absence in this essay not just of any significant reference to time but also of any hesitation to embrace the spatial language of horizons, fields, and realms. Ontology, as Levinas summarizes, relates to a "luminous horizon," an orb of sensory experience that surrounds the ego that beholds the world (*BPW* 10). The face appears on this orb, irrupting and unsettling it; it is a "breach in the horizon" of the ontological field, a crack in the screen. The face is *like* an opening, but even this, Levinas admits, "would be to make it relative to the environing plentitude" (10). These early reflections on the face will give rise to a much more robust treatment ahead. For here, the face of the other appears as a rupture in my realm, something incommensurate with the features of my field of freedom and power. It is a rift deeper than a mere feature of the being it irrupts.

Why does Levinas fail in this essay to correlate this field of freedom with his recent and robust treatment of the instant? In *Existence and Existents* and *Time and the Other* this is precisely how Levinas describes and defines the freedom and power of the subject. The subject, Levinas points out, is master only of the *present instant.* Levinas unexpectedly shifts gears, referring here to an undetected "sphere of relations" rather than the encounter with alterity in time (*BPW* 10). A number of problems will result from this shift, including the configuration of *infinity* according to the spatiality of horizons.

Still, Levinas has not forgotten the temporal edge that he utilized to differentiate between *sight* and *sound* in the 1940s. In "Is Ontology Fundamental?" Levinas extends his preference for auditory encounters over the visible sort. In speech, claims Levinas, one accepts and takes account of the other, even without comprehending the other. Comprehension belongs to the field of objects; however, to speak to the other is not to possess or consume but to acknowledge the very particularity of the other.[10] Interestingly, Levinas here discusses speech in the active sense of invocation.[11] In *Existence and Existents, Time and the Other,* and "The Transcendence of Words" his preference for the spoken word over image had more to do with *hearing* than *speaking.* But in "Is Ontology Fundamental?" he is concerned both with how one addresses the other and with the way the other addresses the self. Likewise, if we read Levinas's discussion of language and speech here alongside his later discussions, we discover a stark contrast in this regard. Here Levinas discusses the way one addresses the other, whereas in later writings *first* speech will belong irrevocably and unambiguously to the other.

Here we see that speech creates a kind of presence in which the self and the other are nonpossessive *partners* in a relation that defies the necessity of an ontological ground.[12] The self and the other have an encounter, Levinas argues, that is not dependent on a first ontology. Ontology, for its part, is almost certain to obscure and condition this encounter and therefore limit moral obligation to ontological structures for the relations between beings. But in describing the way interlocution denies the need for fundamental ontology, Levinas neglected several concepts that he featured in the 1940s and will return to in the 1970s and 1980s.

We might expect Levinas to show some ongoing alignment with his previous understanding of time in his discussion of murder in "Is Ontology Fundamental?" He has, after all, recently reflected on the connection between the alterity of death as it relates to the other (*TO* 69–74, 76–79). But even as he reflects on murder, Levinas outlines the phenomenon in terms of spatiality. The other person, who may *appear* in my ontological field as another object, is clearly *not* like the other objects found under this horizon.[13] I can treat the other as an object, as in slavery, and perhaps deceive myself into deeming this treatment ontologically legitimate. Levinas notices that "the other is the sole being whose negation can only announce itself as total: as murder. The other is the sole being I can wish to kill" (*OS* 9). I can kill another person, as a hunter slays a beast or as a lumberjack fells a tree, but unlike these objects, the other person has never *actually* entered into my field of power. In death, the other person "has escaped me." To encounter a face is to encounter that which only *seems* to enter the interior space of my power and freedom. By killing, I merely affirm the radical exteriority of the other. In death, as in life, the other is beyond a horizon that philosophy has too often neglected. Notably, Levinas has exchanged a deeply time-oriented discussion of death in *Time and the Other* for a spatially configured paradigm here in "Is Ontology Fundamental?"

Levinas is attempting a thorough overhaul of philosophical reflections on being. He is attempting to think *beyond* being and ontology but is constrained by the fact that the philosophical tools at hand are laden with ontological connotations. Since Levinas basically agrees with Heidegger's revolt against intellectualism and his critique of the negligence of traditional philosophy on time, they are both in a similar quandary with regard to philosophical language. In *Time and the Other*, Levinas raises two kinds of metaphors for the separation between the self and the other: temporal and spatial. In both cases, Levinas is attempting to speak of a separation that will not fold into unity. The temporal separation is configured primarily according to the futurity of the other, and the spatial separation is very cautiously framed in the language of exteriority. But in "Is Ontology Fundamental?" Levinas has chosen only to develop the spatial metaphor. Perhaps hindsight can judge this development a misstep, but differentiation from Heidegger is now of supreme importance to Levinas. The spatial imagery becomes handy in this regard given that Levinas is proposing that philosophy must be able to think *outside* the confines of being and ontology.

We cannot accuse Levinas of abandoning his innovative understanding of time, for it will make a substantial reappearance in *Totality and Infinity*, often in support of themes first raised in this essay. But we can conclude that a substantial transition has occurred. If the spatiality of Levinas's ethical philosophy is considered problematic, as Derrida will claim, the lack of caution demonstrated in "Is Ontology Fundamental?" may be near the source of this trouble.

"Freedom and Command"

Levinas begins, in the early 1950s, to escalate his emphasis on the relationship between ontology and violence. This is perceptible in "Is

Ontology Fundamental?" particularly in regard to Levinas's discussion of the "impossibility" of murder. Like many features of Levinas's oeuvre, his worries about the relationship between philosophy and violence increase as the years pass. Even in his earliest writings there is a perceptible concern with violence, which resurfaces and receives significant attention in the 1950s, particularly in the essays "Freedom and Command," "The Ego and the Totality," and "Philosophy and the Idea of Infinity." The writings of the 1940s appear to leave Levinas primed to discuss the nature of violence in light of the violation of the time of the other. This will eventually be the route that Levinas takes, but in these essays he seems to only deepen his commitment to thinking about the relationship with the other through almost exclusively spatial terms. Once again, we do not find here that Levinas explicitly controverts his original understanding of time in the early writings, but our search for the presence and relevance of Levinas's understanding of time must again go underground.

Levinas's growing concerns about the relationship between philosophy and violence in the 1950s lend energy to his many spatial metaphors. War, as Levinas will teach us, is the ultimate act of totality, the evasion of the face, and the exclusion of exteriority.[14] Levinas appears to grow increasingly aware of the inability of ontology to offer anything but *war*. His critique of the way early Heidegger positions ontology as fundamental leads directly to a burgeoning concern that fundamental ontology is defenseless against violence. For Levinas, the science of being is merely the analysis of a struggle in which the other person is inevitably reduced to an *obstacle*.[15]

In "Freedom and Command," Levinas points to the way that violence replicates labor. In violent conflict, opponents view one another "like an engineer measuring the effort needed to demolish the enemy mass" (*CPP* 19). In the midst of this struggle to overcome in battle, adversaries seek to overcome the freedom of the other, but they do so by way of reducing the freedom of the other to "an animal freedom, wild, faceless." This involves a defacing of the other, for to encounter the face of the other is to regard the other person in the height of his or her freedom (19).[16] Against the backdrop of the defacing nature of war, Levinas suggests that the face opposes me, but not in the conflict of being. Instead, I am confronted from *outside* the battlefield of ontology.

Briefly, and perhaps unintentionally, Levinas utilizes the language of time to describe the opposition of the face: "It is an opposition prior to my freedom, which puts my freedom into action" (*CPP* 19). The freedom of the other *precedes* the battle and opposes the bloodshed of war. In war, the face of the other must be converted into a hostile force. This participation in struggle exploits the weak and vulnerable side of the other. The other is "absolute" in his or her alterity, but in the relationship of war, this otherness is reduced to a savage force. In the bodily struggle of being, the alterity of the other is therefore bypassed, and the pacific opposition to violence is obscured. In war, one takes advantage of the weakness of the body of the other. The alterity of the other, "what is strong and absolute in him," is "ambushed" by the exploitation of what "is weak in him" (19).

Violence is the evasion of alterity; it "consists in ignoring this opposition, ignoring the face of a being, avoiding the gaze" (*CPP* 19). Violent action must reduce the freedom of the other to the stubbornness of a tree reluctant to fall or a boulder that is difficult to move. The encounter with the face requires either this reduction to *labor* or it appears as revelation, utterly exterior to even the form and categories in which it first presents itself. The face is either an animal or it is utterly independent in its externality.

The face of the other offers an opposition to my way-of-being, a threat to the stability of my ontological coherence. This face, if confronted, questions the very fundamentals of my freedom. Much easier, and much more commonly, I ignore this fissure in my configuration of being and relegate the alterity of the other to a particularly complex object in my world. The structure of this relation is spatial; in the encounter of war, the self and the other bypass the actual encounter of faces. Opponents subsist internally, each relegating the other to a feature interior to the world of the self. This is the battlefield of ontology; and according to Levinas, ontology can only hope to analyze the way the struggle of being unfolds. On the battlefield, we are either the master or the slave in Hegel's dialectic. Ontology, as framed by Levinas, cannot think outside of these constraints.

Two interesting usages of temporality appear in this essay. First, in his discussion of war Levinas uses the concept of "ambush" to delineate the relationship of war. The term *ambush,* in fact, is for Levinas the very definition of war (*CPP* 19). On the battlefield, the Hegelian master-slave dialectic is vividly embodied—one does not go toe-to-toe with an opponent in war but seeks an advantage, a *priority.* One ambushes an opponent by getting out in front of, or perhaps circling behind, the adversary. Ambush avoids the supposedly neutral collision of "two substances or two intentions" (19); it circumvents the face-to-face encounter like a hunter dropping a net over unsuspecting prey. It is instructive to remember that Sergeant Levinas himself stood among French soldiers encircled by the powerful and clever Nazi forces in June of 1940.[17] To ambush the other is to be abundantly *proactive,* to ambush is to renounce passivity. One ambushes by getting up earlier in the morning than one's opponent, by actively positioning oneself prior to the other. Though Levinas does not configure *ambush* as a temporal term but rather as a kind of spatial evasion, *ambush* also functions as a circumvention of time, positioning oneself *before* the arrival of the other. By surprise, the other is prevented from surprising *me,* and therefore the other remains safely fixed as a feature on my radar. War is a constant game of sight and evasion; to be seen is to be reduced to a fixed quantity, a current situation, or a known commodity.

The second, if fleeting, reference to time appears near the conclusion of "Freedom and Command." Levinas refers to a "meaning" that is possessed by every being "before I constitute this rational world along with them" (*CPP* 22). He briefly refers to the doctrine of creation to refer to this priority of an intelligibility that is radically prior to the constitution of the self. The themes of this paragraph anticipate several developments on the concept of time that will receive attention later in his career: "Creation is the fact that intelligibility precedes me. It is just the contrary of the notion of *Geworfenheit.* This is not a theological thesis; we read the idea of creation out of the experience of a face" (22). The face arises like creation, *ex nihilo,* from a before that is not coincidental or even linear. One does not miss out of the moment of creation because one is too late to observe it but because the very condition by which one is constituted depends on the primordial, temporal priority of the act of creation. Therefore, in the face of the other, one detects a break in the very tapestry of existence on which the face appears. At least for a moment, Levinas's language echoes the philosophy of time that dominated his writings in the 1940s. This serves, at the very least, as a reminder that he is still on the pathway toward his mature expressions of diachrony, even if he remains dominatingly concerned here with exteriority and spatiality. The exteriority of the other is positioned here as radically as Levinas can frame; he even refers to this relationship as "religion, the situation where outside of all dogmas, all speculation about the divine or—God forbid—about the sacred and its violences, one speaks to the other" (23).

"The Ego and the Totality"

Up to this point Levinas has used the term "totality" sporadically, but in "The Ego and the Totality" (*CPP* 25–45), Levinas earmarks totality as a key philosophical theme. Given that the term will appear routinely in the middle stages of Levinas's career, including, of course, in the title of his most well known book *Totality and Infinity*, it is important to demonstrate the way this concept sustains the trend toward spatial configurations of the self-other relation. *Totality and Infinity* will, thus, underscore the internal conflict in that book between the metaphors of time and space. After the 1960s, Levinas will begin resolving this tension by way of increased emphasis on his innovative understanding of time.

In "The Ego and the Totality," Levinas continues his increasingly poignant discussions of the face, and solidifies the technical and practical difference between interiority and exteriority. Levinas also continues to draw liberally from religious language in his attempts to gesture toward that which exceeds ontology. Above all, "The Ego and the Totality" reflects on politics and the ordering of society. The Levinas of the 1950s is clearly worried about the connection between politics, justice, and the relationships between individuals. And among these discussions about money, psychoanalysis, and communism,

Levinas delivers a most strident critique of love.[18] His concerns about love are driven by a fear that intimate societies can replicate the "false totality of the ego," that the exclude the "third person" or the outside, and in doing so they form a kind of autarchy (30–31).

The tension between temporal language and spatial language abounds, even as Levinas attempts to speak of the internal and external function of the ego, the society, and the other. Returning to a theme that first appeared in the 1934 essay "Some Thoughts on the Philosophy of Hitlerism," Levinas considers the way *pardon* functions within intimate societies. Pardon, we recall, was originally configured by Levinas as the *defeat* of time. The genius of Jewish and Christian pardon, Levinas explains, is that it turns the tables on both the power of time and the inevitability of a future bound to past wrongdoings (*UH* 14). The past is rife with exploitation, pain, and suffering, but the past is *undone* by pardon. Pardon cannot be achieved in solitude, which was one of the reasons that Levinas pointed to the tie between time and the other person. The past, the present, and the future are features of intersubjectivity, and the *present,* where the subject is grounded in solitude, is without resources to obtain pardon. Forgiveness therefore must move from the *future,* from the other. This understanding of pardon was utilized in both *Time and the Other* and *Existence and Existents.* Pardon is the redemption of time; it excuses the subject of the weight and consequences of existence. There is no ambiguity in the sustained temporality of this term in Levinas's overall work. Levinas invokes the theme of pardon in "The Ego and the Totality," but in this essay he expresses deep concern about the evasion of responsibility that pardon allows.[19]

Pardon was explicitly configured as a function of time in *Existents and Existence,* where the alterity of the other approaches me in the solitude of my instant and my very being is "pardoned by the alterity of the other" and therefore given a future (*EE* 96–97). The economy of being provides ample opportunity to compensate for our need for *time.* Economic time, which marks the movement of clocks and trains, serves to distract and deceive the subject into an artificial pardon for the weight and guilt of being. In *Existence and Existents,* the labors of economic activity merely provide the illusion of pardon. But labor cannot make full reparations for pain, for being, even if it promises as much. Levinas calls for an eschatological hope, a hope in salvation that must come from a future that is impossible according to the rules that govern world time. Pardon is the future that is unreachable from the present, from the instant, from the now.

In "The Ego and the Totality," we see that pardon retains some of these elements but is now configured in a primarily spatial sense. The self presents an internal world. False pardon therefore becomes the self-delusion that rectifies injustice and fulfills responsibility according to an internal sovereignty (*CPP* 30–31). Yet the ego remains heavy with guilt, for its action creates irreparable damages. When self-delusion fails to address the guilt I carry for intentional and unintentional violence, I need another person to liberate me through *pardon.* "I am shut up in my own portrait," writes Levinas (34). From the privacy of the ego's world, I can construct a private religion where I can grant myself pardon, but this is pure self-deception, like Crusoe creating a society out of the pieces of shipwreck (30). When these delusions fail, the need arises for a face to grant forgiveness. And here Levinas sees the grave danger of insular societies, whose intimacy is forged at the expense of the external world. Modeled after the insularity of the self, the "intimate society" provides a new delusion of pardon. The victimized other, wounded but wrapped into the internal logic of the community, offers forgiveness and absolution. The other's words of pardon imitate the hope of justice and says the word the ego desperately needs to hear. But the insularity of this "violence" and forgiveness is deeply artificial, for it pays no heed to the "third person" who is influenced and wounded but not addressed in the illegitimate pardon of the intimate society. For example, a battered spouse pronounces forgiveness as a child cowers nearby. "The couple is a closed society," writes Levinas, and this enclosure makes the couple "oblivious of the real evil" (31).

Nothing here refutes the earlier treatments of pardon as temporal; one can easily imagine Levinas adding a paragraph about the way insular societies attempt to replicate the self-deluded isolation of the instant. But at the heart of "The Ego and the Totality" is an increasingly active configuration of the self–other relation as spatial, as a matter of internal and external relations. The key contrast in this essay is between a simple existence and a complex one. The simple existence, which Levinas compares to the "immediate" lives of animals, takes in the world around it as "nourishment," as "nutriments" (*CPP* 26). When the world becomes a canvas of possible enjoyments, it is reduced to features of the internal world of the self. "A simply living being is thus in ignorance of the exterior world" (25). The internal society, along with the internal ego, settles for an animalistic existence that never does more than simulate encounters with exteriority. The "simply living being" cannot encounter exteriority:

> A living being as such is then not without consciousness, but it has a consciousness without problems, that is, without exteriority, it has a purely inner world whose center it occupies. This consciousness is not concerned with situating itself relative to an exteriority, does not grasp itself as part of whole (for it precedes all grasping). . . . The inwardness which, for a thinking being, is opposed to exteriority, occurs in a living being as an absence of exteriority. There is nothing mysterious in the identity of a living being throughout its history: it is essentially the same, the same determining every other, without the other ever determining it. (26)

This is a replication of the treatment of the instant in Levinas's early writings, but here the present is converted to interiority. A complex existence acknowledges and attends to that which lies beyond the interiority that produces the natural history of a living being. Conceptually, Levinas continues to refer to the ancient tension between the same and the other, the ego and the totality, and the individual and the world. Simple existence is interiority; the encounter with the other exposes one to exteriority. In terms of time, Levinas has replaced the time-driven metaphor of the instant with the spatially-oriented metaphor of interiority. The shift appears to be somewhat benign, for one can express *excess* in both spatial and temporal terms. But there are consequences, however unintended, for this transition in terminology.

Levinas does not hesitate to utilize religious terminology in this essay, yet the concept of transcendence now takes on a nearly exclusively spatial connotation. Whereas in *Existence and Existents* the religious notion of Messiah was pronounced as characterizing the future that can only be conceived in hope, now we find Levinas invoking a God who is "condescending," "transcendent," and "exterior to this world" (*CPP* 29). Pardon can only come from outside the totality in which the ego languishes. Hence, only God, or a saint, can grant pardon.[20] Pardon cannot emerge from the enclosures of society and being but must come from the exterior.

We need not belabor our review of "The Ego and the Totality," which well represents the transition in Levinas's thought throughout the 1950s. Before exploring the way this transition impacts the key term "infinity," we should also note a softening in Levinas's previously stark pronouncements about images, vision, and sight. This softening seems to be related to the fact that the face, and particularly the colorless pupils of the eyes (cf. *EI* 85), is encountered at a visual level. Levinas comments, in a contemporaneous article about Simone Weil, that "The face of a man is the medium through which the invisible in him becomes visible and enters into commerce with us" (*DF* 140). The *visage* of another person is the way in which the alterity of the other makes itself known in the otherwise airtight world of sight and vision. In order to accommodate his emphasis on the face of the other, Levinas must refine his harsh treatment of images and representation. At least in the case of the face, an image is far more than reality's shadow. The play between the visible and invisible is itself a metaphor of space: the other hides from vision and appears within the field of vision.

"Philosophy and the Idea of Infinity"

In "Philosophy and the Idea of Infinity," Levinas further develops the notions of transcendence, distance, and exteriority, with a particular eye for the way Greek philosophy has positioned the Western tradition. Though Levinas traverses a number of topics in this essay, the concluding pages leave no doubt about the overall thesis: Levinas is chiefly concerned with delivering a clear, resounding refutation of Heidegger's fundamental ontology.[21] If *On Escape* initiated a quiet departure from Heideggerian phenomenology and *Existence and*

Existents named this itinerary explicitly, then "Philosophy and the Idea of Infinity" completes the journey. Levinas remains appreciative of the brilliant innovations of Heidegger, but Levinas fears that the final legacy of *Being and Time* is a confirmation of the worst trends in philosophy. Heidegger's critique of traditional philosophy left him in a position to judge philosophy's reduction of heteronomy to autonomy, particularity to generality, and alterity to familiarity. Instead, Heidegger turns away from this opportunity and consequently ends up "affirming a tradition in which the same dominates the other" (*CPP* 53).

This sharpened critique of Heidegger helps us understand the overall theme of the 1950s for Levinas. The return of *time* in *Totality and Infinity*, especially in the conclusion and the preface, will confirm that Levinas's creative thinking about time has merely gone underground, not disappeared. But throughout this decade Levinas has honed and sharpened his critique of ontology, and Heidegger in particular. He asked the question "Is Ontology Fundamental?" to start the decade, and he clearly doubted it then. By 1957, he proclaims that ontology is essentially atheism, or paganism, and that fundamental ontology is defenseless against violence and the "will to power" (*CPP* 52). Heidegger, for all his brilliance, is more dangerous than Socrates and Plato.[22]

For Levinas to attack Heidegger in this manner requires a rather delicate operation. As he lets us know in *Existence and Existents*, the way forward past Heidegger cannot be a return to something pre-Heideggerian. It remains on the shoulders of *some* Heideggerian themes that Levinas's philosophy stands. But in critical respects, Levinas believes Heidegger does far more harm than good. "Philosophy and the Idea of Infinity" performs a surgical catharsis of the worst in Heidegger, seemingly for the sake of moving toward a philosophy that is post-Heideggerian. The term *infinity*, critical enough to be the key term opposed to *totality* in the 1961 title *Totality and Infinity*, is the instrument Levinas utilizes to exorcise the elements in Heidegger that present such danger for philosophy and ethics. Heidegger, suggests Levinas, is the "apogee" of thinking about the relationship between finitude and infinity (*CPP* 52). If philosophy has gradually forgotten

Plato's respect for the "Good beyond being," Heidegger has made this negligence intentional and constructive.

The thesis of *Being and Time,* as summarized by Levinas, is that "Being is inseparable from the comprehension of Being; Being already invoked subjectivity" (*CPP* 52). This means that the very idea of being is already a feature of Dasein's own being. Levinas now claims that by binding the meaning of being to Dasein's being, Heidegger has refused to think of infinitude. Heidegger's philosophy must settle for thinking about finitude; it cannot think of metaphysics, but only ontology, the science of being. By confining the focus of philosophy to ontology, Heidegger has severed the relationship between being and absolute transcendence. Levinas thinks Heidegger has liberated Dasein, loosening any moorings exterior to Dasein's own analyses of the world. He therefore yields to a "long tradition of pride, heroism, domination and cruelty" (52). Heidegger, as Levinas reads him, confines philosophy to the realm of being, to the natural, to the maternality of the earth; life's meaning is therefore found in a pagan alignment with the familiar landscape of matter and things (53).

According to Levinas, this renders philosophy "ethically indifferent" (*CPP* 53). Guilt, if it even arises, would derive from one's inauthenticity, as a failure to enact autonomous existence. Such a philosophical orientation banishes all "guilt with regard to the other" (53). Heideggerian ontology is therefore narcissism. All meaning, purpose, and fulfillment are found within the identity of the same. In this tendency to reduce the other to the same, Levinas finds Heidegger returning to Socrates and Plato. For their part, the Greeks sought to move beyond the apparent opposition between the familiar and the strange, pushing further into the "beyond" where multiplicity and singularity collapse into unity. Heidegger leads us to believe he has unseated this idealist pursuit of the "good" and redirected philosophy's attention to the "real singularity" of beings (51).

Levinas chooses the terms the *same* and the *other* in this essay to realign Heidegger with the tradition he has rejected (*CPP* 48).[23] For Plato, nothing encountered in the world can be unfamiliar, novel, or unrelated to the ideals toward which every manifestation in being gestures. However deficient and temporary, everything under the sun appears on a metaphysical continuum. Levinas points to the *Phaedo,* where the human "soul" itself is related to the realm of the *eidos* (51).[24] Because of this, the human soul can never encounter anything foreign to it; difference is a matter of perspective. This means reason is "appropriation and power," the act whereby difference and exteriority are shown to be ultimately illusory. Under this all-seeing eye, an idea honed in particular by Aristotle, apparently exterior beings must be generalized, categorized. As Levinas summarizes: "And here every power begins. The surrender of exterior things to human freedom through their generality does not only mean, in all innocence, their comprehension, but also their being taken in hand, their domestication, their possession" (50). This is the way of science, to reduce alterity to a feature of the same. It positions philosophy as atheism, for it can think of no exterior that is not already interior.

Heidegger identifies a fissure in this consideration of being, particularly in the way in which it presents the imposition of formal and disincarnate ideas into the meaning of the existent. But his reversal retains the methodological reduction of alterity to familiarity. Levinas writes, "Let us first observe that this supremacy of the same over the other seems to be integrally maintained in the philosophy of Heidegger" (*CPP* 51). Heidegger has simply replaced the primacy of the transcendent *eidos* with the primacy of man. He has erected a new neuter, some neutral concept around which thought and beings align themselves. Between one face and another, Plato has placed his *eidos,* and Heidegger has merely replaced this third party with his own: the subjective being of Dasein. This is not a new suspicion of Levinas but a repetition of the critiques found in *Time and the Other* and *Existence and Existents.* Humans are gathered together (*Miteinandersein*) facing and confronting a common truth about the subjectivity of existence. This third, this neuter, functions with reductive force. All alterity must be subordinated to the powerful reach of Heideggerian ontology. There is no "outside" this truth, no confrontation with anything exterior to its reach.

Yet it is not enough for Levinas to accuse Heidegger of falling back in line with the Socratic thinkers. Heidegger makes matters *worse,* for he pronounces the end of metaphysics, the end of thinking beyond the confines of being and its obsession with power.[25] Heidegger retains the worst of philosophy's arrogant faults and undermines philosophy's humility. The atheism of philosophy is replaced by Heidegger with sheer paganism. Perhaps Levinas's critique of Heidegger never gets sharper than this: "Heidegger does not only sum up a whole evolution of Western philosophy. He exalts it by showing in the most pathetic way its anti-religious essence become a religion in reverse. The lucid sobriety of those who call themselves friends of truth and enemies of opinion would then have a mysterious prolongation! In Heidegger atheism is a paganism, the presocratic texts anti-Scriptures. Heidegger shows in what intoxication the lucid sobriety of philosophers is steeped" (*CPP* 53).

The paradigmatic experience of the infinite is, for Levinas, the face of another person: "The idea of infinity is the social relationship" (*CPP* 53). All other features of being are highly susceptible to generalization, reduction, and assimilation. The exterior world that arises before the self registers in the realm of finitude as intelligible and sensible. Not so the face, which gazes back at me: "To be sure, the other is exposed to my powers, succumbs to all my ruses, all my crimes. Or he resists me with all his force and all the unpredictable resources of his own freedom. I measure myself against him. But he can also—and here is where he presents me his face—oppose himself to me beyond all measure, with the total uncoveredness and nakedness of his defenseless eyes, the straightforwardness, the absolute frankness of his gaze" (55). The face renders me powerless, not because I cannot play the games of being against the other who stands before me, but because the face arises from a beyond being which does not enter into my world as an object to be appropriated. I can kill the other, but I cannot domesticate his or her face; the face of the human other is utterly other, purely exterior (55–56).

The face of the human other opens to the dimension of infinity, to the beyond that is not susceptible to my manipulation. This is the very beyond that Heidegger cannot consider, for the ontology of *Being and Time* cannot conceive of anything about the fellow human as independent of a common captivity to the neutral truth of being, the sheer fact of a world into which we are all thrown. Levinas believes that Heidegger has forgotten philosophy's humility, if the "atheism" of Socrates, Plato, and Aristotle can be considered humble.

Biblical religion offers another way of thinking about exteriority. The doctrine of *creatio ex nihilo,* to which Levinas returns briefly on several occasions during his career, epitomizes this exteriority. The first of the Ten Commandments underscores the irrevocability of divine exteriority: "You shall have no other gods before me."[26] No human freedom can circumvent this radical exteriority of God; the idea of God, itself a feature of the infinite, introduces a heteronomy that refuses reduction to autonomy. Metaphysics, as opposed here to ontology, is reflection on the beyond.[27] Plato simply failed to sufficiently differentiate between the same and the other, being and being's other. But Levinas wishes to proceed in Plato's general direction, against Heidegger, in pursuit of the Good, which in being is nowhere found.

So what of Levinas's understanding of time? Does it still operate here, amid the discussions of spatial interiority and the infinite transcendence of God and the other? It seems so, if surreptitiously. The idea of *creatio ex nihilo* presents a primordial "before" that is more ancient than any dynamic ontology can consider. But Levinas does not develop the temporality of creation here, as he will in *Totality and Infinity* (*TI* 63, 89, 104). The relationship of creation to being is most clearly a temporal one; *ex nihilo* does not promise God's exteriority to creation but God's *priority.* We can also see in "Philosophy and the Idea of Infinity" that Levinas has a renewed awareness of the weakness of spatial language to indicate the kind of alterity required to refute ontology. Plato thinks of height and transcendence according to *distance,* so Levinas proposes that we go beyond Plato, for "distance alone does not suffice to distinguish transcendence from exteriority" (*CPP* 47). Exteriority may be spatial, but it is spatial only in a loose sense, for Levinas will not let the spatiality of exteriority and interiority be reduced to a mere feature of distance, as though there is a long but perceptible distance between the same and the other. In this sense, he realizes that he must find a way to think about transcendence in ways *less spatial* than Plato.

Perhaps the most important gesture toward time appears near the conclusion of "Philosophy and the Idea of Infinity," where Levinas clearly assigns a temporal element to his analysis of the face: "No movement of freedom could appropriate a face to itself or seem to 'constitute' it. The face has already been there when it was anticipated or constituted; it collaborated in that work, it spoke. A face is pure experience, conceptless experience" (*CPP* 59). Here we see signs of what is to come. No matter how clearly one perceives or categorizes another person, the other escapes the grasp of reason and knowledge. This evasion is a product of *time*, the result of the *priority* of the other. "No movement of freedom could appropriate a face to itself" precisely because the movement is always too late. No *a priori* idea can prepare the self for the experience of the face, because the encounter with the infinite is an encounter with the ultimate unfamiliarity.

The encounter with a face is not an encounter with an object or an idea but "an experience in the strongest sense of the term" (*CPP* 59). The experience of the infinite in the face of another leaves the ego *judged*, shamed for its default blindness to alterity and infinitude. The experience of a face, because it gives way to infinitude, opens beyond the field of assimilation. This means heteronomy stands for more than just the relationship between the world and its creator. The faces of humans are windows to the infinite and, as such, they deny autonomy and the syntheses of reason and generalization. Levinas calls this idea of irreducibility and infinity a "disalignment" of the venerable philosophical tradition that presumed unity beneath or beyond diversity (59).

Though we only see traces of *time* in this essay, they appear at critical junctures and within a framework that is familiar to both the previous and future writings of Levinas about time. "Philosophy and the Idea of Infinity" seems to have completed a pivotal movement in Levinas's independent philosophical development. No one reading this essay could again align his philosophy with Heidegger's for long without running almost immediately into critical and stark differences. In his conclusion, Levinas points to the truly decisive issue at stake in these deliberations: justice (*CPP* 59).

It has become increasingly clear, at this point, that Levinas's central concern throughout the 1950s had been ontology. The dangers of fundamental ontology have become increasingly ethical, and the face has become the locus of transcendence. Levinas undermines Heidegger in the 1940s by challenging Heidegger's constructive statements about time and temporality. Then, in the 1950s, Levinas lays aside this critique, for the most part, and takes a fresh angle in his attack. The face, after all, makes its appearance in being. Even if Levinas is wrong about the fundamental passivity of the instant, and therefore the self, his meditations on the face represent an independent critique of Heidegger. The face, suggests Levinas, appears within being but calls the stability of ontology into question. The critique presented in "Philosophy and the Idea of Infinity," and in Levinas's overall attack on fundamental ontology, does not depend on his philosophy of time. The experience of the other, qua *visage*, reveals the dangerous inadequacy of ontology.

"Lévy-Bruhl and Contemporary Philosophy"

Of the many articles written by Levinas at the end of the 1950s, one essay related to the problem of representation allows us to see how time remains pivotal for Levinas's philosophy, despite its apparent recession to the background of his thinking. "Lévy-Bruhl and Contemporary Philosophy" deepens Levinas's critique of representation he offers in "Reality and Its Shadow." According to Levinas, Lucien Lévy-Bruhl questions the traditional priority of representation over *presentation*. Representation, he argues, privileges thought over experience and prioritizes reason over emotion. For Levinas, this reinforces the dominance of the same over the other and continues the tradition of idealism with its preference for unity and reduction. Levinas detects in Lévy-Bruhl a reversal of the terms and an opening in contemporary philosophy for thinking past the classical conception of "the relationship between subject and object" (*DEH* 115).

In his analysis of the work of Lévy-Bruhl, Levinas sets aside any evaluation of Lévy-Bruhl's sociological or psychological program and focuses instead on the ontological structure of Lévy-Bruhl's assessment of primitive human societies. "Whether accepted or challenged," asks Levinas, "have not Lévy-Bruhl's well-known ideas on primitive mentality marked the orientation of contemporary philosophy?" (*EN* 39). In *How Natives Think,* Lévy-Bruhl claims that the causal and logical categories most familiar to "us" modern humans are of little interest to the mentality of "the primitive" (*EN* 45). Primitive persons, according to Lévy-Bruhl, are absorbed in *experience* and *action.* In contrast, the Greeks prioritized representation over sensation. Sensing, for the Greeks, represents an inferior or incomplete appropriation of the world. Lévy-Bruhl's claim is that primitive persons do not experience "sensing" as an uncertain or "lame thinking," nor as a shortcut. Instead, summarizes Levinas, sensing registers and operates "in another dimension" (46–47).

This dimension is a reversal of the Western manner of thinking about encounters with things and persons. Lévy-Bruhl points to the integration of the supernatural world with the natural world in primitive societies (*EN* 40). This leads to a "permanent possibility of sorcery," undermining any trust in causality or in systematic ways of thinking about time and duration (46). The pervasiveness of supernatural forces offers no firm reduction to a feature of the present. Likewise, sorcery promises no reduction from its permanent priority over persons and things. One is born into, and in every moment faces, a mystical past that continues to operate mysteriously in the present.

Lévy-Bruhl states that such an understanding of existence denies primitive persons any progress toward the great pronouncements of modern philosophy and the Enlightenment. This unrestrained transcendence was encountered at the level of *emotion,* or *participation,* and not at the level of contemplation. Such exteriority is "stripped of the form that guaranteed thought a familiarity with it" (*EN* 47). Such reality is necessarily pluralistic; there can be no domestication of the dark beyond to the present. Exteriority is then experienced as "an exposure to a diffuse threat of sorcery, a presence in a climate, in the darkness of being that is lurking and frightening, and not a presence of things, confronting us *face to face*" (47).

An exchange takes place here that is often unnoticed. Primitive mentality, by Lévy-Bruhl's reading, was forced to embrace a kind of passive reception of an unknowable mystery. Idealist philosophy lays waste to paganism, mythology, and sorcery, and in so doing, renders all sensory ambiguity a merely deficient way of knowing. For Lévy-Bruhl, mystery begs for discovery and explanation. As Levinas claims in "Philosophy and the Idea of Infinity," this renders idealistic philosophy essentially atheistic; there is nothing exterior to the realm of the ideas, including the eternal soul of the thinker (*CPP* 53). Platonic idealism eliminates all vestiges of the supernatural. This leaves behind a "unique, genus-free character of situations and moments, their bare existence" (*EN* 47). A belief in unknowable alterity has therefore been exchanged wholesale for a belief in the opposite, that all mystery and alterity in the end give way to representation, knowledge, and unity.

Yet Levinas detects in contemporary philosophy a revolt against the dominance of Platonic idealism, particularly as idealism regards unchanging, stationary substance as the fundamental character of being: "Ever since Aristotle's metaphysics, substance has represented the ultimate and intimate structure of being; it is the term of the 'analogy of being.' It not only bears an idea of permanence and solidity—but also of a 'polarization' of experience and a mastery exercised by substance over attributes and actions" (*EN* 44–45). Heidegger, for his part, suspects that this default preference for *substance* over action and experience contributes to a forgetfulness of the more primitive nature of being. As Levinas explains, "Being, for example—in Heidegger and the Heideggerians—is not a being, but the being of beings. . . . The condition of every entity, the first revealed one, is not an entity" (45). The possibility arises, at the very least, that the "ultimate and intimate structure of being" may be actions, attributes, and becoming. This is the revolt of contemporary philosophy against idealism, to question its reliance on substance. This revolt appears in a variety of manifestations: as duration in Bergson, as intuition in Husserl, and as ecstasis in Heidegger.

Still, Levinas feels that despite philosophy's "destruction of substance," it has not managed to shake "the logical and grammatical priority of the substantive" (*EN* 45). The thinkers of contemporary philosophy revert to the vocabulary of substantial philosophy, subtly founding philosophy again on the static, stationary nature of substance. We can justifiably accuse Levinas of a similar mistake, especially in *Totality and Infinity*. His rhetoric of distance, exteriority, interiority, and proximity are metaphors of substance and spatiality. But in "Lévy-Bruhl and Contemporary Philosophy," Levinas points to Lévy-Bruhl's discovery of a structure of participation that precedes and overrides the traditional philosophical preference for *timeless*, motionless substance. At the heart of this discovering is a radically different perception of time:

> The being that is about to be is already a being that has traversed you through and through. And, at the same time, this determination and this influence are not causality—since the *I* that is in their grasp decides, is engaged, takes responsibility. The structure is that of a future already sensed in the present, but still leaving a pretext for a decision.... The being of primitive mentality is not general.... Time qua pure form is unknown to primitives; every instant has a different potential, contrary to the homogeneity of form time.... Henceforth, the past has a special format, it is mystical as past; it still acts by virtue of the fact that it was. (48)

Time, for Lévy-Bruhl's primitive, is experienced heterogeneously, off the condensed "single plane" that unifies persons and events into a homogeneous time. Lévy-Bruhl claims that for a modern person to save the life of a "primitive man," to restore him from immanent death, is to fundamentally compromise his life.[28] Paraphrasing Lévy-Bruhl, Levinas notes that the primitive person dwells in a time of "his" own, alive in the "complex network of mystic 'participations' in common with the other members, living and dead, of his social group, with the animal and vegetable groups born of the same soil, with the earth itself" (*EN* 49). So like murder, life saving threatens the alterity of the time of the other; it robs the primitive other of his or her independent moment in a nonlineal experience of time. Levinas compares this way of thinking about time with both Bergson and Heidegger.

To discuss this primitive mentality, Levinas dives beneath, and more importantly *before*, representation. He attempts to describe an experience and a participation that precedes the assimilation necessary for representation. Lévy-Bruhl, claims Levinas, destroys the absoluteness of truth "precisely by showing that representation is not the original gesture of the human soul" (*EN* 49). More original than representation is *participation,* and participation is not something one chooses or upon which one deliberates. The self has no mastery of the exteriority of the spiritual forces; "it is a prey to the events that have already determined it" (47). Whatever the lasting value of Lévy-Bruhl for anthropology, Levinas's investigation of his work on primitive humanity demonstrates that Levinas remains concerned about that which is *prior* to the wisdom of philosophy.

Some Evaluations

I began this chapter promising to explore the ebb in Levinas's philosophical writings about time. It is impossible, of course, to fully resolve the question of this receding tide. A number of possibilities have been left out. For instance, Jean-Paul Sartre was certainly on Levinas's mind during this decade, but he goes virtually unmentioned in Levinas's published writings.[29] This is also the decade in which Levinas began to write with intensity on Jewish sources, publishing a great number of articles on various aspects of Jewish life and thought. Levinas participated in the annual meetings of the Alliance Israélite Universelle, a group of Jewish European intellectuals, and began giving interpretations of the Talmud at these meetings in 1960 (*NT* xi). The issue of time arises in his confessional writings of the 1950s, and it is possible that during this period Levinas's thoughts on time are being incubated in his confessional reflections before reassuming a central place in his published works of the 1960s. Indeed, this practice is standard fare of Levinas's later work; he tends to move ideas into philosophy after experimenting with them first in his confessional works.[30]

In the 1950s, Levinas increases his interest in responsibility and represses his discussions of time. The omission of his formerly serious discussion of time allows us to see that *responsibility to the other,* and not *time,* is the driving force of this philosophical program. This may seem like an obvious conclusion to readers familiar with Levinas's later works, which contain radical expressions of obligation. But the extreme expressions of obligation in his second major work, *Otherwise than Being*, are thoroughly interwoven with Levinas's understanding of time. His 1947 publications also demonstrate an intense concern for time. Furthermore, Levinas's philosophy clearly relates to the projects of Bergson, Husserl, and Heidegger, each of whom was explicitly interested in pressing philosophy to reconsider the concept of time. If the "chicken or the egg" question matters in this investigation, the 1950s appear to side with *responsibility* as the driving force of Levinas's work. Levinas turns toward, away from, and then back to the concept of time in an intense search for language that befits the sense of responsibility he seizes as his central philosophical theme. So in the 1940s, as in the 1970s and 1980s, Levinas uses the concept of time to approximate the indescribable responsibility to the other person. Indeed, in his later work, time is Levinas's principal philosophical tool for thinking about otherness. This is decidedly not the case in the 1950s.

What drives Levinas's receding interest in developing his notion of time? The most compelling explanation for this recession appears to be Levinas's relationship with the work of Heidegger. Levinas's need to distance himself from Heidegger's philosophical agenda was surely reinforced by Heidegger's Nazi affiliation. Levinas recoils, in "Is Ontology Fundamental?" from the whole program of Heidegger's thought, apparently including Heidegger's declaration in *Being and Time* that philosophy needs to rethink *time* from the Presocratics forward.[31] The inclusion of "temporal existence" in Levinas's charge against Heidegger is telling; he is moving against Heidegger at every possible point. It seems compelling to suppose that Levinas suspended his rigorous investigations on time to focus his attention on overcoming ontology. This turn in his thought was already brewing when in 1944, while still in captivity, he pens this line: "An essential element of my philosophy—what makes it different from the philosophy of Heidegger—is the importance of the other."[32] This clearly outranks "time" on Levinas's itinerary during the 1950s.

There is no clear explanation for the disappearance of his *caution* about spatial imagery that was expressed so clearly in *Time and the Other*. The language of exteriority and separation, perhaps gleaned from his reading of Franz Rosenzweig, may be too tempting for Levinas to leave aside. And so the interiority/exteriority binary comes to take a central role in Levinas's deliberations about the self and the other. What seems most tempting about exteriority, in particular, is that it is unthinkable in Heidegger's structure for philosophy. It is not sufficient for Levinas to show, as he did in the 1947 publications, that his understanding of the ecstasis of time is different than Heidegger's. He must demonstrate, too, that Heidegger is only a thinker of interiority and that by binding philosophy to the machinations of being, however temporally configured, Heidegger has settled for a form of paganism.

As we will see in *Totality and Infinity,* the developments from Levinas's work from the 1950s are retained in his unapologetic dependence on spatial terms like interiority and exteriority. There are consequences to this drift into language of spatiality, interiority, and exteriority, for it is spatial imagery that soon allows Levinas to talk about the initial status of the ego in a "dwelling" alongside the enigma of the feminine (*TI* 152–74). Derrida, for his part, will not miss the way these spatial metaphors seem to be irrevocably intertwined with ontology. He will also find problems in Levinas's use of the face, especially given the necessary spatiality and luminosity of the appearance of the face of the other.[33]

FIVE

Between Four Walls

Eschatology institutes a relation with being *beyond the totality* or beyond history.
—Emmanuel Levinas, *Totality and Infinity*

Introduction

Levinas asks in the preface of his own book, "Can one speak of a book as though one had not written it, as though one were its first critic?" (*TI* 29). Answering this question in the affirmative, the preface to *Totality and Infinity* presents both a first reading and an initial critique of the work itself. The preface may also indicate that a transition has occurred between the first lines of the book and the ones written last.

The first sections of *Totality and Infinity* abound with spatial imagery, which tends to relegate the experience of time to the economic time that is still a function of interiority and is designated as "history." However, the last sections of the book renew the use of terms like *messianic time* and *eschatology* while configuring the transcendence of the other increasingly in terms of time. This transition across the very pages of Levinas's book is masked by the use of "eschatology" in the preface. Michael Morgan, for instance, claims that "Levinas does not delay introducing the notion of eschatology in *Totality and Infinity*."[1] Morgan's claim is technically correct; Levinas knows that readers will turn first to his preface. But if our interest lies in the development of Levinas's unique understanding of time, it is significant that the eschatological tone of the preface reflects the *final* portion of *Totality and Infinity*. These are the words written last.

Levinas reinforces the sense of transition with this statement: "How far we are in this preface from the theme of the work announced by its first sentence! Already there is question of so many other things, even in these preliminary lines, which ought to state without detours the intent of the work undertaken. Philosophical research in any case does not answer questions like an interview, an oracle, or wisdom" (*TI* 29). Furthermore, Levinas concludes the preface with a remark about the "very essence of language" in which his prefatory reflections find themselves situated. A preface consists of "unsaying the said" and of "attempting to restate without ceremonies what has already been ill understood in the inevitable ceremonial in which the said delights" (30). It invites us to see the ways that its words already unsettle the book that we, having encountered only the preface, have not yet read. By conceiving of the preface as the final portion of *Totality and Infinity,* as it came from Levinas's hand, we can detect the early swells of the final and enduring wave of Levinas's thinking about time: diachrony.

My examination of *Totality and Infinity* will track the growing tension between the metaphors Levinas invokes to demonstrate alterity. Levinas clearly favored time in his 1947 publications, but in *Totality and Infinity* he favors spatial configurations. This strain will be evident in the surprising limitations that arise from his references to light, darkness, space, time, distance, interiority, and exteriority. The title of this chapter, "Between Four Walls," is taken from Levinas's discussion "The Dwelling," which is the most *spatial* and the most *problematic* section of this book. It is in this section that Levinas utilizes the trope of the feminine to make possible the ethical relation. The pervasive use of spatial language in *Totality and Infinity* leads to the gender-related consequences of this emphasis, yet we find promising solutions and developments in the direction of time that occur in the conclusion and celebrated preface of the book.

The complexity and intricacies of *Totality and Infinity* defy any facile summary. This is a text of such breadth and depth that 50 years after its initial publication it continues to generate intense interest, exegesis, and ongoing scholarship. *Totality and Infinity* established Levinas as an important voice in twentieth-century philosophy, allowing him to transition into official academic positions in the French academy.[2] It is an unusual book in the history of philosophy in that its pages are full of vivid images alongside complicated philosophical formulations. Much of Levinas's style relies on the alignment of philosophy with practical and embodied language that philosophers typically avoid. But this is a difficult book, whose ideas vacillate between subtle and stunning, covert and overt. Levinas draws the reader in with the vivid language of hospitality, welcoming, and the face of the other, and then stretches the patience of his reader with long pages of arid philosophical deliberations. *Totality and Infinity* is a powerful challenge to the history of philosophy, claiming that Western philosophy, and not just governments and regimes, have settled for a philosophy of *totality*. Levinas writes this book to undermine the dominance of totality and to appeal to the infinite.

Focusing on the concept of time allows us to peer under the hood of *Totality and Infinity* and see an ambivalence that is not without consequence. Levinas's inconsistent emphasis on metaphors of time and space creates a unique effect on the arguments delivered. This is not to say that Levinas's work here fails to be "radically original" or "carefully thought through" as John Wild writes; it is both of these things (*TI* 12). Nor is it to say that this book lacks a unifying and forceful theme, which it has abundantly. But the philosophical force of this book relies on two appeals to transcendence, one spatial and the other temporal. For most of *Totality and Infinity*, it is spatial metaphors that reign, and time is treated as a component of history and being. The subtitle Levinas gives this book is, after all, "An Essay on Exteriority."

This fixation on spatial imagery, evident in the title and subtitle, represents an extension of ideas that build across the 1950s in Levinas's thought and reach mature expression here. The use of this language of interiority and exteriority was a concern to Levinas as early as 1947, where he worried that these metaphors might collapse again into "light" and therefore ontology.[3] It is this reliance on spatial imagery that becomes a key component of Jacques Derrida's critique of this work in 1964. The spatial tension between interiority and exteriority, Derrida claims, is a reversion to the "traditional logos governed by the structure of 'inside-outside,' 'interior-exterior'" (VM 88).

Totality and Infinity is a sweeping opus that traverses a host of topics and metaphors; however, focusing on the way time operates in the book, provides a unique perspective to Levinas's work. The tension of metaphors for transcendence in this book may reveal where Levinas's work is both weakest and strongest. In the end, Levinas rests his philosophy on his understanding of time and that there are serious and hurtful consequences for the detour he takes into the language he later admits is still too close to the ontological tradition he wishes to critique and escape (*DF* 295).

Being, Violence, and Totality

The clearest of Levinas's spatial images in *Totality and Infinity* are posed as interiority and exteriority. The exterior, marked by transcendence and height, seem to presume a neutral site from which these geometrical arrangements can be observed. Levinas is exceedingly harsh on this idea of a neutral perspective precisely because he perceived in Heidegger a *neutral* or *neuter* sense of being to which all persons are beheld (*TI* 68).[4] Levinas makes extended and serious attempts to use the idea of infinity to unsettle this neutral perspective and to pronounce an exteriority that is higher than Platonic transcendence.[5]

In this respect, we find the Levinas of *Totality and Infinity* far closer to Hegel than his rhetoric will indicate. Like Søren Kierkegaard, Levinas wishes to attack the Hegelian universalism that sees and knows everything from a perspective that transcends history, being, and the particularity of human life.[6] For Levinas, it is the other that resists the Hegelian system (*TI* 40). But in *Totality and Infinity,* the other disturbs the system primarily by introducing a *distance* or *transcendence* too exterior for assimilation into the system. Hegel's system adapts smoothly and readily to ruptures and distances, using the dissonance and dialectic of difference to press forward toward historical unity—totalization.[7] The agenda of *Totality and Infinity* is complicated and mounts several attacks on such systematic thought, but the core metaphors of totality and infinity function primarily as spatial challenges to the Hegelian dialectic. This constitutes, also, a renewal of the critique of Parmenides. The notion of infinity as developed here by Levinas proposes radical difference as insurmountable *spatial* separation. But in presenting difference in such a manner, Levinas is playing into the hands of the Hegelians. For any difference configured geometrically, however radical or infinite, remains *perceivable* from a vantage that transcends the difference.

It is precisely this concern that worried Levinas in *Time and the Other,* where separation is a function of the isolation of time. Exteriority, as it is discussed in *Time and the Other,* is "neither spatial nor conceptual"; the radical alterity of the other is not "due to any difference the concept would manifest through spatial exteriority" (*TO* 84). Levinas deftly notes that to spatialize the relation with alterity is to reduce that difference to a concept. And that *concept,* even when described as infinite, is therefore thematized. No amount of distance or separation can then strip the philosophical position of conceptual transcendence. Even here in *Totality and Infinity,* where Levinas has adopted the predominantly spatial metaphor of exteriority, Levinas indicates his awareness of this danger by insisting that he is gesturing toward a *radical* exteriority (*TI* 29, 192). For Levinas, *relative* exteriority is a recipe for violence and war.[8]

The issue of violence and war is the driving problem that presses *Totality and Infinity* forward in its investigations. Totalization leads to violence, justifies violence, and lacks the ability to call acts of violence into judgment. This theme, which received increasing attention across the 1950s, now plays a central role in Levinas's philosophy. Levinas opens the preface to *Totality and Infinity* with meditations on war, and the danger of violence is a ubiquitous concern across its pages. If we may take him at his word about letting the preface unread the book before we read it, we might conclude that Levinas wants us to open this book with the problem of war on our minds, and if ever we think we are not reading about violence in the pages that follow, we are misguided.

As with his early essay on Hitlerism, Levinas here associates violence with a self-justifying logic of being: "Does not lucidity, the mind's openness to the true, consist in catching sight of the permanent possibility of war?... We do not need obscure fragments of Heraclitus to prove that being reveals itself as war to philosophical thought, that war does not only affect it as the most patent fact, but as the very patency, or truth, of the real" (*TI* 21). Heraclitus proposed "that war is universal, and that justice is strife" and that "war is father of all, and the king of all."[9] Heraclitus is called the "weeping philosopher," in part for his negative perspective on humankind and hopelessness of being. He also introduces the concept of a controlling *logos,* in accordance with which all things come to be,[10] and this *logos* is apparently one of endless struggle. Here it is Heraclitus's wisdom and the violence of being that interests Levinas; he opens *Totality and Infinity* by confirming a rather pessimistic perspective on the truth and reality of being. He suggests that when one looks with any philosophical clarity at the realm of being, one discovers that "war is produced as the pure experience of pure being" (*TI* 21). This sentiment functions here to position all of Western philosophy within the totality envisioned by Heraclitus. We might already suspect in Levinas confirmation of Heraclitus's pessimism that he has aligned himself closer than he wishes to traditional philosophy and its hubris. War, Levinas tells us, "is fixed in the concept of totality, which dominated Western philosophy" (21).

The way of being, ontology, and all that might be unveiled in the analysis of being will therefore offer no wisdom that may unhinge the foundational status of violence and struggle. Viewed in its totality, peace and war are partners in the only possible dance of being—"The peace of empires issued from war rests on war" (*TI* 22). Within the struggle of being, peace is only false peace that relies on the threat of war. And this struggle gives birth to a self-justifying morality, a codification of the rules of struggle, a morality that dupes us into supposing that the struggle gives rise to a good. Morality is determined in "trial by force" (21). This Nietzschean conclusion is a result of clarity, philosophy's goal. Under proper perspective, the philosopher sees that being is struggle and that struggle defines its own morality. If ontology is fundamental, it can therefore do no more than sanction the morality that is itself derivative of violence. If we fall for this form of morality, we are in danger of being seduced by the introduction of a false infinite. And so morality wears the clothes of an external judge but underneath is nothing but a product of the struggle. The preface to *Totality and Infinity* opens with the important line: "Everyone will readily agree that it is of the highest importance to know whether we are not duped by morality" (21).

Levinas thus accelerates the argument he first articulated in "Is Ontology Fundamental?" If ontology is "first philosophy," then morality can be based on nothing but power and violence. The condition that allows morality to be defined by ontology is *totality*, the inclusion of all difference within a common circumference of truth. Totality denies exteriority, for it must reinscribe all phenomena into the thematizing and generalizing epistemology of the same. Alterity *within* totality is only temporary and provisional, a passing phenomena and a temporary impediment to understanding. Here again, Levinas must press the word *radical* into use to indicate that he is not talking about *relative* distance and *relative* transcendence. Levinas claims that Western philosophy has never truly broken free of Parmenides, let alone Heraclitus, who both insisted on an ultimate and abiding unity.[11] Totality insists that from the proper perspective, all phenomena eventually relate to a common truth. The audacity of Levinas's critique is evident in the vast swath of Western philosophy against which he turns. Traditional philosophy has been "dominated" by totalization (*TI* 21). To totalize is to subordinate all phenomena to the logic of proximity and similarity.

Totality cannot entertain true exteriority, which would require thinking beyond or outside of totality. Levinas thinks that Plato's doctrine of the "Good beyond Being" catches wind of this kind of transcendence but fails to deliver a philosophy that breaks with totality (*TI* 292–93). Following René Descartes, who also glimpsed radical separation, Levinas proposes an idea that, when thoughtfully considered, is radically incommensurate to the machinations of totality. This idea is *the infinite*, an idea that refuses to be confined in a horizon. Levinas appeals to Descartes for the sake of establishing his idea of the infinite. Descartes, Levinas reminds us, "refused to sense data the status of clear and distinct ideas" (130). Our senses, for Descartes, can provide us with enjoyments. They demonstrate the use to which objects can be put, but they are not capable of determining or revealing the truth (135). There is a gap, for Levinas as for Descartes, between what the ego senses and what is known to the mind. Our senses, Descartes urged us to consider, may deceive us, and they must be suspended for a clear discussion of knowledge and the foundations of epistemology.[12] Levinas does not adopt every feature of Cartesian mind-body dualism, but he does wish to show a kind of separation between the ego and the world that the ego enjoys.

How does Cartesian infinitude relate to the concept of time? We can certainly see in Descartes a replication of the traditional preference for the eternal over the temporal. Descartes's infinitude does not seem to point toward any infinitude of time but toward a transcendence that moves beyond the realm of the physical with all of its temporal vicissitudes. For Heidegger, this made Descartes a chief example of the problems with traditional philosophy. Heidegger claims of Descartes that "an attempt has been made to start with spatiality and then to Interpret the Being of the 'world' as *res extensa*. In Descartes we find the most extreme tendency toward such an ontology" (*BT* 94–95).[13]

It seems that Descartes becomes a critical tool in Levinas's endeavor to break with Heidegger. Descartes could conceive of a beyond being; indeed, it is the exteriority of the concept of infinitude that allows Descartes to establish the existence of God. Likewise, with Levinas, the encounter with the face therefore opens the isolated and interior world of the self and reveals that which lies beneath, or more accurately, that which lies *beyond* ontology. This encounter in some ways mirrors the way anxiety opens up to nullity and temporality in Heidegger.[14] But the anxious experiences traced by Heidegger are experiences internal to Dasein. Levinas is proposing an experience of that which is necessarily, entirely, and permanently foreign.

Time seems to be of little or no consideration in this particular argument for the transcendence of the other. It seems that by enlisting Descartes in his moves against Heideggerian ontology, Levinas picks up a version of infinity that has no strong need for time. The emphasis on exteriority in *Totality and Infinity* is decidedly spatial, even as Levinas periodically reminds his readers that the other still remains proximate: "The dimension of the divine opens forth from the human face. A relation with the Transcendent free from all captivation by the Transcendent is a social relation. It is here that the Transcendent, infinitely other, solicits us and appeals to us. The proximity of the other, the proximity of the neighbor, is in being an ineluctable moment of the revelation of an absolute presence (that is, disengaged from every relation), which expresses itself" (*TI* 78). The appearance of transcendence in the interiority of the subject is an opening, a rupture, an epiphany. It appears not as a feature of being but as that which cannot be assimilated into being. Levinas binds the appearance of the infinite, in particular, to the face of suffering: the other's "very epiphany consists in soliciting us by his destitution in the face of the stranger, the widow, and the orphan" (78). But Levinas realizes that it is easy to miss this appearance of the infinite, to treat the other as an object among objects, to forget the for-the-other that is demanded by the face of suffering. The pressure of war and commerce "dupe" us into a morality that is determined by the economy of being.

According to Levinas, there is no way that ontology can speak of anything but struggle, the "will to power," and the dynamics of being and becoming. This permanently lodges ontology in the field of violence: "Ontology as first philosophy is a philosophy of power" (*TI* 46). If it can speak of nothing but being, ontology is confined to the sorting and resorting of power and possessions on earth. Justice is subordinated to a logic of force, an ontology of struggle and autonomous freedom. This relegates ethics to a secondary concern; it "affirms the primacy of freedom over ethics" (45).

Levinas is continuing his tested strategy of building a case against Heidegger by showing the ways in which Heidegger remains similar to traditional philosophy.[15] Heidegger allows the relationship with "Being in general" to reign over the relationship with the other. Whatever his innovations, Heidegger is therefore clumped with "the whole of Western history" (*TI* 45).[16] In Levinas's estimation, Heidegger is defenseless against the claim that he remains a thinker of totality. No pretension toward *ecstasis* or the individual authenticity of Dasein can alter the great scope of Heideggerian thinking; he attempts to situate all beings around a common truth about the existential and temporal nature of being. So despite his destabilization of universalism and idealism, Heidegger still "subordinates the relationship with the Other to the relation with Being in general" and therefore bends his knee in obedience to an anonymous and neutral "truth" (46). Conflict between persons is resolved by resorting again to the foundations of ontology, to the reduction of persons to a common truth, and to a "presence on earth and under the firmament" (46).

Totality and Infinity, claims Levinas, wishes to think about "absolute exteriority," or a relationship with "a being infinitely distant" (*TI* 47). Of course, even as Levinas invokes the concept of infinite distance, the emphasis is placed on the infinitude and *not* on the distance. However, it is possible that the invocation of infinite *distance* to describe the relationship with the other undermines Levinas's thesis. This has consequences, including a drifting away from the sense of alterity that Levinas earlier pinned to time. His later texts, and several key sections of *Totality and Infinity,* help us to think of the other as infinitely *prior* to the self.[17] But here Levinas unapologetically leans on the spatial language that dominates his work in the 1950s. Alterity is outside, exterior, beyond, infinitely transcendent, and infinitely distant.

The use of spatial metaphors is rigorous and thorough, especially in the early sections of *Totality and Infinity.* The other "is not *on the same plane* as myself" (*TI* 101). Foreign to my "dwelling," the other person resides in another *land,* "a land not of our birth" (34). The other dwells on another "shore" (64), a place Levinas refers to as "a yonder" (33). Metaphysics, for Levinas, maintains this distance and separation. Levinas refers to this distance as infinite, so that this transcendence is so extreme that no journey can traverse the distance. "No journey, no change in climate or of scenery could satisfy the desire bent toward" this "elsewhere" or "the other" (33). The other person is "not wholly in my site" (39). These are the images with which Levinas opens *Totality and Infinity.* This is a book about infinitude, but the infinitude is first cast as a feature of distance.

The language of lands, shores, and infinite distance positions the reader above the relation between the self and the other. The omniscience and universalism of this perspective draws closer to Hegel than Levinas may wish, a point not lost on Derrida.[18] How can one think of the distance between two points, even as infinite, without the neutrality of a vantage point? In his examinations of Hegel, Levinas critiques the "synoptic gaze that encompasses" (*TI* 53). But is not the posture of the philosopher, who considers the same and the other according to the metaphor of distance, not forced to adopt just such a "synoptic" gaze?

Despite the ambivalence of his metaphors, it remains clear that Levinas is consistently concerned about the complicity of philosophy in violence and, in particular, the inability of ontology to mount any real judgment of violence. Even peace, ontologically considered, is just the flip side of violence, just a temporary truce. In *On Escape*, an essay already 27 years old when he writes *Totality and Infinity*, Levinas proposes that philosophy must "get out of being on a new path" (*OE* 73). The path he chooses in *Totality and Infinity* reveals a puzzling mixture of spatial and temporal terminology.

The Distance of Time

The primacy of spatial metaphors is never more vivid than in the first pages of Levinas's second section, "Separation and Discourse," where he introduces an updated philosophy of the instant. The distance between the same and the other, claims Levinas, forms an "inner life," similar to the Cartesian *cogito*, which "evinces separation" (*TI* 54). The inner life, or "psychism," is characterized by thought, by recollection and projection, by memory and hope. Levinas speaks of universal time, world time, objective time, and common time to refer to the anonymous, universal passage of time that forces all persons into the totality of history. This economic time remains unbroken by the other, whose face it plainly summons to resign its interiority. History therefore leaves the subject relating to the other as another object among objects. Again, as in *Existence and Existents*, time seems to move with clocks and calendars. But beneath this movement, there is an interiority masked by everyday life. Throughout most of *Totality and Infinity,* time refers to the economy of labor, anxiety, enjoyments, and interiority. In *Existence and Existents,* Levinas pursues a philosophy of time that demonstrates the secondary nature of economic time by establishing a hypostatic self that is separated from all experiences. In such a mode, the ego operates with no awareness of the tenuous nature of the relationship between the subject and the world. All things are reducible to enjoyments, pains, and equipment for the ego in itself; time seems to flow, but really the ego is locked in a world of self-interest and interiority. But the face is the *irreducible* relation; to relate to a face is to relate to that which cannot be assimilated, that whose separation is complete.

For Levinas, the purpose of demonstrating this separation between the ego and the world of enjoyments is the establishment of a fundamental *egoism* that is at the genesis of the story of every person. From birth we find ourselves in a world of sensibility and enjoyments. He explains: "Enjoyment is the very production of a being that is *born,* that breaks the tranquil eternity of its seminal or uterine existence to enclose itself in a person, who in living from the world lives at home with itself" (*TI* 147). To this reflection on enjoyment Levinas adds some thoughts on time, but specifically the time of enjoyment, which is the time of "possession and labor," of economy and "fabrication" (146). In this sense, time is "the nothingness of the future," the uncertainty of tomorrow, a cause for anxiety and evasion. The time of enjoyment is an expansion of the instant explored in 1947; its past and its future are a mere economy of labors and investments. The time of enjoyment, sensation, and labor is for gathering and storing; the future is cursed for the anxiety and uncertainty it brings. Levinas, sounding rather like Heidegger on this point, calls this the "muffled rustling of nothingness" (146).

In the middle sections of *Totality and Infinity,* we can detect a practical, and perhaps less theoretical, invocation of the instant, which was notably important to Levinas in 1947. In *Existence and Existents* and *Time and the Other,* the instant was marked by its absence of time, which it could only receive as a gift. Though the examples of insomnia and the *il y a* pointed to a deeper sense of time than the experience of clocks and calendars, the reader of those early books has to constantly resist the alluring impulse of everyday time to reestablish the sense that time indeed moves on. Whereas in his 1947 works he emphasized the reality of the asocial, frozen instant of insomnia, in *Totality and Infinity,* Levinas discusses time from a complementary angle. Here the focus is on how our interior conceptions of time are exposed as possessive constructions in the face of the other.[19] Time, throughout most of *Totality and Infinity,* takes the form of a history that pressures me to ignore the rupture that is the presence of the face in my field of vision.

Levinas is establishing the roots of a fundamental enjoyment and egoism, which are the conditions of insularity and interiority.[20] In *Totality and Infinity* Levinas refers to time in a mode similar to the descriptions found in his early essay *On Escape,* where time was a cause for insecurity and evasive maneuvers (*OE* 50). The desire to secure one's joys against the impending unknown can be seen as a manifestation of the primordial egoism that Levinas is outlining. The bourgeois, Levinas explained in 1935, is naturally nervous about the future and the pains it may hold. Possessions become capital, invested as "insurance against risks" (50).[21] The past is the remembered past; the future is the anticipated danger of decay and unknown perils.[22] The self, as given over to its natural history is self-obsessed, concerned primarily with its own self-preservation.

Therefore the instant in *Totality and Infinity* remains a moment of internality, soon to be expanded into the notion of dwelling and being "at home with oneself."[23] Interestingly, Levinas cannot resist even here injecting spatial language into his analysis of time and the instant: the instant is defined as "the space of the *cogito*" (*TI* 54). The existence of the internal instant is remarkable, for it pronounces a hiatus that is prior to metaphysics. The instant "articulates the ontological separation between the metaphysician and the metaphysical." This "distance of time" is that which separates the "metaphysician" from history (54–55). The instant is *ontological:* it is the time of being, of interiority. Levinas focuses in *Totality and Infinity* not on the night in which the instant horrifies us with its immobility, but on the false, egoistic, and economic way that we typically face time in the day.

There is certainly a sense in which the interiority of time might appear to be a *freedom* from history, from the other, from the outside.

But for Levinas, this freedom is utterly limited to the instant, to the interior life, to the dimension of *thought*. This time of interiority Levinas also calls "dead time," for it "consists in being between two times" (*TI* 58). The "dead time" of interiority is not the same as a potency that precedes activity or the pause before carrying out a project.

"Dead time" refers to the deep capacity for totality within the inner world (psychism) of the I. But the experience of the other introduces a puzzle, a "secrecy" that historical objectivity cannot understand.[24] The attempt to reduce this secrecy to knowledge is totalization and violence, for it fails to appreciate the radical "pluralism" of human society (*TI* 58). The attempt to envelop the separated time of intersubjectivity into a unified time performs a violence akin to Cronos, who "thinking he swallows a god, swallows but a stone" (58). The time of the other cannot be digested by universal time; it cannot be incorporated into my interior instant. This does not yet seem to mean, for Levinas, the divesting of the subject of its unique temporal moorings in the isolated instant, a move he will develop in the decade ahead. It does, however, indicate a separation between subjects that violence cannot overcome. The time that is *mine* and not subject to history is also safeguarded from violence: "The existence for the other, this Desire of the other, this goodness liberated from the egoist gravitation, nonetheless retains a personal character. The being thus defined has its time at its disposal precisely because it postpones violence, that is, because a meaningful order subsists beyond death, and thus all the possibilities of discourse are not reduced to desperate blows of a head struck against a wall" (236). Thus the role of time in this configuration is the establishment of separation, and this separation places the identity of the other outside of history, in a time that is not available to the forces of history and nature.

Uprooted from History

Levinas levels a steady stream of attacks on Hegel in *Totality and Infinity*. His concerns with Hegel's phenomenology arise at almost every important step in the development of the book, though sometimes Levinas only references Hegel obliquely. As Levinas reads Hegel, the Hegelian I moves forward in history in a possessive, grasping "sojourn" (*TI* 37). Quoting Hegel, Levinas demonstrates the way the Hegelian I encounters alterity: "I, the selfsame being, thrust myself away from myself; but this which is distinguished, which is unlike me, is immediately on its being distinguished no distinction for me" (36–37).[25] Levinas stops his quotation at that point, immediately before Hegel confirms Levinas's suspicion: "Consciousness of an other, of an object in general, is indeed itself necessarily self-consciousness, reflectedness into self, consciousness of self in its otherness."[26]

Here we find Levinas cataloging his deepest concerns about Hegelian phenomenology. Hegel's I moves into a world that cannot be configured otherwise than according to the logic of the self, the same. My "thrust" forward into the world is a movement into a world that is already mine. Levinas summarizes, "Everything is here, everything belongs to me; everything is caught up in advance with the primordial occupying of a site, everything is com-prehended. The possibility of possessing, that is, of suspending the very alterity of what is only at first other, and other relative to me, is the *way* of the same. I am at home with myself in the world because it offers itself to or resists possession" (*TI* 37–38). The encounter with the other offers confirmation or presents a resistance, a correlation or a conflict. Even if the other *resists* possession and assimilation into the world of the self, this too reinforces the identity of the ego: "If the same would establish its identity by simple *opposition to the other,* it would already be a part of a totality encompassing the same and the other" (38). Hegel is for Levinas the paradigm of totality, the heir of Parmenides.

Hegel has described *war,* the primitive origins of violence, but he has not considered the possibility of a "metaphysical relation." Levinas seems to accept that this way of totality is the way of *being;* the problem is not that Hegel's phenomenology has inaccurately described the natural encounter of the self with the objects and persons of the world. The chief problem for Levinas is that Hegel has constricted philosophy to the phenomenology of being and, in so doing, has precluded the possibility of an ethics that might emerge from beyond the totality of being. Hegel, therefore, sanctifies the violence of being and necessarily subordinates ethics to history.

As we examine Levinas's use of time in these portions of *Totality and Infinity,* it is clear that Hegel's understanding of time is a chief target. History, for Hegel, is the final arbiter of truth, and in the light of history, any provisional difference and distinction between the same and the other is collapsed. For Hegel, who is the prototype of modern philosophy in this regard, any "opposition between the I and the non-I disappears, in an impersonal reason" (*TI* 87). Thus Levinas accuses all of Western philosophy, not just Hegel, of substituting "ideas for persons" (88). In Hegel, this occurs through a broad, sweeping reduction of all phenomena to a common movement toward the state, the teleological aim of history. This "alleged integration is cruelty and injustice," precisely because it ignores the other in pursuit of the neutral and universal "truth" of history (52).

No philosophical position could be more directly in opposition to Levinas's own itinerary. Therefore, when Derrida accuses Levinas of a proximity to Hegel, as we shall explore in chapter 6, he attacks the very heart of Levinas's philosophy. At the core of Levinas's unique philosophy of time is a bifurcation between universal time and the time of the other. Philosophy has long instructed us to prefer the universal, the common and measurable time of clocks and calendars. The encounter with the other simply cannot be registered in history, and attempts to appropriate the other in Hegelian fashion are nothing short of violent: "History as a relationship between men ignores the position of the I before the other in which the other remains transcendent with respect to me. Though of myself I am not exterior to history, I do find in the Other a point that is absolute with regard to history—not by amalgamating with the Other, but in speaking with him. History is worked over by the ruptures of history, in which a judgment is borne upon it. When man truly approaches the Other he is uprooted from history" (*TI* 52). Philosophy's concern, claims Levinas, should not be with history but with its *rupture,* and the rupture of history can be detected in various phenomena (52).[27] Grounded in a history of labor, enjoyments, suffering, and egoistic concern, I encounter the face and am "uprooted from history" by the other.

Time, Labor, and Dwelling

Nowhere does the spatial imagery used by Levinas in *Totality and Infinity* become more vivid, and more dangerous, than in his chapter "The Dwelling." This section presents a collision of metaphors and themes, which are worked out across the middle portions of *Totality and Infinity*. Levinas here combines the spatial analogies of interiority and exteriority with his philosophy of the present. He has experimented with the analogy of a habitation before as an illustration of the enclosure of the ego in its interiority. In "The Ego and the Totality" he provides a two-sentence illustration from Charlie Chaplin's film *The Gold Rush* (*CPP* 26). In that film, perhaps Chaplin's finest, Chaplin's character "The Lone Prospector" finds himself trapped in a cabin by a snowstorm that threatens to push his little domicile into the chasm below. There within the cabin, "without windows upon the world," the prospector must recover his relationship with the world (26). Sprawled on the slanted cabin floor, he "gropingly studies the elementary laws of those shaken balances." To rejoin the world, the prospector must *think;* he must gather himself and regain his mastery over the materials at hand and the laws of physics that hold sway even in the privacy of his home. He leaps from the cabin just before it tumbles into the abyss.[28]

This analogy between a home and interiority receives extensive treatment in *Totality and Infinity*. Levinas uses the image of the four-walled habitat to represent the realm of interiority.[29] The most controversial element of this portion of the text involves Levinas's extensive discussion of the feminine in this analogy. Two issues are worthy of discussion here: first, the spatiality of the habitation and its close relationship to the notion of the instant, and second, Levinas's portrayal of the feminine and the consequences it has on the time of the woman.

We can hardly imagine a more spatial and geometric image than that of the domicile, with its walls and interiority. As we turn to examine habitation and the home, we discover that it functions like one of Heidegger's tools, as something already nestled into the complexity of lived experience.[30] For Levinas, to become aware of one's dwelling is to already have the remembrance of a prior dwelling. The consideration of the world *presupposes* the site in which we pull together and assimilate the objects seized by our hands and eyes (*TI* 153). Levinas, of course, does not mean an actual home of boards and nails in the objective world. Nevertheless, he draws heavily from the model of the actual home, with all its spatial overtones. He is attempting to demonstrate that all knowledge and experience of the world occurs within the gathering of oneself in the "private domain" of the "home" and is, in fact, itself a secondary movement, an appropriation of events "*after the event,* after having dwelt in them as a concrete being" (152–53). The fact that my experiences are processed from and within the habitation of my mind denies me the immediacy of experience that the sensual world feigns. All "knowing" occurs in the retraction and retreat into the dwelling. The dwelling is where I sort and classify and *represent* the world of experience. And this "event of dwelling" is always *a posteriori* (153).

Levinas has abandoned his earlier discussions of hypostasis, replacing them, apparently, with the structure of the dwelling. One encounters the world already presupposing this method of enjoyment, though perhaps unaware of the separation between experiences and the way they are appropriated.[31] The dwelling is a "withdrawal from the elements," a hiatus from the world of history and its cruelty (*TI* 153). In the dwelling, the ego is free to sort and stack experiences in isolation from the world. The furnishings are safe; they are my experiences as *I* have retained them.

The ego is driven by hunger and desire back into the world of labor, going "forth outside from an inwardness," from an *intimité* (*TI* 152). The intimacy of the original habitation, which Levinas earlier referred to as "uterine existence," remains configured as a place of safety, gentleness, and warmth.[32] The movement into the privacy of one's habitation is a *break* from the immediacy of experience, though all experience already presumes the ongoing function of this retreat. We are, as Heidegger suggested, thrown into the world, but an analysis of the separation between the ego and the world outside reveals that all experience depends on the gentleness of a prior habitation (154–55). But hunger and thirst and longing press us back into the world. The ego therefore moves laboriously into the world of experience for itself and toward itself, in a journey that already anticipates its own return.

Labor and its goal of "acquisition" is a movement "toward oneself," not toward the other (*TI* 159). The grasping hands of labor operate according to the logic of ontology, which is "pretheoretical" and "spontaneous" (158). There is no *transcendence* in labor, for the work of our hands participates entirely in a "for itself" that has no other aim. Levinas believes that this is innate and natural to "every inhabitant of the earth" (158). Like primitive hunters retreating back to our caves, we leave our habitations, acquire food and possessions, and then retreat again to our abodes. Levinas aligns this possessive labor with comprehension. It is not only material things that we drag back to our caves but the knowledge of the world that we have discovered in our excursion. The dwelling is the place of constitution. For the Greeks, it was understood as a place of origin and stability; for Levinas it is a locus of reconstitution, recovery, memory, and security. Labor and comprehension necessarily aim at "mastery" and "dominion" (161). And in the privacy of my dwelling I am master of my domain; my experiences are mine to do with as I please.

Time plays a role even here, even within the confines of the dwelling. Its function is akin to Levinas's earlier discussions of the instant. In *Existence and Existents,* Levinas outlines the way the ephemeral possession of grasping and sight show themselves to be meaningless in the night of insomnia. Here, in *Totality and Infinity,* economic time is an enemy to the laborer, a threat to the accomplishments of the work. We drag material goods into the "four walls of the home" and tame them as possessions. Labor makes things *present;* it is the way of production and being. Possession, in fact, "masters, suspends, postpones the unforeseeable future" (*TI* 161). The objects or possessions are essentially made into a still life, domesticated into the "now" of the home. The goal is to make the foreign element *present* to the home, to the self. In this sense, "possession neutralizes this being:

as property the thing is an existent that has lost its being" (158). Possession attempts to remove beings from change (160). In taking possession, I affirm my "power over time, over what belongs to nobody—over the future" (160). The goal of possession is security against an uncertain future, which Levinas considers to be the methodology of ontology.

In *Totality and Infinity,* it is clear that it is *ontology* that Levinas wishes to escape. For Levinas, because ontology confines itself to the logic of labor, acquisition, and interiority, it must *fear* and *suffer* the encounter with time and exteriority. The dwelling provides a lapse in this pressure, a break, a temporary escape. Levinas's discussion of dwelling has allowed him to establish a rich exteriority that is found even in the apparent interiority of the home. To have a home, the extension of a body, is to be *from* something other than oneself. Here, freedom is dependence and independence: "To be at home with oneself in something other than oneself, to be oneself while living from something other than oneself, to live from . . . , is concretized in corporeal existence" (*TI* 164). Bodily existence begins in passivity, patience, recollection, and resignation, but to have a body is also to have a home. This tension is the fundamental tension of human life: living *in* the otherness of the world but being nevertheless a distinct instance of dwelling and internality. And this "ambiguity of the body" Levinas calls "consciousness" (165). And "to be conscious," claims Levinas, "is precisely to have time" (166).

So here, where the spatial images of homes and walls reign, Levinas turns to the images of time to stabilize his philosophy of habitation. Levinas provides an intriguing expansion of his earlier discussion of the instant, the present moment. Habitation now plays the roles previously attributed to the hypostasis, the instant, and the present. The dwelling is the secure internal world of comprehension and representation, the place from which labor and effort and possession arise. But Levinas wishes to say more about dwelling than he has articulated in these previous reflections on the instant. He is outlining, in fact, the great temptation to confine philosophy to reflections on *being*. His goal is clearly to demonstrate that Heidegger has set his sights too low, aiming only at a philosophy that describes being and aspiring for nothing beyond the horizon of being. So even the daring and authentic movement of Dasein remains concerned only with interiority. Likewise, attempts to "fore-stall" the expiration of one's goods and one's life, bolstered by traditional philosophy's antipathy to change, are concerned only with the present and its preservation: "To conceive the future is to fore-stall. To labor is to delay its expiration" (*TI* 166).

The goal of labor is self-sufficiency, or more specifically, a security against the needs that might drive the subject out of habitations and into the menacing world. Levinas calls this world of menace the *future.* The ultimate prey, tool, or furnishing in the dwelling would be *time,* except that time "belongs to nobody" (*TI* 160). Prometheus steals fire from the gods, symbolizing "industrious labor" in all its brazen impiety (160). With fire, the dwelling is warm, the food is edible, and *need* is abated. But no sacrificial labor of Prometheus can tame the alterity of time; sooner or later, the food and the wood run out. As hunger and need arise, the dwelling is reminded of its primitive dependence, its original temporality. The dwelling ossifies a present, seizes time like a possession, and hoards for the sake of warding off the threat of decay and the menace of the outside. And this seizing, this "making present," is partly successful. In the very act of consciousness, the subject *has time,* establishing a present "at home with oneself" from which to respond, "not to exceed the present time in the project that anticipates the future, but to have a distance with regard to the present itself, to be related to the element in which one is settled as to what is not yet there. All the freedom of inhabitation depends on the time that, for the inhabitant, still remains" (166).

This sense of staving off the future by *having time* is the very nature of the dwelling, the frozen moment of the present in which one digs in against the ominous future. This "time" that Levinas invokes in this section, and throughout the majority of *Totality and Infinity,* is clock time, the time of labor and economy. The habitation is a *break* from labor; it is the place where I am lord of my own time, where I can methodically enjoy the flavors of a meal and the warmth of a bed. This is not an "inward history," not a time that marches on independently from history (*TI* 231). The time of the dwelling is the present, the instant, the now. The experience of the world occurs as time and history. Time is reason for concern; labor is the process of staving off death for yet another moment. And the other person, encountered within the time of history, is a threat. Violence occurs when the other person, whose face is a break from the ontology of violence, is reduced to some "thing" in the field of my powers (225).

The habitation disposes the ego to reduce life to the sum of pleasures and pains, the resistance of the inexorable approach of death by way of postponement. This is opposed to Heidegger's being-toward-death, as "the 'not yet' which is a way of being against death, a retreat before death in the very midst of its inexorable approach" (*TI* 224). The habitation is a totality; by itself it can only know the "in itself." It knows only possession and war. The ego can broker peace in the world, but even this is just another component of the economy of ontology. The face, in all of its infinitude, calls this mode of encounter into question, irrupting with an appeal to exteriority even within the interior machinations of being. The face arrests the ego, even in its attempts of acquisition and violence. The "You shall not commit murder" written on the face of the other is not a component of the being in which the other is encountered.[33] "You shall not commit murder" is not a rule or regulation native to being; it is the result of the infinitude written in the face of the other. This is a summons that *judges* history, whatever history has instructed the ego with regard to violence and murder. History can, and has, excused murder, but the face is the opening of being. Only the infinite, utterly exterior, can judge history.[34] Time, here, is what the face judges.

Levinas promises to later explain the significance of this prior "relationship with the other, with infinity, metaphysics" (*TI* 166). At the end of the section "The Ethical Relation and Time," which is primarily about the time of violence and economy, Levinas indicates that there is a "primordial effectuation of time" that operates behind and beyond the time of "visible history" (247). Here, Levinas appeals to paternity, as the "primary phenomenon of time." But even here, as Levinas prepares to discuss this originary sense of time, the subject remains the protagonist of the story. Fecundity and paternity are the way that the ego can be bound to a biological life, but live "beyond that life" (247). As Diane Perpich points out, the ego remains the main character in this narrative of sensibility, awakening, awareness, and responsibility.[35]

Levinas seems content in the section on "The Dwelling" to demonstrate that the *time of interiority* is fundamentally a present-making, a preservation of the present. But this time-of-the-same is dependent on a more fundamental sense of time that relates to infinity and metaphysics. Levinas uses this section to show that the internal concerns of the I, whether they operate to secure the present or live authentically toward death, do not yet manage to think beyond totality.

Habitation and the Feminine

Few aspects of Levinas's work inspire as much controversy as his treatment of the feminine, which has already made important appearances in his earlier work and now takes center stage. Careful readers of Levinas have disagreed widely about the implications of his comments about women, the feminine, paternity, and many other gender-related terms that provide the chief set of metaphors in his discussion of "The Dwelling." Some question the relative value of anything Levinas has written about gender. My sense is that feminist critiques of "The Dwelling" have established a serious problem in Levinas's argumentation.[36] There are several aspects of this discussion that weigh heavily on the concept of time, a connection meticulously explored by Tina Chanter in *Time, Death, and the Feminine: Levinas with Heidegger*. By reviewing some of the terms and issues at stake in Levinas's use of gender-laden language, it becomes clear that the use of spatial language throughout *Totality and Infinity* becomes both costly and revealing when applied to Levinas's notions of gender. Following Chanter's analysis, beneath Levinas's problematic utilization of gender there is an internal struggle over the concept of *time,* a struggle that manifests itself tacitly between the covers of *Totality and Infinity* but relates most directly to the next phase in his work.

In an era of heightened awareness about gender, twenty-first century readers might be taken aback by the gender-laden language of "The Dwelling." In bold fashion, Levinas leverages a particular understanding of the woman for the sake of articulating the relationship between internality and externality. The language is so riddled with metaphoric, and perhaps literal, assumptions about the nature and ways of the feminine that more than one feminist has found ample ground to dismiss the whole of *Totality and Infinity* based on the structural chauvinism of "The Dwelling."

Levinas claims that "man" comes from the feminine habitation, a private domain "to which at each moment he can retire" (*TI* 155). The woman is "the other whose presence is discreetly an absence, with which is accomplished the primary hospitable welcome which describes the field of intimacy." She is "the condition for reflection, the interiority of the home;" the woman is inhabitation (155). Levinas himself and a good number of his interpreters have pointed out the metaphorical nature of these references. In other words, he claims that reflections on the femininity of the dwelling need not be directly associated with the actual presence of a female within the home. Levinas attempts to stake out the metaphorical nature of his assertions about hospitality and the feminine, writing, "Need one add that there is no question here of defying ridicule by maintaining the empirical truth or countertruth that every home *in fact* presupposes a woman? The feminine has been encountered in this analysis as one of the cardinal points of the horizon in which the inner life takes place—and the empirical absence of the human being of "feminine sex" in a dwelling nowise affects the dimension of femininity which remains open there, as the very welcome of the dwelling" (157–58).

In hindsight, this disclaimer does indeed need to be added to Levinas's reflections to prevent a justified dismissal of "The Dwelling" as pure sexism. This short passage has been the centerpiece of a defense built by some scholars of Levinas who wish to underscore the purely metaphorical use of gender in Levinas's invocation of the feminine. Richard Cohen draws a distinction between "gendered metaphysics" and "gendered metaphors" to demonstrate the ways that metaphorical language does not extend to the metaphysical.[37] Adriaan Peperzak notes that Levinas frequently binds the ethical relationship to the biblical phrase "the stranger, the widow, and the orphan" (*TI* 215), and then suggests that this invites us to correlate this triplet with "man, woman and child" and to therefore find every human incorporated into the ethical relation.[38]

Nevertheless, neither Levinas's disclaimer nor the interpretations of his more sympathetic readers have successfully insulated Levinas against scathing criticism. Much of the tension originates from a now famous footnote in Simone de Beauvoir's groundbreaking *The Second Sex*, where she meticulously outlines the way traditional Western philosophy has operated on a gender-charged dualism between the self and the other. For de Beauvoir, the concept of otherness has been routinely and consistently presented as a negation, particularly as a negation of the *goodness* of the world of the same. She writes, "Otherness is the same as negation, therefore, Evil."[39] The introduction of the other forges a dualism of positive and negative, good and evil, which de Beauvoir accurately calls a form of Manichaeism.[40] *The Second Sex* is routinely praised for de Beauvoir's successful appraisal of traditional philosophy. She exposes the covert and overt marginalization of the feminine, and therefore real women. Levinas scholar Stella Sandford calls de Beauvoir's 1949 masterpiece a "magnificent and omnivorous study."[41]

The thesis of *The Second Sex* is "to show exactly how the concept of the 'truly feminine' has been fashioned—why woman has been defined as the Other—and what have been the consequences from man's point of view."[42] Early in her exploration of these themes, de Beauvoir claims that Levinas's description of the feminine, "which is intended to be objective, is in fact an assertion of masculine privilege."[43] Since *The Second Sex* appears in 1949, her footnote deals only with the trope of the feminine as Levinas uses it in *Time and the Other* and *Existence and Existents*, which both appeared in 1947. Yet rather than backing away from the charged language of gender, Levinas escalates his usage for the 1961 *Totality and Infinity*. Levinas appears to be taking up a challenge to utilize the metaphor of the feminine to elucidate his account of intersubjectivity. So we may join Alison Ainley in deeming de Beauvoir's footnote about Levinas "dismissive," but only as we remember that de Beauvoir was writing still twelve years before *Totality and Infinity*. And while the reference to Levinas may bypass potentially helpful elements of Levinas's invocation of the feminine, it is incontestable that Levinas in some ways participates in the very patterns named by de Beauvoir in *The Second Sex*.[44]

We should not be surprised that de Beauvoir recoils from the way Levinas makes heavy and sustained use of the same-other pairing, particularly when Levinas overtly aligns the feminine with the other. In this regard, feminists find Levinas in lockstep with a historical bifurcation between the virility of masculine activity and the receptivity and hospitality of feminine passivity.[45] But this straightforward reading overlooks the manner in which Levinas's critique undermines the primacy of masculine virility and individual freedom. The critique of virility is essential to his turn against Heideggerian philosophy. In *Totality and Infinity* Levinas is undermining virility by demonstrating an utter dependence on the feminine other for any appearance of virile activity. Jeffrey Bloechl points out that this can be considered a fairly direct reversal of a similar gender configuration offered by Aristotle in *Nicomachean Ethics*.[46] Virility and activity for the citizens of Aristotle's *polis* rely directly on the literal woman who plays the roles of mother and housekeeper. Bloechl claims that this Aristotelian model relegates women to the margins of society; women are utterly peripheral to the operations of the polis.[47]

In *Totality and Infinity,* Levinas appears to embrace Aristotelian gender roles, but he accords the feminine with a central, foundational, and positive configuration. In addition, Levinas at least attempts to differentiate between the metaphor of the feminine and real, "empirical" women (*TI* 157–58). This *ostensibly* liberates feminine traits from their embodiment in women alone and positions good ways of womanly hospitality at the center of Levinas's ethics. But this strategy nevertheless retains the feminine as a metaphor for certain behaviors and modes. The feminine as metaphor is subject to *representation.*

Levinas vacillates, as Claire Elise Katz has pointed out, between his use of "the feminine" and "woman."[48] This vacillation belies a literal sexism in which Levinas's invocation of the feminine performs more than a philosophical allusion. Stella Sandford challenges Adriaan Peperzak's metaphorical interpretation, claiming that this glosses Levinas's words on this subject. Peperzak admits that the feminine may not be a "good metaphor," but maintains that it must be understood as a metaphor nonetheless.[49] In response, Sandford comments, "Presumably [Peperzak] would not be foolhardy enough to argue that, as a metaphor, the trope of the feminine has no connection whatsoever, no linguistic or cultural reference at all, to empirically existing women, as this would deprive the metaphor not just of its rhetorical force, but of its very sense: of any possibility of it functioning with any intended meaning at all."[50]

Sonia Sikka states it more succinctly: "metaphors matter."[51] Sikka points out that the supposed variation offered by Levinas in his use of the feminine, which intends to express the "ethical dignity of the human in general," is overwhelmed by meanings Levinas may not have intended. She points out that Levinas invokes "the maternal body" as a metaphor, but in doing so "signifies more than the superior spirit it is meant to convey."[52] Good intentions are insufficient to overcome the faultiness of the metaphor of the feminine.

Chanter's strategy, which I am following here, is to peer past the distracting debate over Levinas's controversial treatment of the feminine and to evaluate the intriguing role of *time* in this configuration of gender. Bloechl makes a similar move, pointing out that despite the faultiness of woman as metaphor, we can detect here a return to and reinforcement of Levinas's unique theory of time.[53] Indeed, Levinas has undermined the feminine other precisely by denying her time.

A "Delightful Lapse" in Being

In an early passage from *Time and the Other,* which almost certainly contributed to de Beauvoir's concerns about Levinas's writing, he reflects on the "exceptional position of the feminine in the economy of being" (*TO* 86).[54] By de Beauvoir's reading, this is far from novel; the feminine has been repeatedly configured as the exception to the rule in traditional philosophy. The realm of politics and ethics has been cast as a masculine venture, and women have been relegated to nonessential or exceptional roles in the periphery. This tendency remains palpable when Levinas places much importance on the concept of paternity. It is this concept of paternity that he claims will carry us "back to the primary phenomenon of time" (*TI* 247). In his discussion of the family, Levinas make a pivotal turn in imagery away from space and into the realm of infinite time. If the model of the family rescues Levinas from his dominating spatial metaphors, it also deepens the general concern that he has little substantial place for women in the ethical relation.

In *Totality and Infinity*, we find in the domestic reference to the relation between the father and the son a truly pivotal moment in Levinas's development of the relation between the self and the other. The relation with the other, particularly as configured in the tones of exteriority and distance, hovers close to the pure negation of all knowledge, language, and communication. This investigation has pointed to Levinas's reversion to spatial metaphors and imagery as a chief cause of this misunderstanding of his thoughts on alterity. Paternity, however, undermines the strictly spatial sense of distance and otherness that are so obvious in *Totality and Infinity*. In the son, Levinas tells us, the father "discovers himself" (*TI* 267). But the self that the father discovers in the son escapes him, into a future that is not his own. This relation with the future Levinas calls fecundity.

The dominant language of spatial distance is not unrelated to the controversial gendered language of *Totality and Infinity*. Levinas's problematic use of the feminine has garnered most of the attention of commentators; less attention has been paid to the biological and familial language of paternity. Chanter points out that the relationship between the father and the son provides the ground of similarity or sameness that rescues this philosophy from "all the problems that Derrida has so forcefully articulated."[55] Distance, alterity, and exteriority leave the self-other relation in the vertiginous state of mystery and negation. The other is known only as the non-I or the alter ego. But in the relation between the father and the son, Levinas develops the notions of familiarity, intimacy, and identification. In my son, I identify my movements, my gestures, and a face similar to my own, but though I find myself in my son, I am tangibly and irrevocably absent there.

The son remains a stranger, but in the child one encounters a revealing phenomenon. This other is not just a stranger but also *me:* "My child is a stranger (Isaiah 49), but a stranger who is not only mine, for he *is* me. He is me a stranger to myself" (*TI* 267). In the world of light, the luminous world, I account for the objects around me according to their possibilities, the future I can anticipate for them. But the relationship with the other as son undermines the enclosure that is my sense of the possible. The future of the child comes to pass "from beyond the possible, beyond projects" (267). The child is in some sense my *work*, a work that is as infinitely *mine* as any other. But this child is simultaneously *fecundity*, a future utterly disconnected from the projections and anticipations of the self.

In the child, the father observes "the return of the I to the self" in the face of a person who is utterly strange and utterly familiar (*TI* 177). This return is a teaching, a reconfiguration of the father. Chanter claims that this return "transfigures him from a masterful, powerful subject to an impotent, responsive father."[56] The relation with the child is an open-ended adventure that does not return to any original or projected future anticipated by the father. The child is about the future and about an infinite time whose futurity is qualitatively different from the future known to labors, projects, anticipations, and predictions. But as we shall see, there is a steep cost for resting the infinitude of time on this configuration of the father figure.

It is no coincidence that as Levinas provides the most tangible examples of the self-other relation, he pivots away from the infinitude of space and into the infinitude of time. But before we follow him in this venture, which ends with discussions of eschatology and the preface to *Totality and Infinity*, we must address the complicated contributions of the feminine to this critical development in Levinas's philosophy.

Several layers and issues complicate the investigation of Levinas's use of gender. One issue that appears repeatedly is the question of Levinas's understanding of family, particularly as he is conditioned by various cultural and religious forces to think in particular ways about the feminine. We cannot overlook the cultural influences of Lithuania, Russia, France, and Germany on Levinas's writings. Chanter points out that the concept of dwelling, home, and domestic security should be considered alongside Levinas's reflections on the state of Israel.[57]

Claire Elise Katz's study *Levinas, Judaism, and the Feminine,* outlines the influence of Judaism on Levinas's understanding of the feminine, providing nuance to these discussions of gender in Levinas. Through a number of linguistic and metaphorical references within Levinas's work, Katz demonstrates that his conceptualization of the feminine is deeply informed by the depiction of women in the Hebrew Bible.[58] None of these issues can be dismissed, each potentially coloring and directing the troubling and intriguing discussion of gender here.

In his discussion of the family, of the father, the woman, and the child (gendered masculine), Levinas offers a uniquely frank example of the way an ethical relation functions in everyday life. For Levinas, the relationship between a father and his child provides a paradigm of the ethical tension between the same and the other. In the child, the father finds himself and simultaneously encounters something utterly alterior. This pivotal encounter with the child, with his infinite proximity and infinite alterity, is sponsored by the encounter with the feminine. As Levinas states it, "The encounter with the Other as feminine is required in order that the future of the child come to pass from beyond the possible, beyond projects" (*TI* 267). The hospitality of the feminine, the warmth of the dwelling, is the condition for alterity. The woman makes possible ethics; she is the reason there can be an encounter with the face. But does she have a face?

It seems rather clear that Levinas relegates women to the traditional role of facilitator, even if he continues to think of himself as only working in metaphor. The feminine creates the condition for ethics, the possibility of the encounter with exteriority. Before there is ethics, there must be the welcoming and hospitable domicile. Bloechl summarizes, "Woman, says Levinas, welcomes me; she meets me at my level, and on my terms."[59] The relation with alterity so vividly apparent in the face of the son presupposes a relation with the feminine other, whose existence is required for the son to be. However the desires, needs, and terms of the woman who greets me appear to be literally effaced. Bloechl writes, "Everything that the feminine other does for me by way of rendering my world an inhabitable place is due to neither her possible intentions nor the specific acts she may or may not commit, but her presence alone. The familiarity of the world is a direct expression simply of the fact that I am not alone in it."[60] The encounter with alterity, then, requires a primitive encounter with the feminine, which seems to situate Levinas for strong proclamations about the primacy of the maternal relation. But since Levinas is utilizing the domicile and the feminine as the key metaphor for that which makes the face-to-face relation possible, he seemingly excludes the feminine from the realm where faces meet. Levinas's "woman" seems to provide a *break* from the pressure of being, a "delightful lapse" (*TI* 150, 155), even perhaps a break from the vigilance of everyday life. There is little indication that in this role the woman exists for any purpose other than to ease the pressures of masculine existence, to provide pleasure and distraction.[61]

This question is particularly cogent in light of our investigations on Levinas's use of time. The feminine is *recusal* from time, from danger, from labor, from suffering, from opposition, and from the ravages of time. She is *lapse*. Levinas's invocation of the feminine does not just *imply* the removal of the woman from the alterity of time; he directly declares that she is the absence of all things subject to the forces of time. The woman, metaphorically or otherwise, is lashed to the present of the masculine subject. She is the comfort and delight of *presence;* the feminine provides a hiatus, a fulfillment of the nostalgia for immediacy, pleasure, and happiness.

Levinas therefore binds the woman to the present. Ironically, to "have time" in the middle sections of *Totality and Infinity* is to be free, to stave off death, and to retain the hope in a next moment. The feminine becomes both a component of the habitation and its very condition. Women find themselves excluded by their very *inclusion* in the habitation. The feminine is present either *before* ethics or in attempting to conform to the masculine configuration of Levinasian ethics. Women are either ahead of or behind the game, but they are never quite players in the ethical drama, which remains, as it has always been, "a masculine affair."[62] In this sense, and in ways that de Beauvoir's 1949 footnote could not have anticipated, Levinas does indeed fall into step with the traditional relegation the feminine to a peripheral and secondary gender.

Into the Past

Levinas has, to this point in his philosophical development, focused relatively little attention on the temporality of the *past.*[63] Indeed, Levinas's early works are consumed by considerations about the future. In *On Escape,* Levinas's discussions of time seem chiefly concerned with the future and the tendency of the bourgeois to insulate and protect themselves from the tragic insecurity of time. In *Time and the Other* and *Existence and Existents,* the "time of the other" refers primarily to the way the other comes from the future and dislodges the self from the insular present, the opening of the hypostasis to the future. The fixation on the future is sustained in *Totality and Infinity,* highlighted by the discussion of eschatology in the conclusion and the preface. Eschatology is, after all, concerned with the future exclusively.

Yet in the concept of the feminine, Levinas explores a theme that will eventually come to dominate his reflections on time: the past. Cohen suggests several possible reasons for this gradual shift in Levinas's philosophy.[64] As Levinas reflects on the relation with the other, it becomes increasingly clear that the encounter with an other always occurs within a context, always enabled by a primitive, primordial hospitality. Levinas will eventually focus extensively on the way in which this primordial approach of the other is always prior to assimilation, representation, or encounter. But in *Totality and Infinity,* he has not yet developed a robust philosophy of the primordial past. His discussion of the feminine, in a very tangible sense, has Levinas out ahead of himself, already proposing a way of thinking about the past that he will not fully articulate until *Otherwise than Being.*

Beyond the troubling metaphors that Levinas offers here and the obviously chauvinistic overtones of *Totality and Infinity,* we find that Levinas is again advancing his understanding of time. Before there can be an encounter with the other-of-the-future (the child) there must already be an encounter with the other-of-the-past (the woman). In the idea of diachrony, so pivotal to Levinas's work in the 1970s and 1980s, these others are fused; the encounter with every other becomes an encounter with a past immemorial and the future eschatological. Yet there is an undeniable sense of corporeality and fleshiness in Levinas's reflections in *Totality and Infinity*. The other is not merely absent, but abundantly present. For in the child and in the woman, the father recognizes the *future* that is not his project, as well as a *past* that precedes all remembrance. The child forces the introduction of a "new category: before what is behind the gates of being" (*TI* 266). Levinas's understanding of diachrony is partly born out of his understanding of the family, even as his configuration of the family bears unfortunate emblems of patriarchal models.

The desire to either convict or exonerate Levinas on the charges of chauvinism has obfuscated a truly critical pivot point in his developing understanding of time. Chanter's work has advanced this conversation in critical ways, demonstrating that beneath the questions about gender lies an important advance in Levinas's manner of thinking about alterity and temporality. Among his chief concerns in "The Dwelling" is articulating that which precedes and makes possible representation. One can therefore justifiably read all the reflections on gender in "The Dwelling" as an extended exploration of this *priority*, the ineffable hospitality that makes possible all language, all encounters, all experience.

Still, in *Totality and Infinity*, Levinas makes only small, subtle gestures toward the category of the *priority* of the feminine. The woman is the past, the context for life and ethics. The child is the future, the reason for impossible, eschatological hope. For all the problems raised by these configurations of gender, family, and roles, Levinas has finally provided an incontestably practical example of the alterity he has been struggling to articulate in philosophical language. He has applied his philosophy of otherness to the family, to the phenomena of fatherhood, childhood, and motherhood. Levinas's readers are beckoned here to consider the face of a child, in all its fecundity, and to at least pause in gratitude for the fact that nothing is possible without the feminine. Yet the woman is figured in such a way that renders her ultimately excluded from the ethical relation. Significantly, Levinas has returned in force to his focus on time. "The relation with the child—that is, the relation with the other that is not a power, but a fecundity—establishes relationship with the absolute future, or infinite time" (*TI* 268). The woman, however, appears to be excluded from this relation with the absolute future.

Levinas began his career with a unique philosophy of time, which receded from the forefront of his thought during the 1950s. Time and temporal alterity were replaced, then, for the most part, with the spatial images of height, distance, and transcendence. *Totality and Infinity* clearly takes these spatial metaphors as primary. Yet Levinas never abandons his philosophy of time, even in the largely spatial sections that dominate the first portions of *Totality and Infinity.* But what happens in "The Dwelling" is rather fascinating in light of this transition toward an emphasis on temporality. What could be more *spatial* than the idea of the home, the dwelling, the emblem of a refuge against the external world? But in his reflections on the dwelling, Levinas is forced to acknowledge a relation with that which cannot be thought externally: the feminine. The very idea of "encounter" is in fact undermined by the feminine. Does one ever *encounter* one's mother? The walls of the uterus that are always *before*? Does not the relation with "the feminine other" redirect Levinas toward a kind of otherness that *precedes* the public square, that ethical relationship that is so important to Levinas's thinking?

Because Levinas invokes the feminine other, we cannot help but have *her* in mind as we read the ensuing pages of *Totality and Infinity* and learn about Levinas's "dimension of transcendence."[65] How can *this* other be considered "infinitely transcendent, infinitely foreign?" (*TI* 194). The infinitude of spatial transcendence becomes awkward and ill fitting in light of the feminine other, whom Levinas affords *priority.* This also undermines the disturbingly spatial metaphor of the domicile. In the home, the feminine is inscribed as a necessary component of the process of dwelling. Because the domicile took on many of the features previously allotted to the hypostasis and the instant, the inclusion of the feminine within the isolated *present* of the self carries a negative and dangerous connotation. As preethical, as part of the world of "projects" assembled and arranged in the interiority of the self, the woman's "face fades."[66] She resembles, all too much, the furnishings and accoutrements of the home. The woman becomes a feature of the present, a character in the history and progressing existence of the masculine subject. She is, in this sense, imprisoned in the world of the same. And this is precisely because she is denied *time* in the metaphor of the dwelling. She is denied a past that might irrupt my present, and she is denied a future that might be otherwise than the hospitality of the home. All of this builds a strong case for serious concern. This is chauvinism epitomized.

Toward Diachrony

William Desmond claims that "Levinas's reference to messianic time at the end of *Totality and Infinity* indicates that the work is a truncated book; its real import lies elsewhere."[67] It becomes increasingly clear that Levinas, as he nears the final sections, has begun to again invoke time to indicate transcendence. After an extensive discussion of *eros,* the face, the family, and the feminine, Levinas has arrived at the decisive need to express radical alterity in a temporal register.

In this work, the relationship with totality has been primarily conveyed according to height and transcendence. Yet Levinas sees the problem of totality in terms of separation and pluralism: "Separation opens up between terms that are absolute and yet in relation, that absolve themselves from the relation they maintain, that do not abdicate in it in favor of a totality this relation would sketch out" (*TI* 220). In typical, physical understandings of separation, the very distance between two "terms" already presumes a common totality, a dimension or field in which both terms and their distance imply. Levinas wishes to use the term *separation* in an explicitly metaphysical sense. Given that we are trying to identify an important shift in metaphors, Levinas's struggle to make the spatial metaphor work metaphysically is noteworthy: "The dimension of height from which the Metaphysical comes to the Metaphysician indicates a sort of non-homogeneity of space, such that a radical multiplicity, distinct from numerical multiplicity, can here be produced. Numerical multiplicity remains defenseless against totalization" (220).

The measurable spaces and gaps in being, for Levinas, can only provide an illusion of separation, an illusion that melts back into unity. This reduction of distance to unity is the key feature of *violence* and the very nature of war. War closes not just spatial distance but temporal distance, threatening the time of the other, her postponement of death. The irony of murder is that the other is also absent from the slain corpse. There can be no reduction of the other to the present of the same. The spatial distance can be closed, but the temporal distance is retained and even underscored by the "tension of the instant" (*TI* 225). The other is absent in presence and present in absence; Levinas calls it "a simultaneity of absence and presence" (225). This paradox is difficult to configure in terms of height; how can the other *transcend* the violence of war?

Spatial proximity abounds in the bodily relations of love and violence; the distance between the self and the other is closed by erotic embrace or by violence. These closures, however, conceal the withdrawal of the other from the present. Here we see that the spatial images of alterity are unable to hold up beneath the weight of Levinas's expectations. The relation with the other, Levinas tells us, is *asymmetrical*, a denial of the spatial notion of symmetry. Like asymmetry, the relation with the other "opens time" (*TI* 225).

The other, exposed to violence and captivity by "the hands of forces that break him," remains "unforeseeable" (*TI* 225). Violence aims at the other as she is *present*, oblivious to his or her absence and withdrawal. The resistance of the other to the totality of violence is not physical but moral, and this resistance is evidenced in the face of the other. Hence, says Levinas, "violence can aim only at a face" (225). The relation with the face "subtends war," unsettling the symmetry that marks the interchanges and exchanges of "totality" (225). The encounter with the face of the other solicits a response, a call that destabilizes the logic of war and violence. War's murderous negation of the face is an attempt to reduce this call to silence. But the face is not satisfied with its role within the categories of being and economy and will not be pacified with good intentions and "a benevolence wholly Platonic" (225). War has its moral code, and the face of the other carries the rupture of this totalizing morality. Time, as opened by the face, no longer marches with the universal symmetry of Aristotle nor the existential consistency of Heidegger. Time is unhinged, for Levinas, from the "menace and postponement" of immanent, universal temporality (235).

Violence depends on, and aims at, a reduction to *immanence* and *presence*. These terms become increasingly important to Levinas in the latter stages of *Totality and Infinity* and then particularly critical in the preface. By the time Levinas writes the preface, he is focused clearly on the relationship between war and eschatology and on the relationship between *time* and totality. The structure of the book suggests that Levinas is already beginning a transition in the way he thinks about totality, even as the book draws to a close and he crafts the preface. Totality was at first configured with almost entirely spatial images: exteriority and interiority, proximity and distance, and especially the dwelling and the world. These spatial themes finally give way to a steady escalation in his emphasis on alterity as eschatological time.

Eschatology is a term that Levinas generally avoids across his career, perhaps because the term becomes important to Heidegger.[68] Here in *Totality and Infinity,* however, the eschatological future is the defiance of the logic of violence that dominates the present. This futurity poses itself in an "impossible" relationship with the immanence of the present. Here Levinas claims that "*the time in which being ad infinitum is produced goes beyond the possible*" (*TI* 281). Pardon is impossible; it comes only from outside the field of possibility. The continuity of time must be broken to allow for pardon, which liberates the past from its fixed prison. Eschatological hope is a hope that "instants" will not "link up with one another indifferently" but that they will instead "extend from the Other unto me" (283). Levinas writes, "The future does not come to me from a swarming of indistinguishable possibles which would flow toward my present and which I would grasp; it comes to me across an absolute interval whose other shore the Other absolutely other—though he be my son—is alone capable of marking, and of connecting with the past" (283).

And so we return to the preface, where Levinas becomes the first critic of his own book. The preface announces that history cannot lead to peace: "of peace there can only be an eschatology" (*TI* 24). How surprising to find Levinas resting, in the preface, so heavily on a term that first appears in the final pages of this book? He calls, in the preface, for the pursuit of an eschatology that "institutes a relation with being *beyond the totality* or beyond history" (22). Despite the familiar fixation on the *future,* which has been the primary focus of Levinas's discussions of time, we also see that he has a growing interest in the past. To get *behind* the violence of being means to go *before* being to the "primordial and original relation with being" (22). The immemorial past will become critical to Levinas's pervasive discussions of time in *Otherwise than Being,* as well as in his other late essays and books. As such, these reflections on *Totality and Infinity* have already begun the discussion that will consume the next chapter: what drives the remarkable changes and escalations that are featured in Levinas's second major work?

Six

Time in Transition

> Like language, experience too no longer appears to be made up of isolated elements, somehow lodged in a Euclidean space in which they could be exposed, each on its own, directly visible, and each signifying by itself.
> —Emmanuel Levinas, "Meaning and Sense"

There is a stark difference between *Totality and Infinity* (1961) and Levinas's second major work, *Otherwise than Being* (1974). One need only spend a few minutes with *Otherwise than Being* to discover that obligation is being configured in even more radical and extreme language. Shocking metaphors of "hostage" and "persecution" appear alongside a series of sharp gestures toward a radical, singular obligation to the suffering other. In *Otherwise than Being* the self, the singular I, awakens in a world in which it is already captive, already radically obligated, already *hostage* to the other. In this register, responsibility is like a persecution that cannot be avoided, a captivity that I can only pretend to escape. This notion of responsibility is stated harshly and severely, exceeding what was already a deep and rigorous commitment to ethics as first philosophy. The depth of obligation in *Otherwise than Being* depends, on almost every page, on Levinas's understanding of time as *diachrony*. How and why did Levinas develop these extreme expressions?

Returning to the theme of waves on the shore, at the end (and the preface) of *Totality and Infinity* we have already seen the signs of a new surge gathering offshore. Levinas's new thinking about time is already intimated in his introduction of eschatological time to discuss radical alterity. Time soon provides the primary philosophical support for his expressions of ethical obligation. It is uncontroversial to suggest that *Otherwise than Being* leans heavily on the diachrony of time;

the waves crash high on the shore as Levinas proclaims that the other summons us from a past that is immemorial.

Given the deepening and extreme account of responsibility in *Otherwise than Being,* it is important to consider how Levinas's thought develops during the transitional period between the publication of *Totality and Infinity* and his second major work. Levinas's vast production across the 1960s reveals a broad range of resources and influences. Sketching some themes from a few paradigmatic essays, such as "Transcendence and Height" (1962), "Meaning and Sense" (1964), and "Enigma and Phenomenon" (1965), reveal in Levinas's thought the receding use of spatial imagery, the intriguing renewal and expansion of time as an operative and central theme, and his first usage of the term "diachrony." In addition to these philosophical essays, during this period, Levinas's account of time becomes increasingly informed by Judaism, the Talmud, and Franz Rosenzweig. And finally, one simply cannot ignore the way in which Derrida's critique of Levinas, "Violence and Metaphysics" (1964), influences Levina's language and discussions of time.

Philosophical Developments

After *Totality and Infinity,* Levinas begins a rigorous and productive period of his career. He publishes extensively, presents papers, and interacts with a wide swath of intellectual voices and movements. Levinas was academically active throughout the 1960s in large part because of the interest generated by *Totality and Infinity.* In this decade, we find immediate evidence that Levinas is already working on a few new ideas as well as attempting to answer some of the questions left open in *Totality and Infinity.*

"Transcendence and Height"

Levinas does not appear, immediately after *Totality and Infinity,* to be uncomfortable with the notions of height, transcendence, or distance. In fact, one of his first papers after the publication of *Totality and Infinity* was entitled "Transcendence and Height." This essay and the conversation that followed its delivery reveal the growing tension in Levinas's thought over the questionable capacity of spatial metaphors to carry forward his sense of radical alterity.

In this short essay, Levinas configures obligation in spatial terms. He writes, "The I is not simply conscious of this necessity to respond, as if it were a matter of an obligation or a duty about which a decision could be made; rather the I is, by its *very position,* responsibility through and through. And the structure of responsibility will show how the other, in the face, challenges us from the greatest depth and the highest height—by opening the very dimension of elevation" (*BPW* 17). In the dialogue session that followed this paper, Levinas was presented with a concern that is pivotal to this study. Responding to his essay, Eugene Minkowski challenged Levinas as follows: "When you speak of height, I ask myself if you have not thereby already made a concession that is too static and too spatial. The movement of elevation that you mentioned elsewhere, appears to me much closer to the primitive dynamism of existence than that of height; in the same way that expanding or opening out has a more vital importance than the notion of extension" (27).[1]

As we will see, Levinas's responses to this challenge demonstrate a reluctance to part with the spatial and geometrical notions of transcendence that we have seen throughout *Totality and Infinity.* Levinas insists on the importance of conceiving of the other as "very distant," transcendent in a way that leaves no common feature between the "same" and the "Other" (*BPW* 27). Levinas remains committed to redefining "transcendence" beyond the *relative* transcendence of Neoplatonism, and he continues to consider radical *spatial* transcendence to be the "point of departure" for all "concrete" relations with the other (27).

The other must first be "very distant" before being near, before becoming incarnate in the manifestations of "neighbor" and "fellow human being" (*BPW* 27). The emphasis is on *difference,* and difference is cast as a kind of space too vast for traversal. Levinas essentially dodges Minkowski's question; he does not address the concern that these spatial images create a static situation and potentially undermine the "primitive dynamism" that seems to mark Levinas's work elsewhere (27). These concerns are not, at any rate, sufficient to warrant any alteration in his use of spatial imagery to denote the originary distance between the self and the other. Despite the spatial intonations of the term "transcendence," Levinas will continue to use this term throughout his later works, however, he recasts the term as the "Time of Transcendence."[2] He performs a similar move with the term "proximity," which plays an important role in *Otherwise than Being*. Levinas undermines any sense of space or "Euclidean geometry" in his later invocations of proximity (*OB* 81).[3]

Time and the Trace

"Meaning and Sense" continues some of the patterns of *Totality and Infinity*, moving without caution between the spatial and temporal metaphors for alterity. This essay includes the content of a paper Levinas presented in 1963 called "The Trace of the Other."[4] As is common throughout Levinas's work of the early 1960s, "Meaning and Sense" operates without hesitation in the milieu of spatial metaphors of height, transcendence, journeys, traversals and distance.

Yet near the close of this article, in the portion that is given in the "Trace of the Other" paper, Levinas turns to a heavy focus on time and the trace. This new focus is important for several reasons, as it represents a definitive expansion of the treatments of time considered in his earlier works, as well as in certain portions of *Totality and Infinity*. But in his reflections on the trace, Levinas focuses his attention with great clarity on the question of the past, the past that cannot be recovered as a former present. This focus on the past inaugurates a new phase in Levinas's thinking on time and enables him to renew his case for the coinciding transcendence of both the other and God. More acutely, it is the weight of the past that has never been present that will allow Levinas to turn toward radically escalated notions of obligation. *Otherwise than Being* will depend heavily on this notion of an anarchic past, a past that is not a former now.

Levinas's strategy in "Meaning and Sense" is to trace the consequences of the failure of phenomenology when it comes to the field of language and then to demonstrate that there is a consistency between the way language escapes totalization and the way *experience* exceeds understanding. From the early pages of "Meaning and Sense," we can see a rough outline of *Otherwise than Being,* which is still a decade away: "Like language, experience too no longer appears to be made up of isolated elements, somehow lodged in a Euclidean space in which they could be exposed, each on its own, directly visible, and each signifying itself. They signify on the basis of the "world" and of the position of the one that looks at them" (*BPW* 37). This correlation between the evasiveness of language and the evasiveness of experience now becomes a permanent feature of Levinas's philosophy. Levinas will only increase the strength of this correlation. Reading and relating will soon register on the common theme of *exegesis,* and exegesis will become the key modality or ethics. At this point, Levinas is clearly concerned with the way geometric (Euclidean) space obscures the way we think about experience and language. In geometric space, objects lay exposed and apparent; their meaning and significance are transparent and evidentiary. Western philosophy, reasons Levinas, has grown accustomed to pursuing knowledge in the way Euclid pursued axioms and geometric logic; it has sought to draw knowledge out into the light, to expose, simplify, reduce, and constrain.

Language, however, clearly resists this impulse; that which is read is not the same as that which was written. Experience, like language, no longer seems to function mathematically. The encounter with the other evades the totalizing gaze that would reduce one experience to a universal field of common experiences. Levinas will gradually rely more and more on the correlation between language and the encounter with the other person. He wishes to establish a phenomenological insight about the relationship between sensory experience and meaning. The development of a correlation between language and alterity will provide Levinas with a critical paradigm in his later works. "Experience is a reading," he writes, and "the understanding of meaning an exegesis" (*BPW* 38). The comparison of the relationship with the other person to the reading of a text is anything but a reduction or simplification of experience. For this development, Levinas will return to and advance his unique considerations of time.

When Levinas claims that the "understanding of the Other (*Autrui*) is thus a hermeneutics and an exegesis" (*BPW* 52), he is utilizing the phenomenological analyses of experience given by Maurice

Merleau-Ponty in the 1950s and 1960s.[5] Levinas calls Merleau-Ponty's analyses "remarkable" and uses his work to demonstrate the rigorous complexity of sensation and knowledge. Levinas has continued to moor his reflections in phenomenology and considers himself a phenomenologist to the end. But he is aware that Merleau-Ponty's work remains more clearly embedded in the phenomenological method and in close relationship with the work begun in Husserl. What Merleau-Ponty adds to Levinas's reflections at this point is a robust understanding of the cultural and prereflective postures that tend to predetermine the modes of experience.[6] That is, more than Husserl realizes, the observer of a phenomenon can never quite prepare for the experience of a phenomenon. In "Meaning and Sense," we find Levinas looking for a way to express the failure of phenomenology to appreciate the complex enterprise of human experience.

Language provides a principle example of this need for humility. The very use of a metaphor already shows evidence of this evasiveness; for the "*beyond* which the metaphor produces has a sense that transcends history" (*BPW* 56). Language makes one acutely aware of the "lack of time to turn around" and catch a meaning before it has been altered in the ossification of writing. In terms of the other person, this means Levinas is interested in the role of that which is *left behind* after the experience of the other. More precisely, Levinas becomes interested in the "trace" of the other, the remnant of the presence of the other that lingers after the "epiphany" or "visitation" of the other person (53).

For Levinas in the 1960s, it is becoming increasingly important to demonstrate and highlight the significance of passivity in experience. This plays no small role in his developing understanding of time. It is because of passivity that the observer is simply *too late* for experiences, unable to turn around in time to experience anything in a common now. The fields of math and science incline us to think about experiences from a common, neutral vantage. However, Levinas uses Husserl, and then Merleau-Ponty, to demonstrate that our impressions of phenomena are already heavily loaded, even as we believe ourselves to be having an unfettered experience.[7] Though language becomes, and will remain, an integral way that Levinas talks about time and sensation, he is interested primarily in the enigma of the trace of the other person. At a phenomenological level, this means Levinas is particularly interested in the difference between the object in-itself and the object as it is experienced.

In "Meaning and Sense," before he turns to the richly temporal notion of the trace, Levinas turns loose another set of spatial metaphors. One's initial impression of the other, he says, is "disturbed and jostled by another presence that is abstract (or, more exactly, absolute) and not integrated into the world" (*BPW* 53). This "other presence" intrudes; it makes "an entry" into the world of the self through the phenomenon of the *face*. Again, we find Levinas using spatial imagery associated with the "face," which "breaks through his own plastic essence, like a being who opens the window on which its own visage was already taking form" (53). The face, as Levinas is here conceiving it, is an alien presence that "*enters* into our world from an absolutely foreign sphere." This entry is a complex phenomenon; the face does not enter the world of the self as an immanence, and it does not become assimilated in any way to the world it invades. This lack of reduction will soon, even in this essay, be turned toward a critique of presence and an emphasis on the alterity of the time of the other, but here Levinas uses spatial imagery without reluctance.[8] This spatial language is strikingly similar to that in Levinas's work in the 1950s, where essays like "Is Ontology Fundamental?" relied heavily on the horizon beyond which the face of the other dwells. The spatial imagery functioned then, as it does here, to create a space outside of ontology. The language of windows and openings and faces functions well to distance Levinas from Heidegger, who had described being as "fundamental ontology" (*BT* 34)[9] with Dasein's temporality as the "horizon" (19). But the imagery of space borrows heavily on the *light* and *sight* that Levinas also wishes to critique with severity. Derrida's reading of *Totality and Infinity* will effectively address this remnant of ontology that subsists in Levinas's choice of metaphors.

In his remarkable essay "The Trace of the Other" Levinas thoroughly unsettles spatial imagery by presenting the relationship with the other as a relationship with that which is "before history"

(*BPW* 59). The encounter with the other becomes an encounter with "an utterly bygone past" (60). The call to responsibility, then, is a call to answer for "an irreversible past." The past from which the other summons is prior to any memory; the trace of the other's presence is the remnant of something from the "immemorial past," without also bearing along the conceptualization necessary to convert it into a *sign*. The presence of the other is now a "supreme and irreversible absence," yet the evidence for this absence is the trace of a presence that was never quite present (61). The trace is disclosed as personal and ethical; it is characterized as obligation.

If the trace discloses obligation, however, it does so in a unique manner, without appeals to a system of symbols and signs and laws toward which signs normally appeal. Signs point somewhere; they participate in systems and webs of meaning. Signs and signifiers therefore accede to a sense of the whole, toward the One of Platonism. A sign permits itself to be configured by a reason outside itself, by a system of meaning in which the sign makes sense. But the trace, claims Levinas, is a sign like no other: "But it also plays the role of a sign; it can be taken for a sign. A detective examines, as revealing signs, everything in the area where a crime took place which betokens the voluntary or involuntary work of the criminal; a hunter follows the traces of game, which reflect the activity and the movement of the animal the hunter is after; a historian discovers ancient civilizations as horizons of our world on the basis of the vestiges left by their existence" (*BPW* 61).

Levinas's trace refuses to operate in the reasonable register of other signifiers. The trace does not take its place in a higher order, in a totality that gives sense to the singularity of the trace. Rather, the trace is an exception in its evasion of "every intention of signaling and outside of every project of which it would be the aim" (61). Ontology seeks to know the other by way of the light of being, to seize on the signs given from the other as a detective measures evidence. In ontology, every sign offered by the other person can be constellated with other signs and correlated with other evidence. The other person becomes a science project, a laboratory experiment where every piece of data becomes the possession of the scientist.

Ontology therefore reduces the alterity of the other to a present possession for the now of the self. All things "other" are ultimately "convertible" to the "same" (*BPW* 59). Levinas is fond of hunting analogies; animals leave behind signs, but they do not leave traces. The trace of the other is the passing hint of a presence that never quite coordinated with the present of the self. The trace *does* leave evidence, but it leaves evidence of a very different sort. The evidence of the trace is disruption; the other unsettles and disturbs the system of signs, reason, and knowledge that characterizes the world of the self. This unsettling instigates desire, a desire to recover the past of which the other left behind this trace: "But a real trace disturbs the order of the world. It occurs by overprinting. Its original signifyingness is sketched out in, for example, the fingerprints left by someone who wanted to wipe away his traces and carry out a perfect crime. He who left traces in wiping out his traces did not mean to say or do anything by the traces he left. He disturbed the order in an irreparable way. For he has passed absolutely. To be qua *leaving a trace* is to pass, to depart, to absolve oneself" (62).

The "trace is the presence of that which properly speaking has never been there, of what is always past" (*BPW* 63). The other essentially evades me by dwelling in a time that is not my own, in a time that transcends the cause-and-effect temporality in which I dwell. Levinas's attempts to use analogies in this section are potentially confusing. He does not mean that the other has departed the way that lightning has departed and left behind a forest fire. This establishes a connection between cause and effect, exposing all that is to be known about an event. According to Levinas, we can know the history of a forest fire, but we cannot know anything about the *past.* Levinas deems the trace of the other to be the very force that disrupts history, which calls into judgment history itself. This disruption also disturbs any ability I might have to "suspend my responsibility for the distress" evident in the face of the other (54). Labor for the other requires a work "without recompense," without a gain or a return to oneself. This kind of work Levinas calls "liturgy" (57).

This understanding of the trace allows Levinas to introduce and explore the way the alterity of God performs a similar evasion. He explains that the God of Exodus 33 "shows himself only by his trace" (*BPW* 64). And to go toward God is not to follow a sign, like footprints in the snow, but to turn outside of the economy that the trace of God ruptures. Levinas has begun to answer a critical question left open after *Totality and Infinity.*[10] The trace of the other and the trace of God are similar inasmuch as they call into question the dominance of the present and the efforts of ontology. Important for this study, the alterity of God and the alterity of the human other similarly expose the difference between the time of the other and the time of the self. They arise from "before history" (59). Elsewhere, Levinas seems to set out in "Meaning and Sense" to explore *how* a trace remains despite the immemorial passing of the other.[11] This question is more difficult than Levinas may have anticipated, and it therefore continues to trouble even the pages of *Otherwise than Being* where the concept of the trace will rise to central importance.

Language and Diachrony

The philosophical essays "Enigma and Phenomenon"[12] and "Intentionality and Sensation" both deal extensively with Levinas's understanding of time. Levinas uses the term "diachrony" for the first time in these two 1965 publications, which becomes the linchpin of his understanding of intersubjective time in *Otherwise than Being.*[13] However, it is curious how the word becomes part of Levinas's philosophical vocabulary, and within a matter of just a few years, becomes so important in his work. Indeed, the term, which is completely absent from *Totality and Infinity,* is used more than 60 times in *Otherwise than Being.*[14]

"Diachrony" belongs primarily to the field of linguistics; Levinas borrows the term and transforms it to mean something quite different from the usage it receives among linguists. Used extensively by nineteenth-century Swiss linguist Ferdinand de Saussure, for Saussure, "diachrony" and "synchrony" could be opposed to demonstrate elements of language that evolve and elements of language that are static and unchanging, respectively.[15] Though Saussure told his friends he was working on a manuscript, he died in 1916 and left behind no trace of the book he had been writing. His students pieced together their notes from his lectures and published the book *Course in General Linguistics.* This text deeply influenced linguistics across the twentieth century, though much new light has been shed on Saussure's work in recent years—his lost manuscript, *Writing in General Linguistics,* was discovered in Geneva in 1996 and published in 2002.[16]

The linguistic community took up the version of Saussure's work presented in *Course in General Linguistics.* Linguists used "diachrony" to refer to the amorphous nature of language and sometimes the permanent difference between the sign and the signifier. This is the very first principle of linguistics for Saussure, that "the link between signal and signification is arbitrary."[17] His understanding of linguistics depends on a bifurcation between elements of language that are static and elements of language that are evolutionary: "A language is a system which is intrinsically defenseless against the factors which constantly tend to shift relationships between signal and signification. This is one of the consequences of the arbitrary nature of the linguistic sign."[18]

The terms "diachrony" and "synchrony" in linguistic studies continue to relate to structuralism and its alternatives. Levinas, however, does not seem even slightly interested in Saussure, the father of structuralism, or the linguistic structuralism that develops in the wake of his work. Peperzak summarizes the Levinas's appropriation of diachrony: "Here and in the following pages Levinas alludes to de Saussure's distinction between a diachronic study of language and other systems of cultural expression, and a synchronic study. The latter makes a cross-section across time in order to expose a structure within which the elements are simultaneous. A diachronic study follows down the transformation of structural elements in the course of time" (*CPP* 61n). Peperzak is right; Levinas does seem interested in using this linguistic term as a tool to help readers understand the extreme difference of the time of the other from the time of the self. But Levinas never quite explains his choice to use this term, though he does clearly express a desire to take the implications of diachrony to a moral level. There is, in fact, a quasi moral significance to diachrony, even as it is used among linguists. The linguist and anthropologist

Johannes Fabian considers Saussure's distinction between synchrony and diachrony to be both "famous" and critical for ongoing linguistic theory.[19] Fabian suggests that Saussure already realizes that synchrony is a "proverbial wolf in a lambskin." Synchrony, which presumes a static role for language, or linguistic structure, is always a compromise. From the beginning, argues Fabian, Saussure tells us that "making sense of speaking, understanding others, can be achieved only for a price, namely, ignoring real time and history."[20] Synchrony is a fall, a concession, and a compromise. Levinas does not move with Saussure into the structuralism that arises from Saussure's linguistic theory; however, the bifurcation between synchrony and diachrony suits Levinas's intentions perfectly.

When Levinas experiments with diachrony in 1965, he is attempting to make good on his promise to connect the phenomenon of language to the experience of the other person. His goal, apparently, is to demonstrate that the mode of exegesis is fitting for reading not just texts but all human experience, and particularly the experience of the other person.[21] Levinas demonstrates this correlation here by way of linguistics, attempting to underscore the radical passivity of language. But while linguistics provides a *term* that allows him to divide up the time between the signifier and the signified, linguistics has less interest in the implications of the passivity this conclusion implies. So for Levinas, as he writes later in "Substitution" (1968), diachrony comes to mean the "anarchic passivity of the creature" (*BPW* 89). The trace of the other is experienced in diachrony; the evidence of the "passing" of the other is borne by a different time than the present of the self.[22] Levinas is careful to note that the diachronic time of the other does not mean that the other is just part of a past or a future "in which the overflowings of the present flow back to this present across memory and hope" (67). He concedes that diachrony "maddens" the subject, but in so doing diachrony unveils the utter transcendence of an encounter that is beyond synchrony. The encounter with the other is perceptible as a "signification that would signify in an irreducible disturbance" (67).

For Levinas to begin to call the transcendence of the other "diachronic" is to chase alterity down a new path, the path of language and, most specifically, *time*. The other's transcendence need not be considered analogous to spatial distance, which returns inexorably to ontology. This is what Levinas considers a new "point of departure," the "invisibility which language sets forth" (*BPW* 67). The trace of the other, left by the passing of something older than history, unsettles the very idea of co-presence. The other arises as a "disturbance" that reveals the impossibility of "simultaneity between its terms" (69). The other shames attempts to reduce the trace to a sign, to connect the phenomenon to an "order," or to contextualize the "enigma." The trace of the other signifies the irreversible, unrecoverable, unrepresentable, and immemorial *past*. It manifests itself as a forgetting, or a "forgettingness" (*obliviscence*), not as a distance that is "remoteness" but of a nearness that somehow remains utterly "incognito" (70).

"Enigma and Phenomena" sustains these themes with an extensive discussion of time and temporality. The approach of the other, Levinas now announces, is a "supreme anachronism" that breaks all correlation and denies the common ground of presence for intersubjective relationships and phenomenological experience. He introduces "diachrony" and the dissonance between the "saying" and the "said," all in an attempt to undermine and unsettle the dominance of presence, which seems to always imply a "contemporaneity" that reduces the other to the time of the self (*BPW* 77). We never coexist in the manner that geometry and science imply; we are always too late. The time of the other is *diachronic*, joining God in an "antecedence" to the world "which cannot accommodate him" (77). And here Levinas continues to unfold his understanding of the relationship between the other person and God. Both relations are radically alterior, but more specifically, both the trace of the other and the trace of God make one aware of diachrony.

Renewed Interest in Husserl

These developments draw Levinas close to themes from Husserl's work that have interested him since the 1920s. Levinas, in fact, had already begun to renew his conversations with Husserlian phenomenology even before he published *Totality and Infinity*. In his 1959 article on Edmund Husserl, "The Ruin of Representation," Levinas provides a personal and testimonial introduction to Husserlian phenomenology. He recounts for the reader his youthful exuberance as he encountered Husserl at the age of 20. He notes the mythological proportion to which his young mind elevated the venerated Husserl and retraces the key elements of phenomenology: intentionality, sensuousness, consciousness, and so forth (*DEH* 111–12). But Levinas makes it clear that "The Ruin of Representation" is not designed to be another summary of Husserl's contributions to philosophy and phenomenology—Levinas has done that extensively elsewhere. Instead, he intends to outline "the points that seem to me essential to all post-Husserlian thought, and they constitute the benefit that I, for my small part, have derived from a long acquaintance with Husserl's works" (113).

This essay confirms Levinas's rootedness in Husserl and his transformation of Husserlian sentiments into the vocabulary of his own philosophical program. What is most significant is the sense of temporality that Levinas derives from Husserl's innovative reconsideration of the relationship between subjects and objects. Whether or not temporality is central to Husserl's theories of intuition and consciousness is a debatable question, but for Levinas this was certainly the case. "The Ruin of Representation" provides pivotal insight into the function of time in Levinas's reading of Husserl and into his own developing ideas of time.

Levinas, as we detected in his early work, correlated his understanding of the instant with Husserl's *Urimpression,* which is the initial and primary moment of impression. *Urimpression,* as Husserl scholar Dan Zahavi summarizes, "must be situated in a temporal horizon."[23] This moment, the instant, is already laden with *retention* and *protention,* the experiences of what has been and the anticipation of what will be. These aspects of a phenomenon are trivial if the fundamental meaning of an object is derived from the present. But Husserl teaches us that if the present moment is considered apart from its temporality it is an abstraction that misleadingly reduces an object by way of false objectivity.

Levinas's reading of Husserl plays on the usage of time: "Phenomenology itself is this reversal in which being creates the act that projects it; in which the present of the act — or its actuality — turns into the past, but in which the being of an object is at once *perfected* in the attitude that is taken with respect to it and in which the anteriority of being is again placed in a future" (*DEH* 119). Phenomenology trumps representation by replacing it with the human experience. This is why Levinas finds representation to be a shadowy, ruinous facsimile of reality. For Husserl, the real experiences of the senses are privileged over the representation of experience, and all experience occurs under the direction of *consciousness.* Consciousness is laden with intentionality, protention, and retention. Phenomenology, before Husserl, relied uncritically on this sense of presence and therefore on the philosophy of time as articulated by Plato and Aristotle. This emphasis on the present is evident in a "naïve" sense of the value and reliability of representation (119).

Levinas again utilizes diachrony in his 1965 essay "Intentionality and Sensation." He clarifies his emphasis on the primacy of diachrony over synchrony and explains that the "final secret of the subject's historicity" is a "diachrony stronger than structural synchronism" (*DEH* 148). Levinas, seemingly, concedes the suspicion of linguists that diachrony is *prior* to the synchronizing compromise that makes language and communication possible. For Levinas, however, the implications of this conclusion are pressed far beyond technical questions about linguistics. At stake is the question of activity and passivity. He suggests that there is diachrony within the very intentionality of the subject (143). In this essay, he first uses the term to exploit an ambivalence within Husserl's *Phenomenology of Internal Time-Consciousness.* To what extent, he asks, does the time of the object differ from the time of the subject who intends to analyze it?

This presses Levinas up against the limits of phenomenology as it has been defined in the Husserlian tradition. Levinas's ploy will be to read Husserl against himself. This is the same strategy that Levinas used repeatedly in his reading of Heidegger. Yet he doubles back on Husserl in a different manner than he does in his reading of Heidegger. Levinas has worked steadily to reject most elements of Heidegger's new thinking about time, even as he embraces Heidegger's suspicion that time needs to be deformalized. In his estimation, Heidegger has accurately assessed the problem and pointed helpfully toward an ec-static understanding of time. But the seeds of the solution are not found in *Being and Time,* which liberates Dasein from the universal scope of Western history and temporality, only to relocate truth back in the field of the same.

With Husserl, Levinas uses a very different approach. Levinas is concerned to show that phenomenology can accommodate his escalating sense of the passivity of the subject. This presents an immediate problem: phenomenology has, since Husserl, bound together sensation, intentionality, and consciousness. That which can be described must be ascertained *intentionally,* and intentionality is a property of consciousness. Consciousness *aims,* in a mode that is active, at objects and describes objects by the form that objects make available to conscious life.[24] Yet even in his early works, Levinas was questioning the power of intentionality, particularly as it appears to establish a level of activity and virility in the observer.[25] Husserlian phenomenology has no obvious mode in which to consider sensation that is *immediate,* or prior to intentionality.

To remain a phenomenologist, or to continue to ground himself philosophically in that tradition, Levinas must show that Husserl has provided a mechanism for addressing a passivity that precedes intentionality. It is helpful that Husserl was as least somewhat aware of the problems associated with intentionality and activity. Bettina Bergo notes that the problem of passivity and affectivity was perplexing Husserl as early as 1905 when he was working on his lectures concerning internal time—consciousness.[26] What Levinas notices in Husserl's work on time is a resistance to the reduction of time to an "atemporal gaze" or a "background of preexisting time" (*DEH* 142). For Levinas, Husserl is chasing a manner of thinking about time that begins with the "sensing of sensation" and not with Aristotle's "immobile eternity for a disengaged subject" (142). Time in the Aristotelian sense, even in the Hegelian sense, is an abstraction from the primordial sense of time, which is the flux of retention and protention. In Husserl, as Richard Cohen summarizes, Levinas finds "an inextricable and fruitful bond between knowing and existing."[27]

So Levinas credits Husserl with identifying the need for a division between two different experiences of time: world time and internal time. He writes, "Is not the object of intention already older than the intention? Is there diachrony in intentionality?" (*DEH* 143). Levinas deftly demonstrates that Heidegger's greatest accomplishment, the unhinging of the self-other relation from the totalizing sweep of universal history, is already presented in Husserl. Historical time is secondary, already constituted by the internal time-consciousness that eliminates "the antinomy of spontaneity and passivity" (78). Time is that original sense of freedom, the liberation of the now from the burdens of idealism.[28] Levinas moves forward his understanding of time with a close eye on the tools that Husserl has provided for moving past the borders of phenomenology to explore something deeper than intentionality. He continues, after summarizing Husserl's philosophy of time, noting that Husserl has already contributed "very suggestive views" in his "concrete descriptions of the consciousness of time" (78). These are writings "of a rare subtlety," wrote Levinas, and he notes tellingly that he is particularly interested in Husserl's treatment of time: "I especially have in mind the theory that makes time the very manifestation of freedom and spirituality" (78). Levinas will develop his understanding of time in a manner that turns both with and against Husserl's helpful but ultimately naïve understanding of representation, activity, and passivity.[29]

Husserl, Levinas believes, held obstinately to this model by placing the constitution of the cultural world in the "transcendental and intuitive consciousness" (*BPW* 58). Husserl therefore "rejoins Platonism" by retaining a sense of objectivity with respect to the formation of the cultural conditions that potentially impact the perception of phenomena. But Levinas now believes he has found "another method" by which to pursue "meaning." This method moves by way of a passivity far more radical than anything envisioned by Husserl, toward a "passivity more passive than any passivity" (*CPP* 135).

The problem in Husserl is not that he fails to acknowledge affectivity or passivity in sensations, but that this passivity remains a component for consciousness. The impression that all things perceived have already "passed by," is not therefore an opening to a "beyond" intentionality and sensation. For Husserl, as Lilian Alweiss summarizes, the *Urimpression* manifests itself as a "correlate to consciousness; it is in the fold of consciousness."[30] The fact that those primal intentions remain riveted to the intentional consciousness of the self is deeply problematic for Levinas. He wishes to demonstrate that the preconscious immediacy of sensation is the result of an encounter with the human other. Levinas can agree with Husserl that all sensuous experience is the product of memory, or retention. He can also agree with Husserl that this is the fundamental and original sense in which philosophy can speak about time. But for Levinas, the alterity that arises from being permanently "too late" for intentionality is not a theoretical alterity, which it remains for Husserl.[31] Levinas proposes an alterity that does not return to either self-presence or to co-presence.

When Levinas uses "diachrony" in his 1965 reflections on Husserl, he is pointing to a Husserlian insight that he wishes to retain but to escalate. Husserl knew that phenomenology must accommodate passivity and affectivity, but obstinately insisted on making these experiences only *apparently* alterior to the self. In the end, Husserl admits no "beyond" intentionality and consciousness, and by turning attention in that direction, Levinas may well be positioning himself outside of phenomenology in its strict definition. But Levinas moves in this direction by way of Husserl's understanding of time and the *passive* mode in which time disposes the self in the world. Time is not, for both Levinas and Husserl, some idea that when added to the experience of change makes sense out of ticking clocks and turning planets. Time is the very sense that I am conscious and that my consciousness is both "too late" and incomplete. Kant, Hegel, Husserl, and even Heidegger commonly leave the ultimate sense of time and history riveted to the self, and it is in this sense that Levinas will accuse these thinkers of rejoining the Platonic tradition (*CPP* 59).

To move beyond this storied tradition, Levinas must push further into his innovative understanding of time, radicalizing diachrony to make it indicate the intrusion of a time that is utterly and completely *not* my own. Husserl's work suffices to describe the intricacies of phenomenology and epistemology, but for Levinas, these adventures in knowing fail to attune to the alterity of the other person. The trace of the other is not given to knowledge in the sense of Husserlian phenomenology; the trace of the other person is known, instead, by "the phenomenology it interrupts" (*BPW* 104).

We can certainly conclude that a renewed interest in Husserl is among the many forces and developments that mark the transitional period between Levinas's two major works. Levinas's reflections on Husserl's understanding of time press forward on the underdeveloped notion of passivity in Husserl's phenomenology. This engagement with Husserl opens up both vocabulary and philosophical resources for thinking about diachronic time, which become indispensible tools for Levinas's expression of alterity in *Otherwise than Being*.

God, Messianic Time, and the Talmud

The scholarly field surrounding Levinas's work is deeply divided on the relationship between Levinas's philosophy and his religious writings. Levinas himself is well aware of his position between two discourses, yet this does not prevent him from leaving a perplexing trail of images from both Greek and Jewish sources. The idea of God in Levinas's philosophical works is generally evasive and distanced from any particular religious tradition. Ze'ev Levy claims that in Levinas's philosophical writings, "the word *God* represents a concept that is religiously lucid but philosophically vague."[32] As Levinas draws close to the publication of his second masterwork, *Otherwise than Being*, he is also participating in sustained religious conversations, often dealing directly with messianic time.[33]

In his philosophical texts of this period, there is the sense that the idea of God connotes, perhaps most importantly, a situation in which it is impossible to *hide*.[34] Jeffrey Bloechl notes, "This was the period in which Levinas's philosophy of God emerged, both through an extension of already familiar ideas on infinity and exteriority, and through a renewal of his dialogue with Husserl, now specifically on time and sensibility."[35] For Levinas, God emblemizes exteriority, the beyond, the otherwise. There are some intriguing suggestions that Levinas is saying *more* than he realizes about God, even in *Totality and Infinity*,[36] but in his earlier works it is safe to say that references to God are generally intended to underscore *separation*.

It appears that Levinas's discovery of the Talmud also provided him with critical tools for reconsidering time outside of the confines of Western philosophy, particularly in the deformalized sense of the intersubjective relationship. Both Rosenzweig and the Talmud probably influenced Levinas's philosophy of time that reaches maturity in *Otherwise than Being*.

Double Fidelity: Jew and Greek

Catherine Chalier suggests that Levinas's Judaism functions critically in the development of his philosophy of time and that he explores the concept of time with a "double fidelity" to the Greek and Hebrew conceptions of time.[37] Levinas, following closely on the heels of Rosenzweig, senses that Western philosophy has failed to think about true separation. Transcendence has been a fascination for philosophy from the beginning, serving as an important concept for philosophers since before Plato and Aristotle. In pursuit of wisdom, Western philosophy has sought to bind together all forms of knowledge under common headings. In their own ways, both Plato and Aristotle, along with the tradition they continue to engender, seek to domesticate transcendence by way of demonstrating the reign of rationality at all levels.

The "first philosopher," said to be Thales of Miletos, inaugurated philosophy by demystifying a solar eclipse.[38] Transcendence, whether in the heavens or in the mental heights of the Platonic *eidos*, has been philosophy's fascination from the beginning. Philosophers across the ages have taken innumerable approaches to transcendence, but when Levinas invokes the concept, he makes a significant departure from this tradition. Levinas quickly finds that the tools of Western philosophy are insufficient for thinking through transcendence in a register that allows for a proper account to be taken of the encounter with the other person. Western philosophy takes up transcendence to defeat it, or at least to constellate the transcendent within the metaphysical horizon of epistemology. Levinas will steadily increase his use of the concept of God to designate alterity and transcendence that permanently defy domestication.

For this venture, and to think of God otherwise than by way of Greek versions of transcendence, Levinas moves comfortably throughout his career into crossover language that functions in both philosophical and religious dialogue. For example, Levinas has previously described dialogue as akin to prayer and the experience of the other as an "epiphany," among other very obvious religious terms (*OS* 149). Religious themes become increasingly important as Levinas begins to shift his attention toward *time,* which soon becomes central for Levinas's mature philosophical enterprise. The problems with philosophical transcendence are recalcitrant; no matter how distant or high one configures alterity, the idea of transcendence seems to stubbornly situate the thinker on a neutral ground of observation. The high and the low exist at the same time, in a common present. He will find in Judaism and the messianic another way to conceive transcendence that is less vulnerable to this reduction.

Levinas has perhaps discovered, by trial and error if nothing else, that the distance implied by the spatial language of transcendence inevitably invokes an enclosure in which the poles—no matter how far they are separated—are held by a common vista. Space is inevitably visual; geometry, and the analogies that rest on geometric vocabulary, relies on light. The distance of the other, however extreme or hyperbolic, remains an invitation for Hegelian synthesis, a chasm to be eventually crossed. However lofty the transcendence of the other, or of God, the spatiality of Greek transcendence presumes temporal *presence.* That which transcends me is subtly but firmly tethered to me by temporal synchrony. Hidden within this latent reliance on Aristotelian time and *presence* is a dependence on neutral perspective, on impersonal principles.

When Levinas writes *Totality and Infinity,* he is partly unaware of this recalcitrance of immanence in the images of separation and height, by his own admission. He claims, in his 1963 biographical reflection "Signature" that in *Totality and Infinity* he retained "ontological language" for the purpose of excluding psychological interpretations of his work (*DF* 295). But Levinas determines that the language of "ontology," with its inevitable reversion to immanence and presence, is to be "henceforth avoided" (295). The transitional period between *Totality and Infinity* and *Otherwise than Being* is marked by pointed reflection on this problem. The resolution to this problem moves through time and language: "Time, language, and subjectivity delineate a pluralism and consequently, in the strongest sense of this term, an experience: one being's reception of an absolutely other being. In the place of ontology—of the Heideggerian comprehension of the Being of being—is substituted as primordial the relation of a being to a being, which is none the less not equivalent to a rapport between subject and object, but rather to a proximity, to a relation with the other" (293). For Levinas, the alterity of the other presents a unique phenomenon, something unavailable through Husserlian phenomenology. Western philosophy advocates a kind of "*Logos*" whereby the self can pursue knowledge of external objects (295). But Levinas suggests that the problem with the experience of the other person is that it irrupts from *before* or *beyond* the logic of knowing. The other affects me *anarchically*, from *before* I catch my balance in logic and reason (295).

The tools of Western philosophy lack a method for demonstrating this kind of alterity. It is this mystery of alterity beyond ontology (otherwise than being) that captivates Levinas's interest and drives his intense investigation of diachrony in his later works. His talmudic readings and philosophical works share a common, pervasive antipathy to the immanence of history. History, intoned by Levinas with obviously Hegelian inflection, is unable to judge itself. History and being can do no better than self-definition, allowing for none of the distance or difference required for prophetic judgment. In both his philosophical and confessional writings, Levinas allows religious imagery, and chiefly the idea of God, to establish the distance necessary for moral judgment.

The need for alterity that defies evasion is evident in his repeated invocations of Hebrew texts and ideas, even in the midst of philosophical expositions that he will resolutely call nontheological. Levinas demonstrates that the hopes, the struggles, the suffering, and especially the ethics of Judaism have legitimate application in the field dominated by the language of Greece.

Messianic Time

Alfred Tauber presses for an interpretation of Levinas's understanding of time that is first and foremost rooted in Levinas's Judaism. He claims that the "entire foundation of Levinas's ethics is built from the Judaic understanding of Time as defined by our relation to God."[39] Tauber's claims are exclusionary; he denies that Levinas's philosophical project can be sustained without the Jewish understanding of time that is its main structure.[40] Tauber moves close to Derrida's critique in "Violence and Metaphysics," to which we will turn shortly, by claiming that Levinas essentially compromises his thought with the language of the Greeks.

Tauber's claim fails to account for several of the philosophical moves that Levinas makes in his development of the concept of time, including his deep appreciation for Bergson, Husserl, and even Heidegger's deformalization of time. However, a compelling case can be made for the evidence of a deep dependence on Judaism in Levinas's writing as he begins to establish a temporal way to think about alterity.

The very earliest of Levinas's talmudic reflections concerned the concept of time and questions relating to messianism and the Messiah. Levinas allows us, at times, to see the philosophical imagination that saturates his reading of the Talmud. Though confessional in nature, Levinas outlines the philosophical intent of his readings: "In no way do we wish to exclude from the reading of our texts the religious meaning that guides the reading of the mystic or naive believer, nor the meaning that a theologian would extract. But we none the less begin with the idea that this meaning is not only transposable into a philosophical language, but refers to philosophical problems" (*DF* 68). Levinas then turns to the topic of history. Judaism, he claims, has been aware of "the end of history" since long before Hegel's *Phenomenology of Spirit* (69). For Jews, the question of the end of history ties in closely with the hope for a Messiah, or a messianic future. What are the conditions for the coming of the Messiah? Is the arrival of the Messiah a product of political history, the result of the

Hegelian movement toward a concrete and universal future? Levinas deems the future that is messianic to be particularly absurd; a messianic sensibility includes the rejection of history and political machinations (96). Indeed, Levinas wonders if perhaps the "Emancipation" of the Jews and the subsequent rights to a political share in history, did not indicate an inevitable forfeiture of the messianic sensibility: "Thus messianism in the strong sense of the term has been compromised in the Jewish consciousness since Emancipation, ever since Jews participated in world history" (96). The interplay of forces within being can portend history's end, but they cannot give way to the messianic age. Messianism, in fact, defies assimilation, undermining and unsettling the way that economic success seems to pacify the need for a radically different future.[41] In *Otherwise than Being,* Levinas will come to differentiate between "time" and "temporality" (*OB* 85, 88). In his later texts, the field of temporality will be *history,* the economic time of investments and return. Time, on the other hand, will become infused with some of these messianic overtones.

Levinas draws another way of thinking about time from the Jewish hope in a messianic future. Universal history is "sought by political life and formulated by Aristotle" (*DF* 94). To have a stake in this history may be inevitable, but it is not the way to the Messiah. For Levinas, "Salvation does not stand as an end to History, or act as its conclusion. It remains *at every moment* possible" (84). Even more importantly, messianic time has the alterity necessary to stand in judgment of the temporal flux, of history. The messianic future does not draw near because of history's successes and is not flung any further into the far future by the advance of evil. Messianic time persists on another plane from history; it stands in judgment of the way of violence and the machinations of economics, even if these forces produce an uneasy peace.[42]

These messianic themes arise in Levinas's confessional work about the time he finishes *Totality and Infinity* and sends it off to the publisher. It can hardly be coincidental that Levinas says very little about eschatology or the messianic future in *Totality and Infinity* except in the portions that he wrote last. Levinas is already concerned that he has not distanced himself sufficiently from ontology, and these concerns are rather transparent in his confessional writings, which are remarkably bereft of references to Jewish ritual and worship and surprisingly rife with philosophical references. What Judaism offers is a way of talking about an *absolute* that is the product of neither reason nor history. As Edith Wyschogrod summarizes, "For Levinas, the Hegelian view of cultures is the most formidable threat to the absolute values presented by the ethics of Judaism."[43] By totalizing all of history, Hegel moves "outside of history," and from this lofty perspective, everything can be evaluated, including the very notion of value itself. Judaism, as Levinas takes it up, reverses this gesture toward totality and reinstates "the person as the final source of all values."[44] Messianic time is fundamentally *other* than history, and Levinas turns the alterity of time into his most useful image for describing an alterity that points otherwise than being.

Jewish time is anachronistic, unpredictably and decidedly alterior to the form of time that is announced by politics, cultures, and civilizations. Aristotelian perspectives on time, which reach vivid expression in Hegel, presume all past instances and future events to exist on a continuum of recollection and prediction. The question is, in part, one of hubris and humility. In his 1969 reflection "Judaism and the Present," Levinas offers a sustained reflection on the function of time in the ways that Jews think of their present situation. The great temptation remains, as he expressed it in 1960, to be fooled into thinking that the future, the messianic future, blossoms out of the present trajectory of history. Levinas worries that Jews may fall for the lure of premature messianic claims, which tether the coming of the Messiah to the foreseeable developments of culture and history. He worries that in so doing, "they are thinking that the messianic age is heralded by the events of history as the fruit is by the seed, and that the blossoming of deliverance is as predictable as the harvest of ripe plumbs" (*DF* 214). Levinas sees Judaism as an "anachronism"; it refuses to coincide with history. The temptation to bend Jewish time to the "reasonable time" of history is the very temptation to "erode the rock of Israel" (212). The loss of the anachronism of Jewish history is the defeat of "the Absolute"; such a loss results in both an erosion and an evasion of responsibility and ethics.

In configuring time this way, Levinas is declaring Aristotelian time a "false eternity" and the infinitude of Greek philosophy a "false infinity" (*DF* 213). The time that is bound to the power of reason is also beholden to worldly determination. Aristotelian time makes all time *synchronous,* all events a part of the eternal now. But messianic history safeguards the "primordial anachronism of the human" (213). This second, deeper understanding of time and the human are known to us despite being eternally *anterior* to the mechanisms of reasonable history. The wisdom of the Talmud knows nothing "about antibiotics or nuclear energy," but these aspects of the "human adventure" are posterior to the wisdom that is beyond history (213). The time of the Messiah can stand in judgment on "science and history" because it is *before* these adventures. So even as every contemporary culture tries to conform and contort the eternal time to strangely match its own time, the messianic resists. The time of the Messiah is the time that cannot and will not be constrained or domesticated.

Rosenzweig Reconsidered

These religious reflections and messianic themes pull Levinas strikingly close to Franz Rosenzweig, who we cannot forget is only obliquely cited in *Totality and Infinity.* The cryptic citation of Rosenzweig here is ironic, given that Levinas accuses German philosophers of failing to give Rosenzweig credit for his influence on their thought—Levinas claims that "they never cite him" (*DF* 183)—but elsewhere Levinas does not leave attentive readers in the dark regarding his admiration for Rosenzweig or the way his work is influenced by Rosenzweig's understanding of time. Levinas's reflections on Rosenzweig are sketched in his 1959 paper (published in 1963) "Between Two Worlds: The Way of Franz Rosenzweig." The question of history plays a pivotal role in this essay, and Levinas demonstrates here the close relationship between his own rejection of Hegelianism and the work of Rosenzweig.

We can detect in "Between Two Worlds" a perspective on history that Levinas's finds in Rosenzweig that is both rigorously philosophical and abundantly Jewish. For Levinas, Rosenzweig provides an example and an inspiration of leveraging the Jewish understanding of *separation* against the merge into unity that characterizes both Platonic idealism and Hegelian reason. For Rosenzweig, as Levinas summarizes, the relation between "God, World, and Man" can be "constituted as time" (*DF* 190). But this understanding of time is inseparable from the concrete articulation of time in creation. Time is unhinged from the formal treatments it receives in Western philosophy; it becomes the concretized separation of God and humanity, the possibility of revelation, in fact, the very condition for revelation. And this means that the past, present, and future are not formal aspects of an independent and universal configuration of time and temporality. History in this register is *religious* history, and therefore, the more fundamental understanding of time is the "present" possibility of a revelation that is not bound by a condition to the present. "That is Rosenzweig's anti-Hegelian position" (192).

Though his essay on Rosenzweig is included in Levinas's confessional writings, and it clearly demonstrates the role of Judaism in Rosenzweig's work, Levinas meticulously and repeatedly returns to philosophical themes in the essay. On the question of time, Levinas points out twice that deformalizing time and unhinging it from history draws Rosenzweig remarkably close to Heidegger's understanding of time as ecstasis (*DF* 290, 192). In fact, Levinas seems to see multiple attempts in recent decades to wrestle philosophy free from the dominance of traditional, Aristotelian understandings of time. He sees this happening to a degree in Husserl, certainly in Heidegger, but most definitively in the anti-Hegelian work of Rosenzweig. Levinas remains concerned that Heidegger lapses back into the Greek tradition that privileges the same over the other, which means that Heidegger's time is the time of the self. Dasein's ecstasis is Dasein's facing of the future. The future of Dasein is death, which is Dasein's own-most property. Heidegger's "ecstasies of time" reject Aristotle's eternal temporality but fall into an even more recalcitrant philosophical problem: the dominance of the same. In this sense, Heidegger finds himself again in lockstep with the Platonic tradition. Not so with Rosenzweig. For him, the liberation of time from the formalizations of philosophy is aligned with the release of the Jewish people from the need to win at Hegel's game. Jews are *independent* from history, unhinged from time's flow. As an "eternal people," Jews are captive to a truth that is "anterior" to the universal history of Hegelian history (200–01). When Levinas turns in *Otherwise than Being* to speak of time as diachrony and as the structure of separation that does not merge into unity, he is following closely themes he has known from Rosenzweig for many decades. The reflections on Rosenzweig in the 1960s are noteworthy, but they do not represent a particularly *new* influence on Levinas's thought. There is, however, one aspect of Levinas's Judaism that changes dramatically during the 1950s and 1960s: Levinas met "Mr. Chouchani."

Chouchani and Exegesis

Most of Levinas's confessional reflections take the structure of reflections on the Talmud. Levinas is a noteworthy talmudist; he wrote extensive, unique, and generally well-received interpretations of the Talmud. Despite being born in Kovno, Lithuania, a traditional center of talmudic studies in Europe, Levinas appears to have had little contact with talmudic studies in his youth and comes to know and appreciate the Talmud in the middle of his life.[45] As Levinas puts it, "although educated since early youth in the square letters, we have come late—and on the fringe of purely philosophical studies—to Talmudic texts, which cannot be practiced in amateur fashion with impunity" (*NT* 9). Mid-century, Levinas begins to engage in a serious study of the Talmud and to share some of his reflections, particularly with a group of French, Jewish intellectuals who met to discuss the Talmud (xi). When Levinas first participated in this colloquium, he did not give a presentation, and when he first presented in 1959, he gave a paper on Rosenzweig. Only in 1960 did Levinas begin to tentatively offer exegetical thoughts on the Talmud.[46] Enough of the images from *Otherwise than Being* cross the border from these writings to warrant some investigation about the talmudic resources Levinas brings to the concept of time. Indeed, the fact that Levinas began to publish on the Talmud in the 1960s coincides intriguingly with his increasing interest in time.

From the beginning of these readings, Levinas points out that his reading of the Talmud is deeply influenced by a mysterious twentieth-century Jewish teacher known as Mr. Chouchani. His real name unknown, Chouchani was an eccentric, itinerant genius who also had a profound impact on Elie Wiesel. Chouchani's life is the stuff of legend and mystery; there are amazing stories about his legendary wisdom and peculiarities. Among Chouchani's remarkable expertise in a wide range of disciplines, he was a talmudic exegete with a photographic memory of the entire Babylonian Talmud (*NT* xiii). In his talmudic reflections, Levinas calls him a "master" and refers to him as a prestigious and merciless teacher and exegete (*DF* 291). In fact, Levinas refers to himself as a mere shadow of the "incomparable master" (*NT* 72).

Levinas seems to invoke Chouchani when he is moving into precarious territory in his reading of the Talmud and particularly when he wishes to demonstrate the dangers of a one-for-one correlation between text and explanation. It takes an expert exegete to detect the way an interpretation of the text can slip away from the literal reading and remain a faithful interpretation. A text must be treated gingerly; it is easy to move irresponsibly from a text to a careless, tangential application. Chouchani's skill was to read the Talmud with the whole Babylonian Talmud fixed in his mind, an impossible task for anyone with normal mental capacities. The breadth of his understanding of the Talmud allowed Chouchani to see multiple accurate interpretations of a text, even where a straightforward meaning appeared to be confined and unambiguous.

In one of the only places where Levinas directly conveys an interpretation from Chouchani to his audience, he speaks of a phrase that appears "constantly" in the Pentateuch: "And the Lord said to Moses: 'Say to the people of Israel *lemor*' " (*BV* 80). The Hebrew word *lemor* appears to translate fairly clearly to the phrase "in these terms." Levinas points out that the meaning of this phrase is plain and "devoid of mystery" (80). The phrase seems to be rendered straightforwardly as "Speak to the people of Israel in these terms." But Chouchani, Levinas reports, claimed "to be able to give one hundred and twenty different interpretations of this phrase" (80). Of his vast, memorized commentary on *lemor*, Chouchani only reveals a single one, translating the phrase to mean "so as not to say." This would render the common phrase "Say to the people of Israel so as not to say." Levinas points out that Chouchani's reading intertwines listening and speaking, teaching and learning, and undermines any sense that language and teaching are about activity only. To teach is to be receptive and passive; it is to learn: "True learning consists in receiving the lesson so deeply that it becomes a necessity to give oneself to the other" (80). Levinas suggests one of his own interpretations of *lemor*[47] and then reports that Chouchani left the other 118 significations of the verse to be discovered, carrying their secrets to his tomb.[48]

Through Chouchani Levinas learns that a passage in the Talmud has a complicated relationship with its interpretation. Not *every* interpretation is valid; the interpretation must be held in a tense relationship with the rest of the Talmud. Furthermore, the Talmud almost always presents multiple opinions on every topic and passage it takes up. Expertise on the Talmud is required to offer interpretations, which explains the pervasive humility with which Levinas addresses this book. The seemingly endless, nonarbitrary interpretations of the Talmud present an intriguing parallel to the intersubjective realm, particularly given that Levinas is moving steadily toward a close correlation between exegesis and ethics.

Levinas treats the text as a collection of traces whose implications are as exigent as they are evasive. Hence, Levinas cannot remain silent in the face of the Talmud, which beckons a response even if that response arrives before one can be braced to receive it. Levinas demonstrates his humility even in offering his own translation of the text: "The passage to be commented on has been distributed to you. Perhaps you should not take it with you. The texts of the Oral Law that have been set into writing should never be separated from their living commentary. When the voice of the exegetist no longer sounds—and who would dare believe it reverberates long in the ears of its listeners—the texts return to their immobility, becoming once again enigmatic, strange, sometimes even ridiculously archaic" (*NT* 13–14).[49] His own readings, Levinas fears, ossify quickly into a deformed shadow of their intent. Divorced from the community of face-to-face relations in which they are offered, even the translation given by Levinas becomes dangerous and contorted.

Time in the Talmud

For Levinas, an encounter with the Talmud breaks through the plastic image it presents in a straightforward reading. The in-breaking of meaning and significance in the Talmud is carefully bound up in a *time* that the reader can never presume to share. The text becomes from a time-before-time, and for Levinas, the anarchic nature of the talmudic text is not merely due to the centuries that separate the Babylonian Talmud from the twentieth century.[50] The text is *older* because it positions the reader as responsible, as already summoned, as indebted. Translator Annette Aronowicz aligns the importance of time with Levinas's approach to the Talmud: "For Levinas insists that it is this very willingness to be judged by these sources that have maintained the Jews as an eternal people. It is eternal in that it has not allowed the judgment of history, the judgment of the powers-that-be, to determine the truth or reality of a situation.... Yet what singles out the Jews, what constitutes their particular service to mankind, is that they remind others of a source of truth outside history, of a dimension outside time" (*NT* xxvi).

By leaving one hundred and eighteen meanings unrevealed, Chouchani demonstrates a heavy emphasis on methodology. He does not attempt to leave behind an exhaustive interpretation of the Talmud but to engender a way of reading the Talmud that is fecund. Levinas does not learn all the secrets of the Talmud from Chouchani; he learns *how* to read this sacred text with reverence. Interpretation as taught by Chouchani is not a matter of finding the correct, fixed meaning of text. To interpret the Talmud is to submit to the never-ending process of stumbling into its deeper and deeper meaning. Talmudic study, in this vein, leads not to a definitive or final interpretation, no matter how straightforward a text may appear. To read the Talmud is to be attuned to a time-before-time; interpretations will forever languish in the time of the said. But Chouchani must have taught Levinas to perceive the said as irruptive with meanings and glimpses and traces of that which lies beyond the verse. Cohen summarizes the way Levinas moves this sense of exegesis into ethics: "The time structure of exegesis is thus the very temporality of ethics, of the encounter of one with another, 'I,' and 'you.' It involves a notion Levinas early in his career called 'trace,' and later calls 'diachrony.' Just as the ethical imperative embedded in the disturbing alterity of the other opens up an unanticipated future more future than the projections of the self, so too it bores more deeply into the self than the self's synthesis, however passive; it fissures the self with responsibilities deeper than its recuperative powers of synthesis."[51] Cohen's claim is bold: Levinas's attitude toward the talmudic text is the very " 'theme' of Levinas's philosophy."[52] Exegesis is diachronic, rendering one anarchically passive. No amount of preparation can turn around the relationship to the text and put the reader out in front of the writing.

The text is never liberated from the original intentions of its authors, but neither does Levinas consider the text bound completely by modern understandings of authorial mindset. The text is never liberated from the authority of interpreters past, but neither is it bound by these interpreters. The text is slippery and evasive in a manner that is strikingly similar to Levinas's understanding of the trace of the other (*ITN* 51).[53] We can only guess from whence the text came, or to where it points, but we cannot deny that the text has entered our present and delivered an injunction to justice in the process. There is an intimacy in the text, even as it evades. The *prior*ity of the text means it has been here all along, a part of the Jewish readers who have respectfully taken it up over the centuries. The Talmud is attuned to divine time, to messianic time, which for Levinas is the time of the other.

In a 1969 reflection, "The Name of God according to a Few Talmudic Texts," Levinas invokes the term "diachrony" to refer to the anachronism that is invoked when one speaks of God as the Absolute. God is "a passing beyond all past that can be remembered, a total diachrony" (*BV* 127). To invoke the name of God is to abandon the search for origins in the field of history, to invoke a time that is before being's origin. History is "essentially archaeology," and the diachrony of the Absolute denies any of the thematization that history could attempt in its reductive narration (126–27).

Levinas moves liberally from his confessional reflections to philosophy, and this is particularly evident in his deliberations about time. In his philosophical writings, Levinas credits Bergson, Husserl, Heidegger, Merleau-Ponty, Rosenzweig, and many others with identifying a weakness in traditional philosophy on the concept of time. But none of these resources provides a convincing solution to the problem. Philosophy is attuned to Greek understandings of transcendence, which are strikingly atemporal. Even Rosenzweig's understanding of separation remains vulnerable to the return of an atemporal structure that supports spatial separation. But in the Bible and the Talmud, time is the very condition for the relation with God and with the neighbor. As Chalier summarizes, "Time is the relation with the infinite or this diachrony which, at the heart of every finite life, presents itself and is experienced as a relationship to the irreducible mystery of the otherness of the neighbor; a diachrony which keeps pace with what remains other and which, in the face-to-face with the person, calls me and asks for me; time as vigilance and patience, time as awakening and disturbance."[54]

The importance of time is exemplified in his discussion of Job. Levinas writes in a talmudic reflection in 1966, that it is simply not enough for Job to account for his own past and take responsibility for his own sins. His impeccable past is clean of any infraction to which he can be bound by fault or causation. He is presented to us as utterly righteous in the internality of his being. But to circumscribe responsibility requires a universal vision, a perspective older and higher than the world itself. So Levinas notes that God's response to Job's righteousness is to ask "Where were you when I created the world?" (*NT* 85; cf. Job 38:4). The question demonstrates that Job is responsible for more than his own past, faults, and suffering. Job is neither the author of his own beginning nor his own freedom. He has received freedom and is bound with the others who are with him free: "Your liberty is also fraternity" (85). Here in the Talmud, Levinas incubates an understanding of anarchic priority, messianic hope, and a way of thinking about alterity that is more radical than anything the Aristotelian sense of transcendence can engender.

The Trace and Derrida's "Violence and Metaphysics"

"Violence and Metaphysics" was written by Jacques Derrida and published in two manifestations, the first as a pair of articles in 1964, the second as a substantial revision in the 1968 volume *Writing and Difference*. In this essay, Derrida claims that *Totality and Infinity* is a "work of art, not a treatise" (VM 312n7). The essay has been credited with helping draw attention to *Totality and Infinity,* and it is generally believed to have had a significant impact on Levinas, though neither he nor Derrida ever divulge the nature or extent of this impact.[55] Levinas's biographer, Salomon Malka, summarizes the essay well:

> In well-developed strokes, Derrida went through all of the work's articulations of thought with a fine-tooth comb, full of praise for its originality, its audacity, its style. At the same time, dissecting the concept of alterity, he pointed out a basic weakness. For him, the idea of the irreducible, absolute alterity of the Other was problematic.... The critique was technical—Derrida used Husserl against his own disciple—but it touched the heart of the project, and moreover it led Levinas to modify certain aspects of his thinking on the nature of subjectivity. Was this critique the catalyst for the eventual evolution in Levinas's thought from *Totality and Infinity* to *Otherwise than Being*? There is no proof of this in written form, and the two men themselves never clearly indicated it, but one detects a possible influence by Derrida's essay when studying the two works.[56]

Did Derrida, who gives us the analogy of waves pressing higher and higher on the shore, enhance the intensity of Levinas's final expressions of time and obligation? He offered that metaphor in "Violence and Metaphysics," but he could not have realized how apt the metaphor would be for Levinas's final development.

It seems that Malka's suspicion about Derrida's influence on Levinas is correct, but to a much slighter degree than some scholars have supposed. The constructive foundation of *Otherwise than Being* has already been laid before Levinas even reads Derrida's essay—by the mid-1960s Levinas has renewed a dialogue with Husserl on the topic of time, introduced the linguistic concept of diachrony, and exerted considerable energy on understanding the function of time in

Judaism and the Talmud. He has also shown a steady interest in language and the way that language correlates to the intersubjective relation. Derrida may have had considerable influence on Levinas, but the constructive moves that define the core of *Otherwise than Being* are apparent before Levinas reads "Violence and Metaphysics."

Derrida's reading of Levinas is neither clumsy nor arbitrary. In fact, it may be a tribute to his careful reading of Levinas that the route taken by Levinas in the 1970s and 1980s looks so much like the path outlined by Derrida. Robert Bernasconi identifies several weaknesses of Derrida's essay, but he also notes an "affinity between the course Levinas actually follows and that which Derrida lays out for him."[57] This affinity, reasons Bernasconi, "serves both to establish the rigor of Derrida's reading and to contradict the accusation of arbitrariness which is commonly brought against him."[58] Perhaps it is similarly noteworthy that Derrida appears to reject Levinas's understanding of the trace in "Violence and Metaphysics," only to embrace his own version of this concept in the years to come. This is a complicated essay, neither easily dismissed nor embraced.

Contamination

We must certainly pay some attention to the overall posture taken by Derrida in writing "Violence and Metaphysics." This too has been the subject of no small amount of contention. Bernasconi explores the possibility that calling this essay "Derrida's critique of Levinas" already misunderstands it. Derrida, in fact, "is careful not to present his account of Levinas as a critique."[59] What Bernasconi sees at work in the essay is a classic, early example of Derridean deconstruction. Derrida does not mention the term *deconstruction* in this essay, but this absence is due to the fact that in 1964 he was just developing what Alan Bass calls his "system of deconstruction."[60] If as Bernasconi suggests, Derrida's appraisal of Levinas should be read in light of Derrida's signature deconstructive methodology, then "critique" is indeed the wrong tone in which to read "Violence and Metaphysics." The goal of Derrida's deconstructive method seeks at each turn to demonstrate the seams in arguments that fail to hold together under the pressure of his intense examination. Levinas tries to overcome the Western tradition by lodging himself within it, and Derrida wonders about the wisdom or success of this strategy. In part, Derrida comes to present Levinas's ambition as an impossible pursuit. Derrida accuses Levinas of trying to depart from the very philosophical tradition that has laid the groundwork for "possibility."

Derrida, in the 1960s, was particularly interested in the way philosophical ideas *contaminate* one another. This is the "classic early deconstructive mechanism" that Bernasconi sees at work in "Violence and Metaphysics."[61] *Totality and Infinity* is ripe for such analysis; it claims to move beyond ontology and traditional metaphysics, but at times it seems to exceed traditional philosophy purely by declaration. To effectively deconstruct Levinas's work, Derrida needs to show that Levinas's language betrays him. Derrida sees that even this argument will be difficult, for Levinas already knows that he is seeking to overcome philosophy by embedding himself in it. So Derrida sets out to divide Levinas's language from his thesis and then to read these two elements of *Totality and Infinity* against one another. Levinas's novel attempt to rise above philosophy is particularly vulnerable for such a reading, even if his awareness of these difficulties already softens the impact of this deconstruction.

Considering that "Violence and Metaphysics" is a massive, complex essay, it is important to focus the analysis on how Derrida's reading influences Levinas's escalating use of diachronic time. Bernasconi has pointed out that the question of influence has questionable philosophical value, a point reinforced by Derrida.[62] We are pressed through Derrida's essay not to claim primacy of influence, but to further plumb the relationship between Levinas's two major works and to discover whether the encounter with Derrida resulted in a change of course for Levinas in his thinking about time.

When Derrida and Levinas spoke about the essay, Derrida remembers Levinas saying of the two-part essay: "You anesthetized me in the first paper, then operated on me in the second."[63] Derrida sees that Levinas is attempting something brash and unusual, something that bears a distinct relationship to phenomenology but also dissembles and unravels the project of phenomenology as it has been traditionally understood. In so doing, Derrida deems Levinas to be challenging no less than the "powerful will to explication of the history of Greek speech" (VM 83). Levinas puts in question the very foundations of Greek philosophy, challenging the presupposition that only a Greek source can adjudicate philosophy's primary questions. Derrida realizes that in so doing Levinas is overturning and challenging "the two Greeks named Husserl and Heidegger" (83). By realizing and underscoring the scope of Levinas's daring challenge to philosophy, Derrida sets him on a lofty and vulnerable pedestal.

Derrida does not miss the significance of Levinas's attempt to think the primary questions of philosophy otherwise than through the lens of the Greeks. For while Husserl assumed that Plato was the founder of philosophical reason and Heidegger blamed the forgetting of being on Plato and Aristotle, both remained deeply entrenched in a Hellenic philosophical tradition (VM 83). Levinas has the audacity to address philosophy from otherwise than the language of philosophy and being. He suggests an ethics beneath any morality, without any recourse either before or after ethics to the grander scheme or systems of philosophy. Derrida recognizes that this is unique, daring, and perhaps haphazard. Can something this pure arise from the mind of a twentieth-century thinker, himself so steeped in the phenomenological tradition of Husserl and Heidegger?

Derrida thinks not, though he piles no small amount of praise on Levinas in the process of demonstrating that Levinas's own thought is contaminated by the very philosophy it attempts to usurp. Specifically, Derrida suspects that Levinas has attempted to leverage himself out of Plato's wake with the very tools of Platonic philosophy. This Greco Platonic tradition at which Levinas takes aim has long paid close attention to the puzzle of interiority and exteriority, the "inside-outside" (VM 88). Levinas's chief aim is to establish a kind of radical exteriority, to propose that the other is irreducibly other. But Derrida believes the notion of alterity, particularly when expressed through "the spatial pair inside-outside," uncritically utilizes the notion of Platonic (or as Levinas would name it, *eleatic*) being (88).[64] According to Derrida, Levinas's work remains contaminated by the very thought it opposes.

Light and Space

Derrida claims that Levinas's thesis "can make us tremble" (VM 82), but he also suspects that this trembling is premature. For all the bluster about philosophy beyond the dyad of the same and the other, Levinas retains the ontological language required to think about exteriority in the language of traditional philosophy. He is more Greek than he admits, Derrida contests. And Derrida hinges this claim both on Levinas's attack on light and on his use of the spatial metaphors. One can only think inside/outside in terms of the visible; interiority and exteriority are in fact geometrical images, and geometry itself presumes and relies on light. Levinas speaks of an exteriority that shames the medium of light for its preference for the same and the way light presumes a prior hermeneutic through which illumination and comprehension can be appropriated. So Derrida sees here that Levinas's thought appears to be sickened by the well it poisoned. If light and vision bend us back into being and into ontology, then the geometrical images of interiority and exteriority must guide us there as well.

To reinforce the vulnerability of this seam in Levinas's argument, Derrida points out that Levinas must proceed by way of establishing the "violence of light"; and Derrida credits Levinas for some success in this adventure (VM 84–92). But Levinas is vexed, thinks Derrida, by the medium of phenomenology and the work of Husserl, which depends on light more rigorously than "any other philosophy" (85). Later in the essay, Derrida makes much of this problem as it relates to Levinas's profound emphasis on the face of the other. Must one not presume light to imagine its irruption in the face of the other? For Derrida, this notion of the face is permanently inseparable from its reliance on vision, sight, and space: "If the face of the other was not *also, irreducibly,* spatial exteriority, we would still have to distinguish between soul and body, thought and speech; or better, between true, nonspatial face, and its mask or metaphor, its spatial figure. The entire Metaphysics of the Face would collapse" (115). What Derrida accomplishes in this reading appears both accurate and prescient. His reading of Levinas's *Totality and Infinity* is sufficiently acute to detect a steady reliance on the very ideas Levinas wishes to overcome. He moves, as we have noted before, between spatial and temporal imagery in *Totality and Infinity*. The language of distance, interiority, exteriority, and height gives way to an escalating turn to nonspatial images of alterity in the mode of time. This shift is clearly incomplete in the text. The seams abound, as Levinas sews together the ideas he wishes to *retain* from the ideas he wishes to supersede.

Derrida has detected the problem voiced previously by Eugene Minkowski in 1962 who questioned the spatial metaphors of *Totality and Infinity* (*BPW* 27). But Derrida does more than gesture at this problem; he uses this tension to question whether Levinas has achieved his aim of moving beyond the confines of traditional philosophy. The language of height and exteriority may be unshakeable once taken up, doubling Levinas back onto the ground from which he wished to escape. The alternative imagery, as Derrida seems to be keenly aware, is explicitly temporal.

Derrida is not shy in his admiration for Levinas's earlier writings on time, though he does not spare *Existence and Existence* or *Time and the Other* from his deconstructive reading. It is interesting, however, to note that Derrida credits Levinas with abandoning the notions of exteriority and interiority during the period when he wrote those works: "During the same period, Levinas had expelled the concept of *exteriority*. The latter referred to an enlightened unity of space which neutralized radical alterity: the relation to the other, the relation of Instants to each other, the relation to Death, etc.—all of which are not relations of an Inside to an Outside" (VM 112). Derrida criticizes Levinas for returning to the spatial language of exteriority, but he recognizes, correctly it seems, that Levinas was up to something quite different in the two 1947 publications. And this way of thinking otherwise than the geometrical dyad of interiority and exteriority was indeed abandoned, at least for a while, during *Totality and Infinity*. Derrida notices the subtitle, "An Essay on Exteriority," and points out that Levinas is not merely abundantly employing this spatial image, but that he is also attempting to reconstruct the notion of exteriority beyond its spatial connotations. Levinas, claims Derrida, intends to recycle the very terms he has promised to defeat; he is using the terms he has "used up" and still dwelling in the ruins in which he has just left philosophy (112).[65]

The ploy fails, at least to Derrida, because we cannot be weaned from this language by its use or its neglect. Interiority and exteriority are the very structures of language, "the very heart of conceptuality itself" (VM 113). For Derrida, Levinas has failed because he has not replaced the inside/outside with anything but a negative gesture. Levinas is "unable to designate it otherwise than negatively" (113), a failure that Derrida deems to be systematic. Because Levinas has riveted alterity to spatial terms, he has only confirmed that alterity is inevitably bound to spatiality and exteriority, even if only by its negation. Derrida summarizes: "Henceforth, if I cannot designate the (infinite) irreducible alterity of the Other except through the negation of (finite) spatial exteriority, perhaps the meaning of alterity is finite, and not positively infinite" (114). Derrida believes, then, that if we can only designate radical alterity by way of negating finite spatial exteriority, then alterity can never truly be articulated outside the scope of finitude. So while Levinas claims to introduce the "irreducible alterity of the Other," he in fact only does so by fiat without managing to successfully gesture beyond immanence and presence.

Levinas has certainly left the door open to this criticism, and Derrida finds several ways to investigate this weakness. Derrida does not neglect the promising moments in Levinas's work here; indeed, he *has* proposed another way to configure alterity than the interiority/exteriority dyad. Derrida himself makes use, even in this essay, of the very tools of Levinas's "otherwise": time. We can see, in particular, that Derrida was influenced by the early Levinasian reflections on the future. Derrida's idea of the future is every bit as open and undecided as Levinas's idea of the future, but the difference remains stark. For Levinas the future arises specifically and unequivocally from the other person. For Derrida this possibility, that the other is the future, is neither embraced nor excluded. This is nowhere more evident than in their differing discussions of the very temporal term *trace;* for Levinas the trace is distinctively of the other person, but for Derrida the paradigmatic trace is of the text.[66]

The notion of the trace has a progressive impact on Derrida. The Derrida of the 1960s remains deeply concerned with structuralism and the weakness of structural approaches to language, philosophy, and ethics. His first response to Levinas's trace appears to be mostly negative. Bernasconi claims, "Derrida seems to reject the notion of the trace."[67] Derrida claims that "Violence and Metaphysics" is already in proof form, near publication, when he encounters the two essays that introduce Levinas's concept of the trace, "The Trace of the Other" and "Meaning and Sense" (VM 311n11). Thus, his reactions to these essays is mostly placed in footnotes in the original publication and then migrated into the text for the 1967 reprint in *Writing and Difference.* What gets lost in the transposition of these footnotes to the main text is what Bernasconi calls the "subordinate typographical position of these footnotes."[68] When the 1967 edition of the paper elevates these notes alongside the rest of Derrida's reading, this subordination is lost and Derrida's apparently critical statements about the trace are elevated. Lost is the "more positive attitude toward the trace revealed in other contemporary essays by Derrida."[69]

This reading of Derrida's essay helps mitigate the appearance that he is duplicitous in his treatment of Levinas's trace. In one of these converted footnotes, Derrida writes, "As soon as one attempts to think Infinity as a positive plenitude (one pole of Levinas's non-negative transcendence), the other becomes unthinkable, impossible, unutterable. Perhaps Levinas calls us toward this unthinkable-impossible-unutterable beyond (tradition's) Being and Logos. But it must not be possible either to think or state this call (VM 114). For all appearances, Derrida seems to think in his 1964 footnotes that any positive access to the infinite will make no sense philosophically. The trace, for which Levinas has already used terms similar to Derrida's "unthinkable-impossible-unutterable," cannot be meaningfully expressed. But elsewhere Derrida appears more inclined to pursue Levinas's sense of the trace, albeit in his own fashion.[70] By 1972, in *Margins of Philosophy,* Derrida will write in much more sympathetic terms about Levinas's understanding of the trace, which becomes embedded in his definition of *différance:*

> The alterity of the "unconscious" makes us concerned not with horizons of modified—past or future—presents, but with a "past" that has never been present, and which never will be, whose future to come will never be a *production* or a reproduction in the form of presence.... One cannot think the trace—and therefore, difference—on the basis of the present, or of the present of the present. A past that has never been present: this formula is the one that Emmanuel Levinas uses, although certainly in a nonpsychoanalytic way, to qualify the trace and enigma of absolute alterity: the Other. Within these limits, and from this point of view at least, the thought of *différance* implies the entire critique of classical ontology undertaken by Levinas.[71]

The trace, which Derrida realizes points to a "past that has never been present," is in both Levinas and Derrida unequivocally attuned to a time-beyond-presence, a permanent denial of immanence and presence. Here, in the nature and notion of the trace, we find Levinas and Derrida moving relatively close to one another, and both very much in the wake of Heidegger's grand critique of the history of philosophy and its obsession with presence.

What Derrida identifies and exposes in Levinas's work is a kind of *drift* of Levinas's work away from his early attentiveness to Heidegger's critique of presence. Here, once again, we can see that Derrida's reading is insightful. In response to the declining role that time and temporality played in Levinas's work across the 1950s and into the first-written sections of *Totality and Infinity,* Derrida is concerned with pushing Levinas back toward the posture of the earlier works, whose resistance to spatial language Derrida cites approvingly.[72]

Alter Ego?

Among Derrida's many engagements with Levinas in this essay is Derrida's "alter ego" argument. Derrida contends that Levinas's claim to asymmetrical relations is undermined by the fact that "the other... would not be what he is (my fellow man as a foreigner) if he were not alter ego" (VM 127). This argument is also concerned with the necessity of originary violence. Levinas's contention of radical passivity is undermined by the fact that "I am also essentially the other's other, and that I know I am" (128). The sense that I am aware of being the other's other means I am positioned in a "strange symmetry whose trace appears nowhere in Levinas's descriptions" (128). Furthermore, this symmetry is the very condition of "dissymmetry" that is so important to Levinas's argument. Here again we find Derrida exposing the weakness of Levinas's spatial language. Is it possible to imagine asymmetry without first imagining symmetry? Does not the spatiality of this image of symmetry default to a metaphysics of presence, or as Levinas said in 1964, the "Euclidean space in which they could be exposed, each on its own, directly visible, and each signifying by itself" (*CPP* 37)?[73]

Before moving to some concluding remarks about the question of violence and passivity, it is worth underscoring the way the lens of time allows us to reexamine the relationship between Derrida and Levinas. The first half of "Violence and Metaphysics" points repeatedly to the inevitable fate of a nonspatial philosophy that is "rooted in space, which cannot conceive of separation and absolute exteriority" (VM 116). Derrida supposes that "history" supersedes Levinas's supposed "absolute" exteriority: "History is not the totality transcended by eschatology, metaphysics, or speech. It is transcendence itself. . . . Metaphysics is *economy:* violence against violence, light against light: philosophy (in general). . . . This becoming is war. The polemic is language itself. Its inscription" (117). Derrida's initial conclusion, that Levinas has fallen back into the economy of violence, is instructive for understanding the larger purpose of this essay, which Geoffrey Bennington deems to be deeply political. Derrida's interest in the second half of the essay is to establish the inescapability of violence and the "primordiality of an 'economy of violence.'"[74] So while Derrida is pressing Levinas to turn his attention more directly toward Heidegger's critique of presence, he is already also exploring the relationship between Levinas's claim to fundamental passivity and his own developing philosophy of violence. For Derrida, as he sharpens his differences with Levinas, light and its otherwise are both aggressions and forms of violence. Philosophy's role, Derrida suggests, is to do gentle violence, to "speak and write within this war of light, a war in which [the philosopher] always already knows himself to be engaged; a war which he knows is inescapable, except by denying discourse, that is, by risking the worst violence" (VM 117). Levinas's work is intensely interesting and perhaps somewhat threatening for Derrida; he proposes a philosophy whose origins are irenic, gentle, and passive.

The second half of "Violence and Metaphysics" capitalizes on the vulnerability of Levinas's language to undermine his suggestions about passivity. Levinas, in the mode of Kant or Rousseau, points to an original human condition that is passive and nonviolent. For Derrida, the Kantian gesture toward original passivity is already a "founding" violence.[75] The *establishment* of original, preoriginal, or teleological passivity is itself a move of violence. Derrida is consistently interested in demonstrating that in every establishment of peace, however ancient or futuric, peace is conditioned on violence. The very establishment of the "law," however cautious or democratic, is an act of force, and therefore conditioned on a preoriginal violence that no peace or passivity can precede. Derrida wants to press for a best possible violence, a way of fighting the war that avoids "bellicosity" (VM 117).

For Levinas this originary peace is not endemic to being, and we cannot expect it to be found anywhere within being or its violent ways. This makes Levinas's claim to original passivity a unique challenge to Derrida's philosophy of originary violence. Derrida has routinely dismantled claims to original or eventual peace as they appear in Kant, Rousseau, Hobbes, Hegel, Marx, and others.[76] His method has been to demonstrate the traces of violence retained even in these proposals to original or eventual peace. His ultimate fear is that Levinas's "avowal of the war within discourse" makes him the "best accomplice" with bellicosity (VM 117). Levinas locates philosophy outside of the war for peace, whose origins and condition is violence. And in doing so, Derrida fears, Levinas abandons the pursuit of peace and becomes an accomplice of war. Who, wonders Derrida, "has shown this better than Hegel?" (117).

Levinas elicits a particularly complicated and extended response from Derrida, precisely because Levinas has threatened the tools whereby Derrida typically identifies and exploits the contamination of violence. After all, Derrida notes in admiration, Levinas has staked his claim for ethics and metaphysics on "nothing other than themselves"

(VM 83). Derrida has made a steady practice of demonstrating the way ideas flow into one another, but Levinas declares a nonviolent relationship to the infinite as infinitely other "without making them flow into other streams at their source" (83). At first glance, this seems to position Levinas outside the game of deconstruction, or at least beyond the violent interplay of light-against-light. Derrida finds Levinas's opposition of totality and infinity to neglect the *history*, or the contortion of history's meaning and reality.[77] Derrida proposes that history *is* transcendence and violence. There is no "before" violence, no "after" violence, and one should not hope or philosophize as though violence can be overcome. Derrida deems violence to be the permanent condition for peace, and he believes that the sooner philosophy and politics come to grips with this, the more effectively they can harken to their calling to engage in nonbellicose violence. Philosophy must live in the presence of the present, which is "originally and forever violent" (133).

It is intriguing that Derrida makes much of the notion of eschatology in this essay, perhaps more than Levinas has used the term himself. Derrida does not, in fact, seem concerned to tie his reflections on "messianic eschatology" to Levinas's own use of this notion in the preface to *Totality and Infinity.* For Derrida, this theme designates "a space or a hollow within naked experience where this eschatology can be understood and where it must resonate" (VM 83). In the analysis of this conversation as it relates to time, the decidedly spatial rendering he gives to eschatology is surprising and puzzling: "This hollow space is not an opening among others. It is opening itself, the opening of openings, that which can be enclosed within no category or totality, that is, everything within experience which can no longer be described by traditional concepts, and which resists every philosopheme" (83). Derrida uses this spatial rendering of Levinas's appeal to "messianic eschatology" (83, 103, 144) to show that Levinas's attempt to break from history is itself only one more transition in Hegelian history: "Is not the beyond-history of eschatology the other name of the transition to a more profound history, to History itself?" (149). In other words, Derrida uses spatial imagery to reinscribe Levinasian alterity into the dynamics of Hegel's history.

After "Violence and Metaphysics"

In 1973 Levinas wrote "Jacques Derrida: Wholly Otherwise," in which he addresses portions of Derrida's critique, perhaps the portions that irked him the most on the eve of the publication of his second major work. Levinas points, unsurprisingly, to the concept of time. He also seems happy to credit Derrida with the insightful critique of time usually attributed to Heidegger: "Derrida's critique—which frees time from its subordination to the present, which no longer takes the past and the future as modes, modifications, or modulations of presence, which arrests a thinking that reasons upon signs as if upon signifieds—thinks through to the end of Bergson's critique of being and Kant's critique of metaphysics" (*PN* 60). Derrida has indeed pressed Levinas and philosophy itself in these directions, past a "gathering synchrony" that constantly reabsorbs the past and future into the present (56–57). Levinas sees in Derrida's work a valiant attempt to awaken philosophy from a long slumber, "a delirium in which, since Plato, the discourse of Western metaphysics is conducted." This seems like a succinct summary of Heidegger's accusation of Western philosophy's negligence on the concept of time, but Levinas here deftly credits the insight to Derrida.[78]

We can certainly detect the residue of Derrida's reading of Levinas's use of "exteriority" in the language of *Otherwise than Being*. Throughout *Otherwise than Being*, we find that Levinas uses the idea of exteriority with intense care. Levinas invokes the term several times in that work and frequently couches it within the language of time and anarchic passivity or within the provision that this exteriority is not "objective or spatial."[79]

Something shifts in Derrida's approach to Levinas across the decades between "Violence and Metaphysics" and Derrida's death in 2004. Even by the end of the 1960s, Derrida is grouping Levinas with the most important contributors to a philosophy that can think of *différance*.[80] Wyschogrod, for example, points to a kind of conversion to Levinasian understandings of alterity in the 1980s.[81] We can certainly see a progressively changing attitude toward Levinas in the decades after the 1960s and perhaps a declining interest in Levinas's work. Derrida credits Levinas with invoking the trace but develops his sense of time and alterity in another direction.

The ongoing developments in the philosophical relationship between Derrida and Levinas must be reserved for another study, but they never move far from the important dialogue over time as it must be reconsidered after Heidegger. In an intriguing exploration of the philosophical relationship between Levinas and Derrida, Paola Marrati points to the concept of time as what indicates Derrida's closest "rapprochement to the thought of Levinas."[82] Their thought is similar from the beginning, in that both embrace a "passivity of time" and find in writing the opening and judgment of our ordinary understandings of time. In his later texts, especially *The Gift of Death* and *Adieu to Emmanuel Levinas,* Derrida seems increasingly comfortable with what Levinas calls the diachrony of time. Marrati summarizes Derrida's "more positive account" as follows:

> The disjunction of the relation to the other is the very disjunction of time. Only a dimension of time irreducible to the synchrony of the present, to the horizon of anticipation and to that of remembrance, makes possible a thought of justice defined in these terms—and of the only justice Derrida considers to be indeconstructible. Disjunctive time, time as anachrony and diachrony, is the opening of what Derrida does not hesitate to call "infinite promise," a "messianic opening," or again, "an eschatological relation" to the coming of an event, of a singularity, an alterity that cannot be anticipated.[83]

At this point we can see with more clarity the trend of Levinas's developing understanding of time across his lengthy career. Levinas moves early and forcibly to explore the implications of Heidegger's critique of time in Western philosophy. He moves quickly against Heidegger, without ever questioning the basic critique of philosophy offered in *Being and Time.* Perhaps because of Heidegger's Nazism, Levinas takes a strong interest in the 1950s in differentiating his work from that of Heidegger. In those years, Levinas offers a sustained critique of the way that Heidegger's philosophy remains beholden to ontology and therefore never manages to escape from the philosophical matrix of Parmenidean and Platonic *unity.*

Totality and Infinity pronounces the differentiation from ontology and traditional philosophy complete, but Derrida thinks that Levinas has overlooked the need to express alterity in a register that does not fall back into the totality of being. Derrida painstakingly shows that spatial metaphors, the stuff of light and space and distance, cannot be disentangled from the metaphysics of presence that give them meaning. But Levinas has already sensed the insufficiency of spatial metaphors, and in the early 1960s, he begins to rethink alterity as time. This movement is demonstrated in his renewed conversations with Husserl's thought on passivity and in his intensifying use of language from the Bible and the Talmud. This is not to insinuate that Derrida does not play a role in Levinas's development; Bernasconi is surely right about the way Derrida refocuses Levinas on the basically Heideggerian quest for a philosophy of time that is not beholden to the presumption of presence in Greek philosophies of being.[84] Furthermore, Levinas's language after the mid-1960s reveals a new and distinct caution about using the spatial binaries of interior/exterior, inside/outside, symmetry/asymmetry, and so forth. Although the constructive directions in which Levinas develops his later philosophy seem to be well underway before Levinas's reading of Derrida's critique, the surgical work of Derrida sharpens and hastens this development.

Many of the essays Levinas writes in the late 1960s and early 1970s bear remarkable similarity to *Otherwise than Being*. In fact, key components of *Otherwise than Being* were published as early as 1968.[85] Between 1968 and 1974, Levinas begins to introduce terms like "hostage," "substitution," and "persecution," which had already played pivotal roles in his confessional writings,[86] terms that are laden with intonations of time and diachrony. It may be the case that harsh terms like "persecution" and "hostage" can only be properly read in the diachronic register that Levinas delivers them. At any rate, Bernasconi sees a transition underway in Levinas's use of these terms; he is now making a concerted effort to "avoid the traditional language of ontology, something which he conceded that he had failed to do in *Totality and Infinity*" (*BPW* 79).

By 1974, Levinas is prepared to deliver a book with as much force as *Totality and Infinity.* His first major work was criticized, and not just by Derrida, for its brazen attempt to think against the entirety of Western philosophy. Rather than retreat from his severe articulation of alterity in 1961, Levinas again escalates the alterity of the other person. And by 1974, alterity is established in Levinas's unique understanding of time as diachrony. It is this diachrony that allows Levinas to gesture toward an ethics that precedes all philosophy and ontology, a responsibility that arises otherwise than being. As Levinas turns his attention to the diachrony of the other person, he is carrying his readers into the depths of the difference and priority of the other person. Levinas is using tools from various corners of philosophy and religion; from Heidegger, from Husserl, from Judaism, and now finally with critical encouragement from Derrida, he finds that rethinking alterity means further reconsidering and deformalizing *time.* And so we turn to Levinas's final works and their implications for thinking about time *after* Levinas.

SEVEN

Diachrony and Narration

> This diachrony . . . is responsibility for others.
> —Emmanuel Levinas, *Otherwise than Being*

We have seen the next wave of Levinas's thought approach from afar, moving through the signs of its beginnings in the works that lead up to *Otherwise than Being*. Having devoted so much attention to the beginnings of this movement, one would think that Levinas's second major book could no longer catch us off guard. Yet even a book that has traced the pattern of Levinas's development cannot fully prepare us for the stunning language of *Otherwise than Being*. Levinas opens his second major work with an unrelenting emphasis on the priority of the other. And "priority" for Levinas does not mean a ranking or ordering of relative obligations. For Levinas, now, the priority of the other is a matter of "transcending diachrony" (*OB* 9).

This is a peculiar sort of transcendence; it is the time of the other. It is a transcendence that is accompanied by a proximity too close for spatial representation. And when we speak of the time of the other, we are not just speaking of someone who is asynchronous with the time of the subject. Nor are we speaking of the time of the other as merely disjunctive, which would leave open the question of whether my needs or the needs of the other take priority. The transcending diachrony of time indicates that the other's time is not just separate, but superior. The summons that addresses me from the face of the other precedes any moment of recollection, recovery, or synthesis. And because the other confronts me before I take up these tools, these defenses and evasions, I am responsible before I am conscious.

On what does Levinas base such radical claims? Is this still philosophy? Outlining just a few of the philosophical moves in *Otherwise than Being*, especially with the way Levinas resituates the self according to the diachrony of the other, will demonstrate their rich dependence on Levinas's unique (and late) way of thinking about time. This reflection will help prepare the way for further reflection on some other themes of *Otherwise than Being* in the final chapter, particularly the questions of God, holiness, and Levinas's late reflections on the feminine.

KEY ARGUMENTS IN *OTHERWISE THAN BEING*

The face, particularly the face of suffering, is not like a face I can observe in a picture or painting, whose colors and lines are already *re*-presented to me. The fleshed face of the other presses against me with an immediacy that precedes my very ability to assimilate these colors and lines into the memories and images of my past. The face is already, even in the instant of a smile or a tear, *before* me. The temporal lapse, the diachronic chasm that separates the words or expressions of the other from my ability to incorporate them, is permanent and beyond traversal. The very attempt to make it present, to *synchronize* the alterity of this experience with some known memory of mine, robs the face of what is fundamental to its presentation: its nonpresence. The nonpresent appears in the face as "invisible, separated (or sacred) and thus a non-origin, an-archical" (*OB* 11). Synchrony, for Levinas, is a denial of the preoriginality that is revealed in every face.[1] The face of another human being can summon me to the interhuman substitution of one person's suffering for another person's suffering, an "expiation" (15). The food on its way to my mouth can be turned over to another. The warmth of my place in the sun can be forfeited to warm the skin of another.[2] This is the extreme outpost of humanism, where the diachronous transcendence of the other pins me to myself and asks these things uniquely of me. The self to which I am pinned turns out to be already not my own. The other is prior to me, and therefore my priority. To think of the other person as transcendent in this register requires a yet more radical invocation of time.

Levinas proposes exactly this strategy in the opening section of *Otherwise than Being*. He knows that to think of the other as diachrony requires that we "unravel other intrigues of time than that of simple succession of presents" (*OB* 10). Our normal mode of thinking about the past relies on "a linear regressive movement, a retrospective back along the temporal series toward a very remote past" (10). This movement will never arrive at the other person, whose past is not subject to my recuperation. Any mode in which I might *represent* the suffering of the other, even to myself, is already a recuperation of a past that is not mine.[3] Levinas understands that in order for me to hand the other bread, or to stand aside for the other to have my place in the sun, I must listen to and appropriate and consider the needs of the other. What Levinas wishes to establish is that the summons to responsibility is *primary*, precisely because the call of the other bypasses my present (11). The attempt to feed or warm the other and allay the suffering in his or her face cannot be reduced to the conjunction of supply and demand. The other is incommensurable; his or her need evades the reduction to my present.

The *present* is the site of my purported freedoms and the moment where I recover my senses and make my decisions. In the present, I select one object from a collection of objects, one tool among many. This is the mode of thinking that Levinas considers the default understanding of the Western self; we are blind to the central egoism of our attachment to the freedom of every present. Levinas now presses to demonstrate that our present is already haunted, already overwhelmed by the other, who is too proximate for assimilation.[4] Such proximity is not a spatial closeness but an anarchic presence of the other in the interiority of the self. This makes freedom, the sense that I choose like a consumer among free options, a secondary sensation (*OB* 10).[5] More primary than any freedom is the past of the other and its exigency. If I could *know* the past of the other person, reconstruct it as a timeline without mystery or evasion, I could address the other in freedom. The compulsion to action, which is evident in the human face, combined with the irrevocable and diachronous priority of the other make freedom an abstraction and a fall from obligation. I am always the second one "on the scene," at best; the other was here first (87).

The situation with regard to the face is one in which I am initially and irrevocably bound. Still, this denial of initial freedom is not a concession to determinism. That I am free to act as I please constitutes a freedom of sorts, but the very condition of freedom is *responsibility.* The freedom to be for-oneself is already an egoism, a choice for the self, a reversal of the for-the-other in which the self has always already been established. To find oneself is to lose oneself in this past that precedes the self (*OB* 9). The presence of the other in the self is prior to my "consciousness"; it precedes the gathering together of the "conscious" self (16). But the content of this proximity is evasive, evidenced by traces that will not be assembled into a theme. To become conscious of the priority of the other is already a secondary movement. This is the movement *into* essence. And the participation in essence, or being, is not a "fall from a higher order or disorder" (16). This is a question not about whether being is evil or if it should be treated with "disdain." Levinas wants to establish the manner in which one is attuned to being. He claims that the logic internal to being is self-justifying and self-judging, and no matter how intensely we scrutinize being, it will never reveal morality or justice.

In my present, my consciousness, I see in the face of the other hunger, loneliness, joy, hope, sadness, and coldness. These needs are addressed in *being,* in essence, in history. The food and sunlight that I bequeath to the other are components of being. Levinas contests that these acts of responsibility attune themselves to the Good, *à-Dieu,* to God. He explains, "Being must be understood on the basis of being's other" (*OB* 16). But they are offered to the other not as an economy of hunger-meets-food; they are offered in the mode of a prayer, as gifts before a transcendent need that does not submit to a reduction to my understanding of the present. Like prayer and other acts of liturgy, the presumption is that these words and activities rest on the very one to whom they are offered; I give to the other, from the other. Levinas can call this attunement to responsibility a religion precisely because it attunes the activities in being to that which transcends being. But this is no theology; Levinas is not offering a doctrine of God to support certain activities within the world of being. Prayer demonstrates the mode in which one speaks to the other who transcends; therefore, acts of service to one's neighbor are *liturgy.* The biblical reference to a God "whose origins are from of old, from ancient times" (Mic. 5:2 NIV), establishes the antecedence not just of God but of the neighbor in whose face "God comes to mind."[6]

Levinas therefore offers what Jeffrey Bloechl calls "the religion of responsibility."[7] Just as prayer is the proper mode of discourse, acts of responsibility are like liturgy, which resist the economics of investment and return. Levinas dreams about, but spends relatively little time discussing, what this might look like in relationship to the third party, to politics. In a world that is lasting peace, and not a hiatus between wars, peace results from a form or reason that is issued from this "order of peace," from the religion that is responsibility (*OB* 16). Such a future is what Levinas sometimes calls "the kingdom of God." This is not an optimism for history, but a logic that depends on a "reason" that is oriented by being's other, forever embracing the order of passivity and patience.[8] If such a politic seems unlikely or impossible this is hardly a deterrent for Levinas, who wishes to place this kingdom in an eon that is beyond history, beyond essence. Levinas attempts, especially in the final pages of *Otherwise than Being*, to further discuss the way the third person disturbs and interrupts the relationship with the other. This interruption is the call for justice, the summons to a responsibility that breaks open the closure of any face-to-face relation and that hopes for justice for all humanity (see *OB* 131–71, esp. 153–71).

Levinas is less concerned about the political problem, in *Otherwise than Being*, than he is about the *otherwise* toward which both ethics and justice should be oriented. The attunement to this otherwise is no easy philosophical task, and it presses philosophy beyond its typical limits. Levinas attempts to reorient philosophy yet again, and this time he radicalizes his earlier suggestions about alterity and transcendence. *Otherwise than Being* scandalizes even the other-centered ego proposed in *Totality and Infinity*. In this wave of Levinas's thought, the for-the-other is already underway before the ego awakens to itself; it is already older than old. The primary question in *Otherwise than Being* is not whether I participate in acts of hospitality, but the way the other already makes this act possible by constituting me in a world in which I can be hospitable. The difference, Levinas proposes, is between thanking God for something, and thanking God for the capacity to thank God. The second act presumes that which it is grateful for in the very act of thanking (*OB* 10). Now ethics rests not on some dynamic that arises from the face-to-face relation but on something that in the face of the other one discovers to be already anarchically prior to this relation.

Otherwise than Being proceeds by way of a "triple signification," as Richard Cohen has indicated.[9] Levinas proceeds with a series of tightly interwoven investigations of *time*, *language*, and *ethics*. These are the chief philosophical tools that Levinas uses to build a case for unprecedented responsibility. His strategy is direct, but it does not follow a particular structure. As Diane Perpich notices, Levinas delivers the principle content of his book in the first section and spends the rest of the book working out the implications of his claims.[10] Thus, with regard to the Levinas's considerations of time, which is invoked on nearly every page of *Otherwise than Being*, it will be helpful to investigate a few key components of his arguments as they are considered across the pages of this volume.

Plot Revoked

Totality and Infinity is still haunted with ghosts of a self that seizes and embodies a right to exist, an ego that claims a story uniquely as its own. In *Totality and Infinity* Levinas borrows the term *conatus essendi* from Baruch Spinoza to refer to this "natural right to self-survival" (DEL 24). And though *Totality and Infinity* purports to move past this *conatus essendi* by undermining the primacy of intentionality and consciousness, we can still detect the remnants of a plot whose main character remains central and stable despite the hyperbolic obligation implied in the face of the other.

In *Totality and Infinity* Levinas first positions the ego in original enjoyment and insularity, and then he narrates the opening of the interior subject to the infinite, which happens in the encounter with the face. The analogy of the dwelling reinforces this narrative structure; I am born and nourished in the feminine interiority of the dwelling, the precondition of my enjoyment, and make forages into the world of experience. The real event, the ethical event in the life of the ego, is the encounter with the face. This encounter is the apex of the narrative of the ego. However, Perpich sees in the transition between 1961 and 1974 an abandonment of this narrative structure. Her thesis offers an assessment of this transition, and it appears to support the suspicion that Levinas's final moves complete the utter abandonment of the time of the self. She summarizes,

> *Totality and Infinity* engaged in an extended narrative that purported to show how a separated and atheist ego could nonetheless come to be commanded by and responsible for the other. If the ego had not been separate, if it were but a dependent moment of the ethical relation, its becoming ethical would be an unremarkable achievement.... In *Totality and Infinity* the narrative form (in conflict, at times, with its content) leads one to expect an answer to the skeptic. You are responsible whether you know it or not, says this text; but the narrative form implies that one could in fact be *brought* to know, that a narrative could be produced that would show the ego to itself in the right light, despite its own attempt to position responsibility outside cognition and intentionality.[11]

In chapter four of *Otherwise than Being*, which Levinas calls the book's "centerpiece" (*OB* xlvii), Levinas rejects the narrative structure that Perpich identifies as critical to the structure of his own argument in *Totality and Infinity*. Here Levinas declares that responsibility is not some option for the ego to measure and consider alongside other options. Responsibility arises *before* the ego has its footing, on the "hither side" of the establishment of any self-identity.[12] The problem, it seems, with the narrative structure of *Totality and Infinity* is that it still locates the ego as a protagonist in the drama of coming-to-responsibility. There is an evangelistic urgency in *Totality and Infinity*, a concerted effort to convert the skeptic to radical responsibility. In this effort, Levinas may at times, perhaps in spite of himself, credit the ego with a "freedom of consciousness" (114). Whether this is the remnant of a stylistic effort to convince readers or a fundamental piece to the arguments of *Totality and Infinity* is debatable. Perpich appears to be on track, however, in her assessment of the text's latent "protagonist" and the (probably unintended) impression that radical responsibility is *optional*.

No such status, or story, is afforded to the ego in the heart of *Otherwise than Being*. Here the ego awakens to a responsibility that is already overwhelmingly ancient. In a sense, the narrative structure of the ego is completely dissipated as the ego takes its place in a time that is not its own. The awakening to responsibility is neither an event in my story nor a component of my existence; it is the very structure and substance of my existence. There are moments, of course, where I come to an awareness of a responsibility that is anarchically prior to my identity, but these are not "events that happen to an empirical ego" (*OB* 115). To think of awakening to responsibility in this sense is to resituate the ego at the center of the drama and, in so doing, to quiet the madness of responsibility by domesticating it to an event in *my* time. In *Otherwise than Being*, one awakens to find oneself already in a plot that one did not begin, but in which one is cast in the role of responsibility.

If responsibility is something I can measure and consider and not an anarchic condition for my identity, then I turn to the question of responsibility as "an ego already posited and fully identified," a Hegelian self whose trials lead me toward a more acute consciousness (*OB* 115–16). If these events are components of history, then Levinas wishes to point toward the "prehistory of the ego" (117). If these are events of consciousness, then Levinas wants to speak of that which is "prior to activity and passivity" (116). If these are debts of which we become aware, they are always anachronistic, like a "debt preceding the loan" (112). The list goes on; Levinas takes up a long series of images and examples of the events in the world of the ego that turn out not to be events in *my* story but summons that are infinitely older than I am. The revocation of the egoistic plot is the *loss of time*.[13]

This unseats the ego from the center of its own story. No longer the protagonist of my narrative, even the normally self-referential term "I" is transfigured: "The word *I* means *here I am*, answering for everything and for everyone" (*OB* 114). This is essentially what makes me "irreplaceable" in the summons to substitution. The summonses are *anarchy;* they precede any architecture of the self, any story, any assembly or configuration, however ancient. The events of discovery and the moments where I become aware of my debt and my configuration as for the other, are trivial. The most important component of my dawning awareness is the simple fact that I am too late; I am already cast in the drama of diachronic time of the other. My role is to be my brother's keeper; the question asked by Cain already misunderstands the situation. Levinas does not expound, but one might wonder if this is not why the biblical narrative leaves Cain's question dangling in the wind. To ask, "Am I my brother's keeper?" is already to "suppose that the ego is concerned only with itself" (117). As Robert Bernasconi concludes in his essay "To Which Question is 'Substitution' the Answer?," the conditions for substitution are not the point; "Substitution happens. In some sense it has already happened."[14] The question, Cain's question, is already too late.

Ethics is silent in the face of Cain's question; to answer Cain is to concede to immanence and to domesticate responsibility to faults, blames, and limitations. Reflecting on this passage, Steven Smith points out that Cain plays the part of the "moral skeptic."[15] In Perpich's analysis, the skeptic has a key role in *Totality and Infinity*, as the reader who is to imagine his or her ego moving through the drama of awakening. But in *Otherwise than Being*, the right of the skeptic to protest is presumed, and the very words that are spoken in skepticism become a conviction, a revelation of my obligation. "Am I my brother's keeper?" is the skeptic's query, but in the utterance, the question is already answered. If Cain wants a reasonable delineation of his responsibility to Abel, he finds none. Likewise, *Otherwise than Being* makes no attempt to forge a reasonable responsibility. Levinas purports, instead, to abandon the construction of an ego that bears any independence from the other. Gone are the remnants of a lonely instant or a hypostasis in which one operates in the prison of internal freedom.[16] For Levinas, I only awaken to a story that is already not my own. And in the story in which I find myself, the fact that I am my brother's keeper is both the reason for and fact of my existence. All of this rests on *diachrony*, on the way the other evades the present and moves within the subject from an anarchic past.

This interpretation of the transition between *Totality and Infinity* and *Otherwise than Being* is supported by a prior reading by Fabio Ciaramelli.[17] In his analysis of the changes between the two books,

Ciaramelli suggests that *Totality and Infinity* portrays an understanding of the self that first emerges from the gentleness of elemental enjoyment of the habitation.[18] The emerging I then encounters the face or opens the door for the other person. Ciaramelli questions the *genesis* of this emerging self.[19] In *Totality and Infinity*, the I is first of all unified in its sensuous enjoyment, as master of the elemental world and the habitation (including, one would presume, the feminine "other"). Bettina Bergo claims that this subject is for Levinas essentially derived from Husserl: "Levinas's subject has always been fundamentally Husserlian: it comes *out* of itself, remains *in* itself, and gets shaken from its solidity, before experiencing itself as thrown into a world."[20] The Husserlian ego is, by definition, primitively objective, necessarily neutral, nonviolent, and premoral.

This gentle origin is then disturbed by the encounter with the face, an experience that introduces radical vulnerability to the I, which before the encounter existed as preethical. The problem, Ciaramelli suggests, is resolved when Levinas makes passivity utterly *initial*.[21] I do not come to be responsible or deliberate on whether I want to open the door to the other; rather, it is my responsibility to the other that individuates me, that supplies me with myself. The very source of the self becomes "ethicized" and the vulnerability becomes my origin. This enacts a destruction of the initial singularity that is required for any narrative to have cohesion, for any tale of the self to be told.

The theses of Ciaramelli and Perpich concerning Levinas's use and abandonment of a narrated core are compelling, particularly when aligned with the progressive development of Levinas's thinking about time. Levinas uses *Otherwise than Being* to unseat the time of the ego, even to obliterate it. Diachrony is the rupture of the time of the self, and its rupture arises from within and before the establishment of the ego. In Levinas's early discussions of time, the internal time of the self is carefully examined, particularly in regard to his treatment of the instant. This sense of internal time can easily be traced out of Levinas's readings of Bergson, Husserl, and Heidegger. Bergson pulls time free from mathematics and physics, but it remains a function of a single person's consciousness.[22] Husserl relocates the original sense of time in the internal consciousness of the ego.[23] And Heidegger (of the 1920s) appreciates this deformalization of time but presses Husserl's discovery out into the fabric of being itself, which he then takes to be irremissibly temporal.[24] In each case, time is definitively returned to the primacy of the self, as complicated and as ecstatic as Heidegger's temporality remains.

So there is reason to consider that Levinas's understanding of the ego in *Totality and Infinity* may have retained remnants of the narrative structure of the deformalization of time as appropriated in Husserl, Heidegger, and Bergson. These remnants, it should be added, persist in spite of several contrasting themes stated explicitly in *Totality and Infinity*. What Perpich and Ciaramelli have detected appears to support the theory that Levinas is thinking in progressively radical terms about the implications of rethinking alterity through an unsettling of the ego as the stable center for time's flow. *Otherwise than Being* appears to deny every construction of a narrative for the ego, as Perpich and Ciaramelli argue. This stripping of the narrative center for the ego allows Levinas to renarrate the self in the plot of the other person.

Perhaps this is as far away as one can move from Hegel and his concept of history. Levinas is here refusing to entertain any semblance of narrative, refusing to allow the "otherwise than being" to be any mere negation of that which is identified in being. Derrida wondered in "Violence and Metaphysics" if Levinas had succeeded in breaking away from Hegel or perhaps only *confirmed* him (VM 120). Levinas now resists Hegel by questioning the very freedom of self-consciousness. Hegel's temporality depends on the "return to self involved in consciousness and time" (*OB* 53). For Levinas this produces a tale, an emblem of the said, and the ego remains the central character in the story. Narration is invariably an act of synchrony.[25]

Stories synchronize, and not just in the sense that they are linguistic and are therefore subject to the betrayal of language. Stories often move sequentially and depend on time as it appears in the moving hands of clocks and calendars. Sometimes a narrator may jumble the time sequence of a story, but this is only a delay in synthesis, forcing the reader to labor to find the temporal synthesis beneath the disjunctive chronology of the story. Narratives, however they are told, are attuned to an eternal now, moving forward by way of connecting one instant to the next, aligning characters and events between pages of books, between titles and credits. Narration conquers diachrony by way of synthesizing and synchronizing plurality into the singularity of a page, a screen, a stage.[26]

This does not mean, for Levinas, an outright rejection of narratives and stories. It means that storytelling, like nourishments, are a component of being, but they do not "escape the same order" (*OB* 8). They participate in the said, which always retains traces of a saying that cannot be reassembled.[27] History "assembles events into an *epos* and synchronizes them, revealing their immanence and order" (8). Narration is plastic and forged; it is artificial. Narrative stories are synchrony, the domestication of diachrony. Narrative *is* the time of memory; morality summons from before any "once upon a time" takes its bearings.

Already in *Totality and Infinity,* again especially in the preface and in the conclusion, Levinas worries about inflecting the subject with a *drama* that speaks of a subject having "active interventions realizing projects" (*TI* 28n2). Such a configuration makes the encounter with the other about *disclosure,* about aims and intentions. Levinas's concern in *Totality and Infinity* involves the reduction of the encounter with the other to an event within being. When Levinas turns his attention to what he calls "beyond the face" at the end of *Totality and Infinity,* he is already concerned with breaking the I free of its central place in the philosophical order. But Levinas's chief example of this is fecundity, or specifically paternity: "The I breaks free from itself in paternity without thereby ceasing to be an I, for the I *is* its son" (278). The relation between father and son allows Levinas to consider a relation that is both "a rupture and a recourse at the same time" (278). In the son, the father encounters an alterity that is nonetheless unique and proximate. "*To be* one's son means to be I in one's son, to be substantially in him, yet without being maintained there in identity" (278–79). Despite these efforts to undermine the drama that positions the ego at the center of awakening and fecundity, *Totality and Infinity* relies on the metaphor of the family in which the father is the center of action and relationship.

Otherwise than Being proposes a hyperbolic sense of responsibility that outstrips the narrative bearings the self might have. *Totality and Infinity*'s narrative bears the seeds of its own deconstruction; Levinas struggles in the preface and in the concluding section to dislodge the ego as protagonist. Perpich knows this, pointing out that "this tale of 'I meets Other' was not meant by Levinas to be a tale at all, and its narrative structure is importantly at odds with the most original impulses of the work."[28] Embracing the basic thesis suggested by both Perpich and Ciaramelli about the philosophical transformation between these two texts, we may explore the recasting of the subject in *Otherwise than Being*. Here, by way of diachrony the ego finds itself without a story of its own and is therefore caught up in the time of the other. To be *in* the time of the other, however, is to abide in a time whose structure forever evades knowledge and consciousness. It is to live in a time with beginnings and ends that radically evade the present as it is known to me.

In the Time of the Other

Narration, as a said, falls prey to the problems of representation and consciousness. The story I tell is the story I control. But there are indications that the revocation of an ego-centered story does not mean a loss of narrative. *Otherwise than Being* does not dismiss history or the flesh and blood and food that are found within being. But no investigation into being will reveal the *Good;* this must be given to being by being's other. For Levinas, the primary problem with history, even with Hegel, is not the narration of being but the audacity that supposes that being's ways are meaningful, that they contain the truth, that they reveal the Good.

This seems like a helpful angle to use when considering Levinas's critical but sometimes confusing distinction between the saying and the said. To tell a story is to perform a classic example of representation. A storyteller crafts a tale with a beginning, an end, and a collection of moments in between. The details of the plot are ordered and assembled; all the unruliness and diachrony of these moments are hidden and suppressed in a said that masks the saying. The other who is a character in a narrative is enclosed and totalized by the plot. This is the nature of the said, to be thematized and ossified into a concretion that hides the sincerity and purity of the saying.[29]

A child, Levinas points out in an interview, has not yet learned "to dissemble, to deceive, to be insincere" (DEL 29). The words, or the cries, of a child arise as *almost* pure saying. In the realm of being, however, they are nonetheless exposed to the betrayal of language and its instabilities. Words that are given over to the other are already converted into a said, a component of a past that has been assimilated and sorted. To pursue the saying, still resonating as a trace in the said, is to find oneself already bound to the other, who is already proximate even in the evasiveness of my own saying. This is diachrony in language; the time of the other is already embedded in my every utterance. A child, therefore, responds without duplicity, without reserve, and without representation. For human beings, once adults, even cries of pain are communication, inflected, and offered as a language, encrusted with themes and synchrony. But behind these cries or words there is the intersubjective encounter. Stories are a paradigmatic said, but this does not make them evil, no more than it makes them good. Stories are dangerous, however, for they make us forget the diachrony that divides the saying from the said. Every story is a betrayal of a saying, a compromise that is necessary for communication, for community.

So the problem with narrative, and history, is our failure to perceive it as a said. The said, Levinas mentions in a late interview, "is an ontological closure to the other" (DEL 29). In narration, one presumes simultaneity, either as a moment in the story, or even in the moment when the story is being announced. It is this sense of story that is visibly absent in *Otherwise than Being*. What Perpich identifies is the complete absence of an independent narrative structure for the ego in *Otherwise than Being*. The ego has been renarrated; stripped of the heroism of its own epic adventure, the self now finds itself in a story not its own. The self is renarrated as the one who holds the bread for which the other hungers. There is a story, well under way, with beginnings that radically precede me and a future that escapes any anticipation. The loss of a narrative structure follows the complete obliteration of any isolated or independent subject. By demonstrating that my words already betray my primordial obligation, Levinas has unseated me from the center of my story.

This unsettling of the narrative of the ego has been underway since Levinas's early writings. Whatever we say of story or history after the invocation of radical diachrony will be awkward and disjointed. How might we think anew about narration and stories, which until now have been about assessments and representations? The renarration of the self will invoke a story whose "once upon a time" and "happily ever after" are components of the beyond essence. If there is a narrative to *Otherwise than Being*, it is the story of the endless tardiness of the ego, which finds itself situated in a story that is not its own. This is a puzzling, irruptive, awkward story, a story whose plot is always uncertain and whose moments are disconnected. This is the exasperating story of one who is always already late for a drama that has already unfolded.[30] The plot itself happens to me from a passivity that is not just affectivity;[31] I receive the plot as the very context of my identity, the mode in which I become.

This fundamental passivity, which is infinitely prior to the passivity of receptivity, gives Levinas the ability to move past some of the thorny biological issues raised by *Totality and Infinity*. In that work, Levinas entitles a chapter "The Ethical Relation and Time" and hinges his argument on the notions of fecundity and paternity. Paternity becomes the "primordial effectuation of time" that allows for what happens in the "biological life" of the ego to be "lived beyond that life" (*TI* 246–47). Thus, time rests on the infinitude normally reserved for God, who alone can judge history from beyond history.[32] The son emblemizes that the father's life is unhinged from its biological confinements. The son is invoked for his relationship to the father, the subject. The feminine and the maternal are invoked to sustain the necessary components of the story of the masculine self.[33] But Perpich's critique echoes here again; it is hard to see the invocation of the woman, or the son, apart from their apparent role in the accusative case. Their roles are affixed to what appears to be the "personal singularity of a unique exemplar" (*AE* 72; my translation; cf. *OB* 56). "Child" and "woman" are "tropes" which have their ultimate reference in the, apparently, masculine ego.[34] The story of the ego is therefore still a synchronization of these other characters, and not yet diachrony.

In the very term *diachrony,* delivered and explained dozens of times in *Otherwise than Being,* we cannot help but detect a retention of *chronos.* Levinas has not fled from the concept of time, but radically resituated its implications. Not just Cronos and Hegel, but all *history* that privileges recollections and anticipations has been stripped of the capacity to give way to truth and meaning. History can only refer to itself, can only settle for what has been said, blind to the primordial saying that vibrates with the otherwise than being. The ego of *Otherwise than Being* is not just robbed of its own free rights to a private history. Levinas goes an enormous step further and claims that the ego is returned to time without any say over its bearings. For Levinas, diachrony situates me as captive to time, to the time of the other, a time in which I find myself already caught up. Levinas borrows from theology the term "election" to connote this sense. I am elected and cast in a story that is not my story but *is* nevertheless a story.[35] The point is not my awakening to the diachronous time of this story, nor my ability to *re*-present this story in any coherent manner. Obligation *happens,* or rather *has already happened,* and all that is left to do is to scramble to *respond* to the call that is as indescribable as it is exigent.

The ego's role is cast in a plot that is both utterly exigent but maddeningly elusive. The *new* story that is being told is of a drama that escapes the analysis of being and essence. It is an anterior history, the primordial time of the other, the story whose beginnings are *ex nihilo* and whose ends are "an unimaginable future" (*OB* 89). The time of the ego is always already saturated by the time of the other, but in "immanence" it is too intense, too "proximate," to be reduced to what Levinas now calls "essence."[36] The abiding, temporal proximity of the other is taken up as the ego's participation in a story whose markers are illusive traces, the reverberations of the saying left behind in the said. Aware or unaware of its plight, the ego is already abundantly captive to this transcendent history, the history that never becomes reduced to *my* "once upon a time" (56).

To be taken up in the story of the other is to be burdened by an external history, and Levinas turns without hesitation to the harshest images of this burden: "The subject of responsibility, like the unity of transcendental apperception, is not the individual singularity of a unique exemplar, such as it manifests itself to be in the said, in tales; 'once upon a time...' Here uniqueness means the impossibility of slipping away and being replaced, in which the very recurrence of the I is effected. The uniqueness of the chosen or required one, who is not a chooser, is a passivity not being converted into spontaneity. This uniqueness not assumed, not subsumed, is traumatic; it is an election to persecution" (*OB* 56).[37] Levinas's "individual singularity of a unique exemplar" sounds very much like Perpich's protagonist and Ciaramelli's preemergent "self."[38]

The loss of a core narrative for the ego might seem to forsake the other by abandoning any position from which my story might intersect the story of the other. But for Levinas, the reverse is true. The lack of an independent narrative, crafted to glue together my identity in a synchronized plot, is the very reason for my unlimited responsibility. Because the other's story is forever before me, and because I have no story that is *mine* before this encounter, I have no way to escape the burden of the other.[39] This is why diachrony means trauma and persecution. Levinas is not interested in masochism; the point is not that the subject seeks constant pain and suffering. The story of the other is also, potentially, *jouissance:* "I can enjoy and suffer by the other only because I am for-the-other" (*OB* 90). The original sensibility of anarchic passivity has what Bettina Bergo calls "two aspects, that of pleasure and enjoyment and that of pain and ageing."[40] These aspects are unified for Levinas: "It is the passivity of being-for-another, which is possible only in the form of giving the very bread I eat. But for this one has to first enjoy one's bread, not in order to have the merit of giving it, but in order to give it with one's heart, to give oneself in giving it. Enjoyment is an ineluctable moment of sensibility" (72).[41]

The importance of *jouissance* is approached hesitantly by Levinas, because of concerns about the knot that typically binds enjoyment with concupiscence. Nevertheless, the enjoyment of the bread in my hands is in part the condition of my responsibility to give it away. To enjoy here is for Levinas primitive and pure, the converse of the trauma that strikes me in the pain of the other. Levinas refers to the enjoyment of one's bread right on the heels of calling passivity "a vulnerability and a paining exhausting themselves like a hemorrhage, denuding even the aspect that its nudity takes on" (*OB* 72). To expect only joy, or only suffering, in the diachronic time of the other is to already anticipate and bind the "encounter" with the other again into the time of the same. Both *jouissance* and suffering happen as trauma, for they embed the unsayable in my very speech. If one awakens to find oneself a part of an utterly alterior history, there can be no adjudication regarding the joys and pains in which one finds oneself cast. Of the notion of enjoyment in *Otherwise than Being*, Kenneth Reinhard writes, "For Levinas, enjoyment is not simply renounced by the subject of responsibility, but remains its intimate and ongoing condition: 'only a subject that eats can be for-the-other, or can signify' (OTB 74). Levinas articulates the responsibility of 'for-the-other' as a substitution that determines not one meaning among others, but rather opens the field of signification as such. Like Lacan's substitutive love, Levinasian responsibility institutes the process of metaphorization without abandoning *jouissance,* which indeed depends on the primal signification of substitution."[42] To find oneself enmeshed in a history that is anarchically prior to the self is to be positioned *within* the joys and pains of the other's diachronic time. I am not replaceable here because I do not choose the histories into which I find myself always already summoned, invoked, accused, and riveted. As Ciarimelli points out, the responsibility that individuates me and disturbs the "time of my existence for-myself" nevertheless does not "annihilate my singularity."[43] The opposition is true; I am not obliterated but riveted to a history, and thus there is a sense that my place in this evasive, anarchic history is irreplaceably singular.

What are we to do with the old forms of the story of the ego, of birth and death and the anxieties in between these poles? It is this problem that gives Levinas pause in *Otherwise than Being:* "But then we have this problem: is not diachrony characterizable only negatively? Is it pure loss? Has it no signification? For such a signification what is signified would not be a 'something' identified in the theme of the said, a 'this as that' illumination in the memorable time of essence" (*OB* 38). If we are no longer to understand life as a series of chronological moments between the poles of birth and death, what does one do with all the residue of the egocentric narrative whose drama was still narrated in *Totality and Infinity*? Is the time that was once configured as *mine* "pure loss"?

Levinas takes up these questions in his somewhat surprising reflections on "ageing" and "senescence" (*OB* 52–53). He understands that our lives are full of events and sequences; we are chronological creatures in the fields of essence and being. What his understanding of diachrony performs is a denuding of the meaning and significance of these chronologies. My "I am" has been converted to the "*here I am*" (114), my identity converted to my responsibility. Not just the pronoun "I," but the whole of my existence is converted to the accusative case (142). Even my remaining and surviving is converted to a patience and a waiting for the other. I am not absorbed with anticipation of my death, in fact, remaining alive is *obedience* to the existence in which I stand accused: "Thus the passage of the *Phaedo* which condemns suicide (*Phaedo* 61c–62c) is meaningful, and not only pious. Being for death is patience, non-anticipation, a duration despite oneself, a form of obedience. Here the temporality of a time is an obedience. The subject as a *one* discernible from the other, as an entity, is a pure abstraction if it is separated from this assignation" (52).

These discussions reveal a deep struggle in *Otherwise than Being* to think through the idea of an ego after the defeat of synchronous time. The ego still is born; it still synthesizes events and traces and impressions into itself. "Time passes (*se passe*). This synthesis which occurs *patiently*, called with profundity passive synthesis, is ageing" (*OB* 52). To age is therefore to subsist within the time of the other, and *aging* is emblematic of the irreplaceability of the ego that is riveted to the diachronic time in which it finds itself assigned. I age because I am assigned to age, in patience, in "the passive synthesis of life," because I am affixed and held to the diachronic time of the other. As in the theological doctrine of creation *ex nihilo*, to which

Levinas remained philosophically committed throughout his career, the subject is affixed to a story whose origins and future are *prior* to any "essence."[44]

In *Otherwise than Being*, Levinas overturns any remnant of the centrality of the ego as protagonist, but he also pauses to outline a new way to think about everyday life in light of the loss of the plot of the ego. To age is to age *for the other*, a senescence that is obedience. "All human experience does in fact take on a temporal form," Levinas would say fifteen years later (*EN* 232). The realm of the sensuous, of skin, of pleasure, of pain, of suffering, and of ageing is a function of a fundamental passivity rooted in the diachrony of the other. Asked about his major philosophical preoccupations in 1988, Levinas said: "The essential theme of my research is the deformalization of the notion of time" (232; cf. 175–77). Levinas was not done pressing time to *less* formal manifestations, *further* from Cronos and Hegel and even Bergson, Heidegger, and Husserl. To press forward with Levinas's final movements is to find out what it means to further deformalize time.

Ricoeur claims that *Otherwise than Being* performs a mere plot reversal, one that surreptitiously unravels the reversal performed in *Totality and Infinity*.[45] As Ricoeur reads them, the first book reversed the direction of concern from the "same" to the "other." This reversal removed what Ricoeur calls "attestation" from the original event in the intersubjective encounter. But suddenly, in *Otherwise than Being*, Levinas has reversed his reversal. Now it seems like the first word is testimony, the saying. However, Ricoeur misunderstands this movement, mistaking an escalation for a reversal. In the saying, the ego is stripped of itself; it is utterly given over to the other. The story that glues the moments of the self together has been given over as well. Levinas has now replaced all remnants of the establishment of the self with a complete for-the-other. Ricoeur's critique, in the tenth study of *Oneself as Another*, hinges on the alleged similarity between his understanding of "attestation" and Levinas's understanding of saying, "testimony."[46] So Ricoeur stands back from both the moves in *Totality and Infinity* and the more radical moves of *Otherwise than Being*, sticking with his original position, which begins in the attestation to the self. Levinas has not reversed himself; he has radically escalated the manner in which the self is cast in the world.[47]

Levinas attempts to question the primacy of the self in *Totality and Infinity*, succeeding in many regards and failing in others. *Otherwise than Being* exceeds the failings of its prequel by scandalizing all that might remain of an isolated ego. Here, the primacy of the other is not a matter of who spoke first or of whose speaking is most important. For the final wave of Levinas's thought, all speech is already saturated from before time with the other. And to support this claim, Levinas turns, perhaps surprisingly, once again to the model of the family and the woman, this time as *mother*. In search of a metaphor for a past that is prior to representation and consciousness, Levinas again turns to biology, and in maternity Levinas finds the emblem of an "irrecuperable, pre-ontological past" (*OB* 78).

Irreplaceable

For Levinas, the maternal origin of every human life demonstrates the diachrony of sensibility. The mother is the model of the passivity beyond passivity because she is the preoriginal hospitality that makes possible all human beginnings. Maternity is election, the election by the Good beyond being, in which the subject is elected to labor for another. Catherine Chalier writes,

> As the maternal body answers for the other and makes room for him or her inside itself, it is evicted from its harbor and disturbed so far as to be out of breath, and this is precisely the signification of subjectivity. It is the ethical signification of the maternal saying. The "pre-original not resting on oneself" of the maternal body entails anxiety and listening but lacks free choice. It is the time of an inalienable mercy for the other, an infinite patience when facing an election that gives birth to the self in the very moment that interrupts its essence. The maternal body is ruled by the Good beyond being; it has not chosen the Good but the Good has elected it.[48]

The maternal model offers much to Levinas's articulation of the diachrony of sensibility and speech. The mother is *hostage* to the other inside her; she is *persecuted* by the pains of pregnancy, and she gives herself over to pain for the sake of hospitality. In the next chapter I will revisit Levinas's use of the feminine in his later works, which remains somewhat problematic. But here Levinas absolutely revolutionizes his earlier reflections on the feminine. He introduces the maternal as the "complete 'being-for-the-other'" and the ultimate model of primordial responsibility (*OB* 108). Philosophy has reflected too little on the field of intersubjectivity that is opened by the diachrony between the mother and the fetus inside her. Maternal hospitality also recognizes the sheer irreplaceability of the self in the responsibility of the other. Labor pains cannot be shared; the labor that precedes birth is a labor that falls uniquely and irreplaceably on the mother.

This is the sense of irreplaceable responsibility that Levinas wants to suggest irrupts from before every encounter. In the face of the other, I am bound as irrevocably to this other as a mother to her nearly born child. This is most clear in the case of suffering. When the other suffers, I am bound by that suffering from before I gather any resource to unravel myself from the other. I am already bound and persecuted by the suffering that confronts me in the face, as one whose suffering is *before* my awareness, no matter how ready I am for it. The suffering of the other is *mine*, and the evidence of the mineness of the other's suffering is already a trace in the very word I speak to her. This is suffering for the other from a unique form of passivity, a passivity that is the very reception of my own identity from the other. I am not interchangeable: the face that suffers is the one who has given me my identity in his very suffering.

Among the problems with suffering as expressed in story, as a said, is that it delivers itself to the synchrony where pain is understandable. From the lofty perspective of history, my blood is as red as my neighbors and just as painfully spilt. If substitution occurs, it happens optionally, as a freedom to choose precedes the encounter with the suffering other. This is the choice that Shelly Kagan calls "agent-centered options."[49] By means of narration, the suffering occurs in context, in a common present. As such, the suffering other cries out to all humankind, *all* stand accused, which also may mean *none*.

Levinas takes up this theme of the meaninglessness of suffering and the irreplaceable election to suffer for the other in his 1982 essay "Useless Suffering." This essay reflects on how philosophy and Jews should think of God after Auschwitz, and it speaks powerfully against the reduction of suffering to meaning and use. Levinas ponders the way theodicy inevitably participates in the "outrage" of justifying "my neighbor's suffering" (*EN* 98).[50] In history, suffering is part of the plot and a component of the drama. Suffering is exchanged for comfort or exploited for gain. Suffering, in history, is justified and quantified and made meaningful by lofty goals and high aspirations. But Levinas resists this, perhaps in no small part due to a constant reluctance to speak directly about the murder of his family members by the Nazi regime.

Levinas's resistance of synchrony and insistence on diachrony undermines the ability of the ego to narrate away the suffering of the other, to evade the responsibility inherent in the face of suffering, and to shift responsibility to another. The other suffers in a time immemorial to me and strips me of the tools whereby I might excuse myself. In the summons, before I even hear them, I am already "responsible as one assigned or elected from the outside, assigned as irreplaceable" (*OB* 106). This suffering, this face *before* me, has already elected me for substitution. This is clearly burdensome beyond comprehension. Wherever there is suffering, it is *my* suffering, *my* responsibility. This is the product of abandoning synchrony and embracing the diachrony of the suffering of the other. I simply am too late to excuse myself from the responsibility for suffering, both suffering that has passed and even suffering that has not yet happened. This leaves me "answerable for everyone else and for everything," a line Levinas quotes repeatedly from an intriguing passage of Fyodor Dostoyevsky's *Brothers Karamazov*.[51]

Dostoyevsky's *The Brothers Karamazov* contains a short story about Markel, a young man of seventeen whose health is failing and whose death seems imminent.[52] Despite his impending death, Markel refuses the comforts of both church and family. He will not give confession, take communion, or fast during Lent. He even mocks the comforts that others received from such rituals: "All that is delirium . . . because there is no such thing as God."[53] However, as Markel's illness intensifies, he begins to soften to the comforts offered by church and family. His body is driven by suffering and sickness into a passivity in which he begins to reconsider his life and relationships. He is clear about his reasons for engaging in religious rituals: he wants to comfort his widowed mother, who will soon lose her son.

As time passes, and Easter arrives, Markel seems to undergo a radical transformation. Suddenly his disposition is not colored by rudeness or by resignation to the suffering and uncertainty of his life. To his nurse, who he once forbade to light a prayer candle, he says: "Go ahead, dear nanny, light it. I was a monster before not to let you light it. For that's your way of praying to God, and watching you makes me happy and in my happiness I pray for you too, which means that both of us are praying to the same God."[54] Markel does far more than just reverse a bitter disposition. He does not merely exchange his egoism for altruism; he begins to speak of himself as being responsible for the suffering in the world, and not just the suffering he caused his mother, his brother, and his nurse. He embraces responsibility for all sins. He longs not just to release the household slaves who attend him, but also to become their servant. Markel proclaims an infinite and unreasonable responsibility; he, more than others, is responsible for the tremendous suffering and pain of the world.

Markel declares, not long before his death, "Every one of us has sinned against all men, and I more than others."[55] His mother figures that this excessive confession must be a product of his illness. Had not this boy recently refused to appear at church and confess the tangible and obvious sins of his young life? Yet now he confesses these sins and much more. He expresses remorse for his prior wretchedness and now embraces responsibility for *all* suffering. His concerned mother can only smile and weep, simultaneously, asking him how he could consider himself more responsible than the murderers and thieves who have so obviously caused endless suffering in the world. But, Markel replies, "every one of us is answerable for everyone else and for everything." As Levinas writes in *Otherwise than Being*, "to support the universe is a crushing charge, but a divine discomfort" (*OB* 122).

It is not difficult to see why Levinas was attracted to this little narrative. Markel becomes aware of debt when it is far too late to repay even a fraction of the suffering he has directly created in the lives of his nurse and his family. He longs to climb out of bed and begin to reverse these injustices, but he cannot. Markel has quite enough debt of his own to embrace; he hardly needs to embrace guilt for suffering outside of his own wake. But he does. And his conversion, which Dostoevsky cleverly crafts as only quasi religious, positions Markel as utterly passive in the face of the guilt for everyone and everything. Markel's passivity is literal; he is stuck in bed and near death. But his waiting for death is an abiding patience. This patience marks his life, each breath, as a movement for-the-other. His body becomes his obedience, the site of the pain he suffers, smiling, for his loved ones and for the world. For all appearances, Markel has abandoned even the self-loathing that his past sins might induce; even these would-be remnants of the for-myself he has discarded.

Markel's senescence is no longer an attribute of his own existence, no longer a component of his story. This leaves him defenseless against the weight of the world's suffering, which he embraces so completely devoid of a narrative of his own that he does not even shudder under the "divine discomfort" of this burden. To live another day is not to add a page to his story, but to add a page to the story of the other, with a beginning that precedes him and an end that exceeds him. The said, which is the narrative of his life, cracks; the very fabric that constitutes his existence shows itself to be already a gift of the other.

The overwhelming nature of the "divine discomfort" may direct our attention toward a tone or mood that inhabits Levinas's later work: that of *lament*. This theme, which I will discuss in the final chapter, does not receive direct attention in Levinas's writings, but may provide helpful insight into the way these texts should be read and extended.

The Betrayal of Language

Levinas has, by the time he publishes *Otherwise than Being*, been considering for some time the relationship between language and intersubjectivity. In "Meaning and Sense," Levinas points to the ways that experience and language perform similar evasions of the typical ways that we think about signification in geometry and science. The power of ontology over philosophy is comparable to the "hold the *Said* has over the *Saying*" (*OB* 5). By 1974, Levinas uses language as a critical tool for introducing diachrony. In language and discourse, one is presented with the very riddle and mystery of ethics.[56]

It is important to demonstrate a difference between the more general problems of language and the distinction Levinas is drawing between the saying and the said.[57] Language is imperfect; it lacks the ability to communicate ideas without loss or remainder. This imperfection of language is less apparent in simple conversation than it is when people try to communicate complex ideas, but it is always there, undermining the clarity of language and letting meanings slip into and out of play without warning or intention. This unruliness of language was of significant fascination to Derrida; he emphasized that there is a sense in which my words betray their author. But this is not Levinas's primary concern in *Otherwise than Being*. In fact, Levinas worries that this level of skepticism may become a play where the gravity of language is obscured. The instability between objects, signs, and signifiers is not yet a diachrony; it may be a mere *game*.[58] Meaning is disjunctive, but the disjunctive nature of language can be of arbitrary concern, or more often, of economic concern. We try to select the least ambiguous language when ordering food so that we do not end up with a meal that we do not wish. The efforts to communicate in this regard are troubled by the instability of language, but these are problems within essence. This is the ambiguity of the said, and not necessarily traces of diachrony, of the saying.

The disjunctive nature of writing and speaking is not trivial; it poses a serious problem for language and communication. But this is merely the "difference amounting to the relationship between the known and the unknown" (*OB* 154). For Levinas, the real intrigue of language is not the never-ending vacillation between "clarity and obscurity, or distinctness and confusion" (154). These are problems of *consciousness,* the struggle between what is revealed and what is hidden to my consciousness. Language framed in this way is an act of "free commitment" (153). The self encounters and reads and listens to language in apparent freedom. I play with words and endeavor to understand and express them better but always from the firm foundation of my present. For play or for profit, I begin from myself and return to myself, experimenting with the effectiveness of my words against the ambiguity of being. Language in this register operates out of a prior commitment, a foundation of awareness, and on the basis of self-possession. This freedom is most obvious when I am observing artistic expression, to which *response* is typically unnecessary. Art leaves the viewer with all the power and freedom; the viewer has the power to critique, to ignore, to dismiss, or to burn. I can put down the book, quit reading the poem, and drop the game of attempting to assimilate new impressions into my foundational consciousness.

This is not the case, however, in dialogue. In human discourse, the other has already spoken the first word, and I am pressed to speak. We should remember these moves are not, for Levinas, necessarily verbal.[59] My words arise from within me but also from before me. The other is already before me and within me, summoning me to discourse. The other "orders me before being recognized" (*OB* 87). My speech is not arbitrary; it is not a game played to experiment with the effectiveness of a code. As Bernhard Waldenfels summarizes, "The other takes words precisely out of my mouth—my speaking begins elsewhere, outside of myself."[60]

Levinas gave the first word to the other person in *Totality and Infinity*.[61] But here in *Otherwise than Being*, even my response is not my possession, not my composition. The word that first spoke to me gave genesis to my speech, but not as an event in my past that I can recover and analyze. This is not a play of ambiguous meanings, between signifiers and the signified. The invocation to speak, and the very power of speech, is already from the other, even as they are absolutely and thoroughly within me. Diachrony refers to this interplay between my time and the time of the other; for Levinas, diachrony is not merely linguistics, but exigency. Diachrony is not just the sense that the other precedes me anarchically but that the precedence of the other is saturated into every word I might speak. This is proximity; the other who has called me is already closer to me than I am to myself.

So I speak. The words I annunciate and the cries that I exclaim are both a *saying* and a *said*. More precisely, my speech originates as a saying that is already a response to a former said and then it instantly breaks up once again. The saying, the verbal act of speech, does not perfectly correlate with the words and noises that fill the air. There is a *lapse* between the act of saying and the words that are ossified in a said, which is the inevitable cost of language. But in these words and noises, and despite the way they congeal into the said, the authority of the other announces itself.[62] This is the "move into language," which creates a correlation between the act of saying and that which is said (*OB* 6). This correlation is not without remainder, however, for through language, the saying is inevitably subordinated to the said, hidden within it. That which I speak is immediately thematized, converted, and assimilated to the common field of space and being and bodies, but it still carries traces of the saying.

Every word and every expression engages in this interplay. There is no pure saying, and neither is there a pure said. Everything that has been said still resonates with a saying, however faintly. Every saying must find its way into being through language, where it is thematized as a said. This is what Levinas calls the necessary betrayal performed by language. However carefully I craft my words, language betrays me: "In language qua Said everything is conveyed before us, be it at the price of betrayal" (6).

In this betrayal, a central component of the saying is lost, an "ancillary of angelic variant" that will not submit to the transmission of any message. But the interplay between the saying and the said, the necessary interplay of all discourse, reveals something not about linguistics or communication but about the immemorial relationship with the other. In the ossification of language as the said, there is evidence of a diachrony that divides my words from their author, which reveals their source in a "before" that is immemorial. The word that I say to the other, therefore, already carries the declaration that I am in the mode of response and that my response is already dependent on the other. This is the passivity of childhood, of a baby who cries for milk but whose very cry is already a reconfiguration of feedings and nourishments immemorial. Language betrays this dependency; the saying becomes obfuscated by the egoism of the said. I forget that what I have said is not my creation, not my act of fiat on a passive world that receives me as actor. The drama is already underway, and I am late for it; my fumbling for words is a haphazard but exigent responsibility to speak *from* the other, *to* the other, and *for* the other.

And here we see that *dialogue* has a peculiar relationship with being. The word that I speak becomes a said, but as a said it contains an epiphany, a revelation of something that is prior to me. Levinas is driving the wedge of alterity into the very lapse between my words and the I who speaks them. The other is already there, in the diachrony that divides my speech between the saying and the said. The saying resists its synchronization, for it moves from behind and before the said and never becomes fully encapsulated in the said. As Adriaan Peperzak summarizes, "The activity of speaking robs the subject of its central position; it is the depositing of a subject without refuge."[63] Levinas points out that it is ironic that this *activity* should be thought of as radical passivity.[64] But the so-called activity of the saying indicates a primordial passivity that shows there is no *beginning* to my story that is my own. I come to be, for the other, in the very act of saying.[65] My supposed self-conscious, independent, self-directed speech is divided from me upon its very utterance. In the said, in the present moment where the saying takes its form in themes and ideas and categories, the saying still echoes its immemorial diachronic origins.

The responsibility that arises from diachrony carries a paradox; I am bound in obligation to a responsibility that did not begin in me, "as though an order slipped into my consciousness like a thief, smuggled itself in" (*OB* 13). I find responsibility for the other deeper within me than my consciousness can trace; my obligation comes from another time, from the time of the other, which is "immemorial, unrepresentable, invisible, the past that bypasses the present" (11). Responsibility is reconfigured by Levinas to be a more primordial category than consciousness; my coming-to-awareness of the exigency of a call is secondary to the fact that my very identity is established by the call, on the face of the other.[66] The so-called freedom that allows me to embrace or to denounce hospitality, that allows me to waiver before I forfeit my place in the sun, is merely part of the way I configure my *present*. The gap between my present and the past for which I am obligated is a "gratuitous lapse" that defies recuperation or reminiscence (11). This defiance is not a product of the "remoteness" of the other or of the saying, "but because of its incommensurability with the present" (13).

The betrayal of language is not entirely bad news, for it is not just the said that betrays the saying by forcing it into categories. The said is also betrayed by the saying: the seams that fasten the said to the saying show themselves, and in this betrayal, we find the inadequacy of synchrony. The said gives the saying a "present" from which it can show itself.[67] That which is *before* being, otherwise than being, shows itself within being, within the said. Language is therefore connected to bodies, skin, faces, and needs. Bodies and stories and the said inhabit being, existing in a spatial present that can be marked by clocks. Perhaps even more concretely, we can mark time in the present of the said, in the rhythmic and predictable interchange between hunger and satiation. The said is *necessary* for the exchange of food; the saying is liberated only through the unsaying of what has been said.[68] This makes language an affirmation of being, not as the site of the Good, but as that which can make way for goodness, for hospitality that exceeds any economy or theme that finds itself said.

There is an unfathomable gravity to this claim; every word I speak comes from and to the other. Levinas suggests that we must "awaken in the said the saying which is absorbed in it" (*OB* 43). On close examination, language demonstrates that diachrony robs me of the chronology that I once considered *my* time. I do not approach or encounter the other, offering speech as one who has caught my breath and taken my bearings.[69] Dialogue reveals this to be the self-delusion of an egoism; language issues from a diachrony that lays claim to the very constitution of a self in consciousness.

This diachrony of language is of the utmost importance for the final wave of Levinas's overall thought, and it is utterly dependent on the manner in which Levinas liberates time without turning time into the chaos of asynchrony. The fact that I can never get out in front of my own words or my own consciousness demonstrates the raw and unprotected passivity from which responsibility has already irrupted. I am hostage to the other because I cannot wrest the past back into a former present. I am bound, from a time of passivity that precedes all affectivity, to this anarchic proximity of the other. Evasions are possible; I can deny that my language betrays my egoism or act as one who is not already always possessed by obligation. I can cling to the for-oneself that is the logic of being and economy, which presupposes a synchrony in which my resources and the other's needs make sense. I am hostage nonetheless.

A note is needed about the nature of persecution in Levinas and about the way that I find myself hostage to the other. These themes do not presume any *power* or physical authority in being; I am not a literal hostage held by a captor who can deny me the freedom of evasion. I am hostage to the other because of his or her "august command," which precedes and obsesses me, but which allows me the freedom of evasion. I can look away from the face, stuff the bread into my mouth, and elbow my way into the sunlight. The persecution of the other is a function of the utter exigency of a proximity that cannot be evaded; but *responsibility* is up to me. The other, Levinas claims, "does not constrain or dominate and leaves me outside of any correlation with its source" (*OB* 150). This is the liberty to say no to *myself*, to deny the alterity that judges me from my own saying.

The other is my *priority* because he or she is preoriginal to my very being. This priority is a debt; I am too late, arriving on the scene as his or her suffering is already, from time-before-time, underway. And because that suffering arises from a time that I can never recover, from a "past more ancient than every representable origin" (*OB* 9), the other's suffering never becomes understandable to me. I can never recover my balance; I never assimilate the other's suffering with other forms of suffering familiar to my memories and other representations of suffering. Levinas goes so far as to say that my "debt increases in the measures that it is paid" (12).[70] In this sense, the other's priority is utterly overwhelming. I am caught, from the beginning, not because the other has evidence against me that binds me to his life and not because of any past transgression or debt that holds me responsible. To encounter an other is to find oneself already guilty for his suffering.

I am hostage to a debt I have not chosen and called to *substitute* my suffering for this unsayable suffering that confronts me in the face of the other. If such disturbing ideas were not enough, Levinas presses even further. Because I am too late to sort through or appropriate the situation in which I find myself responsible, my responsibility escalates even further beyond my capacity to understand and enclose. I am responsible for the irresponsibility of the other, even for the irresponsibility of the other toward me, and even when I am not at all to blame. I am responsible for my own persecution.[71]

Can these themes of diachronous time, radical responsibility, and the unraveling of *any* independent ego-narrative be assimilated with the history of philosophy? Need they be?

Diachrony and the History of Philosophy

Levinas's later essays often seek to demonstrate how this diachronic understanding of time resonates against various philosophical voices, especially Aristotle, Plato, Kant, Descartes, Hegel, Bergson, Husserl, and Heidegger, but also a host of other figures. This is a critical component of Levinas's late development; he wants to demonstrate that all is not lost for philosophy and that even if philosophers have generally failed to think beyond being, their language still can be turned over to gesture toward the unsayable. For Levinas, it is critical that his radical philosophy of responsibility remain rigorously philosophical and in clear relationships with the work done by other philosophers. He is concerned to show that the "ambiguity of being" (*OB* 9) has been glimpsed at many moments over the course of philosophical history. If Derrida is right, and "philosophy died yesterday," then Levinas remains intensely interested in the way philosophy "wanders toward the meaning of its own death" (VM 79). Perhaps unsurprisingly, Levinas finds repeatedly that philosophy has failed because it has been negligent in its thinking about time; thus, he uses his mature understanding of time to reread the history of philosophy. Levinas demonstrates this throughout *Otherwise than Being* and in dozens of essays and lectures from the late 1970s and across the 1980s.

Levinas was clearly concerned with remaining in relationship with the history of philosophy as he penned *Otherwise than Being*. As he wrote in the early pages of *Otherwise than Being*, "The history of philosophy, during some flashes, has known this subjectivity that, as in an extreme youth, breaks with essence. From Plato's One without being to Husserl's pure ego, transcendent in immanence, it has known the metaphysical extraction from being, even if, betrayed by the Said, as by the effect of an oracle, the exception restored to the essence and to fate immediately fell back into the rules and led only to worlds behind the scenes" (*OB* 8). These moments, however brief their flash, provide Levinas with points of departure and places of intersection. Levinas does not wish to offer a new philosophical system, but to demonstrate that philosophy's work is far from done, that philosophy must pass through the abandonment of itself to find its primordial, foreign master in ethics. Levinas is interested in demonstrating the way philosophy has yearned for this "death," has shirked away from this "end" in fear and trembling, and has in brief moments given voice to the "break up of a fate that reigns in essence" (8).

Philosophy has, from the beginning, borne traces of its own demise, its own inability to fulfill its transcendental purpose. These traces are the "ambiguity of being," indications that the "beyond being" is not to be conceived by way of modifying or tinkering with some ontological concept apropos to being. Levinas writes, "The philosopher finds language again in the abuses of language of the history of philosophy, in which the unsayable and what is beyond being and conveyed before us. But negatively, still correlative with being, will not be enough to signify the *other than being*" (9).

Though Levinas's philosophical path is unique and unprecedented, he does not strike out on a path that leaves behind the rest of philosophy. As Pierre Hayat notes, "Between the concept of the infinite, between totality and transcendence, Levinas's philosophy traces a singular path, one that runs into that of masters of Western philosophy."[72] Levinas's 1989 essay "Philosophy and Transcendence" reveals just how far Levinas has moved away from spatial configurations for proximity and alterity. The other, in *Totality and Infinity*, comes "from another shore" and teaches "transcendence itself" (*TI* 171). But by configuring transcendence as diachrony, as he does in this late essay, Levinas resituates the transcendence of the other as a proximity that is too proximate for assimilation. Levinas's intent is to align this novel understanding of transcendence with key figures from philosophical history. In "Philosophy and Transcendence," he moves methodically through a series of conversations with philosophical masters, all in the interest of constellating his own understanding of transcendence with these other figures in philosophy. Along the way, Levinas sharpens some of his usual arguments about representation, presence, and death. He also demonstrates how much his understanding of transcendence has been transformed by his developed understanding of time.

Levinas is acutely interested in demonstrating throughout "Philosophy and Transcendence" the various ways that philosophy has become intoxicated by the idea of "transcendence" that begins and ends in "cognitive thought" (*AT* 5). The modern person is "more indubitably planted in his *cogito* than his feet on the ground." As such, the idea of transcendence has just become a feature of immanence. The future and past remain just another manifestation of the present of the cogito. And the cogito becomes "drunk with being *in himself* and *for himself* in the presence—or the modernity—that he unveils" (5). Levinas aligns this drunkenness with an inability to hear "a silent voice in which God comes to the mind." This original locus of infinitude has been overlooked, neglected, and set aside, and Levinas methodically demonstrates how this happens by examining some of the major thinkers of Western philosophy. In nearly every case, as we have come to expect in an essay as late as 1989, Levinas pins the problem on an inability to consider the importance and significance of diachronic time.

Philosophers never tire of discussing transcendence, but Levinas has long suspected that most gestures toward transcendence amount to little more than an elevation of some form of immanence. His reading of Plato and Neoplatonism is instructive in this regard. Often in his writings, Levinas has admired the relationship between Platonism and being, especially Plato's effort to think beyond or without being. Nevertheless, Levinas will constantly play with and against Plato, like rattling a tree to force fruit to fall. The opening lines of *Otherwise than Being* point to Plato for the sake of moving beyond him: "Since the *Republic* there had been a question of what is beyond essence" (*OB* 3).

In "Philosophy and Transcendence," he turns his attention to Plato and the notion of transcendence engendered by Plato and fulfilled in the Neoplatonic tradition. The Platonic celebration of unity, of the One, weds philosophy with the idea of fusion. Neoplatonism demonstrates this in the literal "ecstatic itinerary" toward the One (*AT* 8–9). This makes transcendence the very nature of philosophy; its endeavors and characteristics are all cast in the mode of ascension and transcendence into union. Sociality, including relations with other persons, is a function of immanence (8). The Greek Titan Cronos must satiate himself by consuming his children, for "all eternity is worth more than love" (10).

Levinas's reference to Cronos and Plato's *Symposium* invites closer examination. Levinas has found a seam in Greek mythology and philosophy, a deference to the eternal and a consequent subordination of time and bodies. Levinas cites Plotinus's invocation of Cronos in *The Enneads*, who suspends time's flow by consuming his offspring.[73] In this way, change and the fecundity of differentiation is halted and prevented. Love is less important than eternity and bliss; unity is preferred over diversity. Cronos fears his children; he must "absorb his offspring that, full within himself, he may be also an Intellectual-Principle manifest in some product of his plenty."[74] Time, for Plotinus, is undoubtedly the undifferentiated bliss of the eternal. Diversity and sociality are emblems of a fall from the unity of unchanging identity. Elsewhere, Plotinus writes that Cronos turns toward his own father (Ourano, the Absolute, the One), "ignoring the lower world," and leaves the inferior world of sociality to his son Zeus.[75] The god Zeus had only escaped Cronos's belly by way of trickery.

Levinas thinks that Plato is more nuanced but that he still sides with the defeat of time on behalf of eternity. In his later years, Levinas reconfigures the concept of love to befit his understanding of alterity and responsibility. "Love of one's neighbor," which he also called a "love without concupiscence," is a "non-transferable responsibility" (*AT* 29, 30, 129). This love that Levinas invokes resists the merging of unity and totality. To demonstrate this difference Levinas refers to Diotima's speech in Plato's *Symposium*, in which she suggests that love is only a demigod. As for Cronos so for Plato: "all eternity is worth more than love, still subject to time" (10). Love's curse is that it binds one to one's neighbor. For Diotima, who appears in Socrates's speech but does not drink around the table with the other guests, the love for other persons is a component of the ascent. Diotima charms Socrates by guiding his eyes upward toward the beauty and unity of wisdom. The *Symposium,* interestingly, swings its way forward in a narrative escalation, appreciating lower forms of love but aiming steadily away from sexual and fraternal love and toward the asexual passion for beauty and wisdom. The highest admiration and passion, however, must be reserved for what Diotima calls "beauty itself, absolute, pure, unmixed, not cluttered up with human flesh and colours and a great mass of mortal rubbish."[76] In Plato, Levinas has long admired the symptoms of a dissatisfaction with being, a hunger unmingled with concupiscence and the self-interestedness that dominates the economy of being. But Plato's "turn" beyond being, or outside of being, is not humanism. It turns both from *time* and from the "mortal rubbish" of "human flesh." Plato provides a glimpse, a moment where philosophy knows itself to be wounded by the "beyond essence," but he quickly recovers and reestablishes ontology.[77] Eternity trumps flesh; the face of the other is exchanged for a "world behind the scenes." Philosophy moves beyond itself, only to again fall back on itself. As Levinas writes in his 1986 collection *Of God who Comes to Mind,*

> A philosophy that has been handed down to us could not fail to name the paradox of this non-ontological significance; even though, immediately, it turned back to being as to the ultimate foundation of the reason it named. The placing of the Idea of the Infinite within the finite, surpassing its capacity, as taught by Descartes, is one of the most remarkable expressions of transcendence.... Under different terms, this relation of transcendence is shown—if only for an instant in its purity—in the philosophies of knowledge. It is the beyond being in Plato. It is the entry, "through the door," of the agent intellect in Aristotle. It is the exaltation of theoretical reason into practical reason in Kant. It is the search for recognition in the *other man* in Hegel, himself. It is the renewal of duration in Bergson, who grasped there, perhaps, rather than in his conception of the integral conservation of

> the past, the very diachrony of time. It is the sobering of reason in Heidegger. (*GCM* 119)

Plato's weakness, as Heidegger detected, is to prefer eternity to time and to forget that eternity rests uneasily on Cronos's full belly. In Plotinus's interpretation of the Cronos mythology and in Diotima's preference for the unity and singular beauty of the absolute good, Levinas sees a defeat of love-without-concupiscence alongside a defeat of time (*AT* 29). The defeat of time, in these and other examples, has a very different outcome than the anxiety that troubles Heidegger. Levinas appreciates the way time is partially "deformalized" in Bergson, Husserl, Heidegger, and Rosenzweig, but he is explicitly interested in correlating overly formal renderings of time to the danger of violence. It is the exigency of ethics, driven certainly by the consequences of twentieth-century negligence, which drives Levinas's breathless movement toward a radical and diachronic philosophy of time.

As Levinas closes the essay "Philosophy and Transcendence," he makes a new turn with his use of diachrony, inflecting the diachrony of time toward the *future*. He first outlines and reinforces the time of the other as a diachronic "past irreducible to the present" (*AT* 32). Diachrony originates as "an immemorial past" and as "my participation in the history of humanity, in the past of others, who are my business" (32). As if completing the moves against Hegel and Parmenides that he initiated in the 1940s and even earlier, Levinas here declares that the diachrony of time means that philosophy must cope with the loss of Plato's "unity of the One." But after repeating his familiar invocation of the anarchic past, which summons me from before myself to forfeit my "place in the sun," Levinas ponders the "diachrony of the future" (36). Packed within the history of philosophy, from Descartes's notion of the infinite to Kant's categorical imperative, lies a subtle diachrony that, when exposed, harkens not only to a past that precedes it but to a future of prophecy (36). The diachrony of the future is manifested as hope for a future that exceeds the foreseeable, a future that transcends both anticipation and possibilities.

Levinas hesitates, in his later writings, to speak with much specificity about the diachrony of the future, for it is difficult to disentangle the prophetic "hope" of the future from the return of "consolations" (*GCM* 96). In "Questions and Answers,"[78] Levinas says of hope, "I do not know if one may speak here of hope, which has wings and does not resemble the patience in which the intentionality still so alive in hope is engulfed, in order to turn back into ethics. . . . The anticipation of the future is very short. There is virtually no anticipation. The future is blocked from the outset; it is unknown from the outset and, consequently, toward it time is always diachrony" (96). Yet just as the past is too ancient for recovery and nevertheless proximate by the exigency of my own words, the future has me bound without any constraint by the present in which hope is already a hope *for*. Whatever we may say about the future, it must move beyond "the fecundity of the I," and even beyond the fecundity of the father displaced in the child (*TI* 277). The future is the future of the other; hope itself is a said that carries but traces of the prophetic saying.

Diachrony, we may conclude, unsettles all things one might collect in the present. All thoughts, recollections, hopes, memories, ideas, and words are already inscribed with the proximity of the other. The act of speech reveals this, as my words turn around mid-air and flash traces of the other who was already always before me, even as I formed the words from what I thought was freedom. The result is that all the accumulations in my dwelling, in the habitation of my consciousness, are the traces of a debt that I cannot contest. My story is ruptured from before I have it put together.[79] Levinas writes, "Proximity is a disturbance of the remembered time. One can call that apocalyptically the break-up of time. But it is a matter of an effaced but untameable diachrony of non-historical, non-said time, which cannot be synchronized in a present by memory and historiography, where the present is but the trace of an immemorial past" (*OB* 89). The encounter with the other is predicated on this debt, and so the encounter situates me irreplaceably before another person into whose story I find myself situated from before time. The self loses and regains a plot, but the plot in which I dwell is now the time of the other, the time of diachrony. Levinas sees that philosophy has caught glimpses of this primordial diachrony, even if it settles repeatedly for the said over the primitive saying that the said betrays. Time as diachrony is offered here as the philosophical structure of responsibility. I am bound to the other because the other is anarchically *prior*, my priority. It is in responsibility that my identity is founded. Responsibility is the "very fact of finding oneself by losing oneself" (11).

EIGHT

The Time of Restoration

Time is pure hope. It is even the birthplace of hope.
—Emmanuel Levinas, *God, Death, and Time*

Despite the radical reconsiderations of time that are evident in *Otherwise than Being* and other late essays by Levinas, we have ample reason to suppose that he still felt there was significant work to be accomplished on the concept of time. The deformalization of time, which Levinas saw underway in Bergson, Husserl, Heidegger, and Rosenzweig, is not yet complete in his final works. When Levinas died in 1995, he left behind a number of clues that might open us to the future development of the philosophy of time after Levinas. Adriaan Peperzak even suggested in 1993 that Levinas was at work on a "new and independent book on time as diachrony, which seems to be a focal topic in Levinas's recent efforts."[1]

Given that Levinas was working on a more deformalized[2] understanding of time, one might expect that he was moving toward the configuration of diachronic time and of the saying and the said in more user-friendly terminology, without the complexity of philosophical jargon. Indeed, it may be the case that the complexity of Levinas's writings contributes to many of the misunderstandings and misinterpretations of Levinas's work. The widely respected philosopher Paul Ricoeur remained a reader of Levinas's work to the end, and yet his critiques show puzzling signs of ignoring key components of Levinas's work.[3] Further research and more lucid expressions of Levinas's philosophy of time might help address, or confirm, critical readings of his work. Much could be gained, in clarity and in application, by pursuing a less theoretical, more informal application of Levinas's understanding of time. His desire to further deformalize time appears to press us in that direction, though these developments exceed the parameters of this study.

We also might detect applications for diachrony in Levinas's ongoing attempt to find the traces of this idea across philosophical history. This was clearly a late fascination of Levinas's, as evidenced by his engagements of philosophical history in, to name just a few: "God and Philosophy" (1975), "The Old and the New" (1980), "The Thinking of Being" (1986), and "Philosophy and Transcendence" (1989). Strong evidence that Levinas was busy thinking through the relationship between his thought and philosophical history also arises in *God, Death, and Time,* a lecture series given by Levinas in the 1975–1976 academic year (*GDT* 1). In his final writings, Levinas appears increasingly inclined to appreciate the moments when the philosophers of the past "could not fail to name the paradox" of the diachrony of time, even if they repeatedly failed to follow this idea beyond being (*GCM* 119).[4]

There are intriguing implications for the idea of "God who comes to mind" and the notion of *holiness,* which Levinas found captivating to the end. Levinas invokes holiness already in *Totality and Infinity,* and the concept is also featured in *Otherwise than Being.* Levinas's late interest in this topic is succinctly reported by Derrida in his funeral oration for Levinas, when he claims that Levinas told him: "You know, one often speaks of ethics to describe what I do, but what really interests me is not ethics, not ethics alone, but the holy, the holiness of the holy."[5] Given the evidence in Levinas's late work that he felt much was left to be said about the concepts of God, holiness, religion, and the implications of these concepts for ethics, it is important to discuss some of these developments, especially as they relate to a pressing and lingering problem with regard to Levinas's understanding of time.

In each stage of Levinas's development he moved forward with a surging wave of innovative claims, only to find that components of his claims would later require alterations, retractions, and especially escalations. At the conclusion of this chronological study, what then, is *missing* in Levinas's expressions of time, diachrony, and passivity? The strategy may not require a more radical expression of time's alterity, which hardly seems conceivable after *Otherwise than Being*. Perhaps the strategy, when one thinks of time after Levinas, ought to lead us toward more precise and careful ways of expressing the radical responsibility delivered to us by the diachrony of time. As Michael Kigel points out, Levinas's method "isn't too harsh, it is too blunt; and so its action really causes needless pain. Its cutting edge needs to be *sharpened*."[6]

This methodology of reading Levinas against himself continues the spirit of his commentary on his own work, which repeatedly found ways to overturn his earlier efforts with a new wave of ideas that scandalize and hone even his own prior expressions. Images and metaphors continue to haunt Levinas's works, necessary as they may be. Further, some of his language still needs to be questioned for the way it has settled for the said and obscured the saying; the language that Levinas uses betrays him.

In evaluating Levinas's account of time and considering possible further developments, there are a number of avenues to be explored, particularly with regard to Levinas's invocation of maternity, the role of the feminine, and his use of religious language as a way to move past his troubling reliance on gender. We can also identify a "fourth person" underlying Levinas's account that not only eludes ethics by slipping behind Levinas's less visible reliance on the feminine in his later work but practically embodies Levinas's philosophy with a vision of hope and restoration in the future. And further, we might read Levinas's late work as a form of philosophical lament, which might help the interpretation and extension of Levinas's thought.

Developments on Levinas's Thoughts about Gender

Following Tina Chanter and other exegetes of Levinas's late work, the next step in thinking about Levinasian diachrony requires another consideration of his thoughts about gender.[7] Chanter claims that scholarship around Levinas's work has failed to engage with sufficient seriousness the relationship between the feminine and Levinas's understanding of time: "Even those readers who have provided interpretations of Levinas which take up some of the earlier themes in order to follow how they are recast in the later work have neglected, in my view, to explain the importance of the feminine and corporeity in Levinas's work, or to expand sufficiently on the relation between time and the instant."[8] Given the failure in chronological studies on Levinas to trace to completion problems relating to the feminine, corporeity, time, and the instant, I have noted at length the concept of the instant and the perilous way that Levinas appears to enclose the feminine in the instant or "lapse of being" that is configured as habitation in *Totality and Infinity*. Though Levinas treats the feminine differently in his later works, some troubling and underattended problems remain.

There do seem to be various tendencies in Levinasian research in regard to his invocation of the feminine. As Chanter points out, scholars tend to treat the concept of the feminine as subordinate to the larger themes introduced by Levinas. This allows the obvious and serious problems to be dismissed or resolved "in a footnote or in an aside, or merely neglecting the theme altogether."[9] But Levinas's use of gender demonstrates a deeper problem with his philosophical discourse that is most perceptible in terms of *time*. Hand in hand, Levinas falters on the concepts of time and the feminine. The woman is captive to the four walls of the instant, the present, and the delightful lapse of time that allows her to be the condition of ethics. Levinas, using what Donna Brody has called a "Maternal Method,"[10] has from the beginning explicated his philosophy by way of gendered bodies. The apex of hospitality, as configured in *Otherwise than Being*, is the mother, who is elected irreplaceably to suffer for the other from before the freedoms we normally associate with ethics (*OB* 75).[11] A mother nourishes a fetus before she is even aware of its existence; no one can shoulder for her the burden of her pregnancy or the pains and anxieties it involves. Levinas turns to the maternal as the exemplar of the ethical subject, who from time-before-her-time is bound to the defenseless other, to substitute, to be persecuted, to be hostage, to be responsible.

This is a significant alteration in Levinas's usage of the feminine. Craig Vasey even claims that Levinas "makes a more legitimate use of feminine imagery" and arrives at a "theoretical position that is practically indistinguishable from feminism."[12] Sonia Sikka puzzles at the ease with which Vasey names what are to be considered fundamentals of feminism, particularly in light of the fact that feminists rarely agree on this point.[13] Furthermore, Sikka accurately questions the wisdom of resting philosophy on a metaphoric reliance of the mother "as trope."[14] Diane Perpich points out that Levinas's problem does not stem from failing to represent the feminine properly, but "precisely from ceding to the temptation to representation."[15] It has been a signature of Levinas's philosophy to recoil from representing the other, but he has made a habit, even in *Otherwise than Being,* of falling to constrictive representations of the feminine. There is an undeniable appeal to Levinas's invocation of the maternal as the paradigm of welcoming and bearing the other. Yet I suspect that what we discover under the surface is yet another instance where the feminine has been removed from Levinas's otherwise all-inclusive understanding of diachrony and time.

Rather than repeating this critique in full,[16] I simply wish to examine closely a significant component of Levinas's understanding of diachrony and passivity that finds itself hidden by a subtle patriarchy that still inhabits his later thoughts. Here there is a seam in Levinas's argument where he again uses feminine imagery and in doing so occludes a critical aspect of both the diachronic future and the diachronic past. In order to more faithfully develop Levinas's dual concepts of responsibility and diachrony, we need not a *better* representation of the feminine, but a more rigorous examination of the consequences of these representations, especially as they remain more quietly in place even in his late writings.

When Levinas utilizes the feminine in *Totality and Infinity,* he demonstrates that the "formal" structure of his understanding of time is still a poor fit for the corporeal world of gendered bodies. This is not a trivial failing, for even as he causes his readers "consternation and pain"[17] through his patriarchal approach to gender, he also demonstrates how easily the face of the other can be reduced again to presence. In the home, within its four walls, the discrete other falls back into the present, the instant, the preethical time of *freedom*.[18] This creates obvious and well-documented problems for the feminine and for gendered societies that attempt to embody Levinas's philosophy of responsibility. The ironic essentializing of women in *Totality and Infinity* coincides with and is enabled by Levinas's ongoing employment of spatial and ontological language. In the habitation, in the instant, the relation with the feminine other takes its bearing from *being*.

This is precisely the problem Levinas finds in Kant, Husserl, Heidegger, Descartes, and others; they detect the Good beyond being but then revert "back to being as the ultimate foundation of the reason it named" (*GCM* 119). Levinas's argument about the face rests on the capacity for an encounter with an other who is utterly separate and external to the self, but he founds this encounter on a prior relation with the feminine. The feminine intimacy of the home is necessary to establish gentleness, not violence, as the original disposition of the ego. Both as mother and as wife, the feminine enables the face-to-face encounter. The woman is cast as a component of interiority, barely (if at all) separable from the tranquility of the objects earned in the world of labor and returned to the home. She is inscribed in being.

One can see why Vasey claims that Levinas corrects these problems in *Otherwise than Being*. Levinas invokes diachrony to declare that the absolute other is always already before me. The "woman" in *Totality and Infinity* is the quiet presence that enables every word and makes possibile the encounter. In *Otherwise than Being*, these traits are transferred to the ego. If the woman could be considered hostage to the masculine presence in *Totality and Infinity*, it is now the subject itself that is subjected to this primordial position in the stories of others. If the "woman" was denied her own, distinct story or time in Levinas's earlier work, it is now the I that finds itself similarly destabilized.

The maternal, in *Otherwise than Being*, is "bearing par excellence" (*OB* 75). Levinas finds that the image of the maternal offers an intriguing and captivating metaphor for bearing, suffering, substitution, and obsession. By some analyses, this could ostensibly constitute an elevation and celebration of the values often attributed to the feminine. Levinas is intent on demonstrating the diachrony of the other-in-the-same; for this reason, one can see the appeal of the maternal. No metaphor rivals this one; a mother carries and sustains a fetus in every conceivable way and yet finds herself beset with obligation relative to the fetus, in passivity and election. Here *seems* to be the ultimate model of diachrony; the maternal body is already social, already plural, already transcendence-in-proximity. One has always, already been *mothered,* and the maternal is the precondition for all speech, all action, all supposed freedoms. It is easy to see why Levinas turned to the maternal for an example of the themes he developed in *Otherwise than Being.* There are, however, many consequences to this move.

Donna Brody wonders if this is not a kind of exploitation of the female capacity for childbearing. She questions whether this reduces the alterity of the feminine to the most eminent and bodily of meanings.[19] The problem of representation is also sharp; how can Levinas insist so clearly that the other is beyond representation and then so thoroughly represent particularly feminine others? Does this constitute a colonization of maternal generosity? All of these are legitimate problems that cannot be explored here except to point out that Levinas, even in *Otherwise than Being,* falters on the notion of *maternity* precisely because he still does not fully extend diachronic time to the feminine. The trouble begins not on the periphery but at the center of Levinas's philosophy, at the point where diachrony intersects with the flesh and blood of history. This intersection makes possible the trace of the saying in the said, the appearance of the Good in being, and the very possibility of hospitality and responsibility. Something must make being capable of this opening to diachrony, to transcendence. And as in *Totality and Infinity,* Levinas requires this of the maternal. In order to demonstrate these moves, and eventually present some suggestions for the resolution of the problems they create, I first return to *Totality and Infinity* to seek clues about Levinas's understanding of the maternal in his earlier work.

Maternity in *Totality and Infinity*

Much has been written about the invocation of maternity in *Otherwise than Being* and the relationship between the use of the feminine in that book and *Totality and Infinity.* Less attention has been paid to the appearance of maternity in *Totality and Infinity,* mostly because it plays a very small role alongside the "feminine alterity" upon which Levinas rests so much of his argument. But the problems with Levinas's use of the trope of maternity in *Totality and Infinity* do not disappear in *Otherwise than Being,* even if they are less evident. Levinas *must* continue to rely on the feminine, because he has not developed any other way to retain the thread of contact within being that makes possible both history and its other. By continuing to represent the feminine, Levinas has robbed the feminine, including literal women and mothers, of diachronous *time.* A consideration of time after Levinas must feature a way to think about the relationship to bodies, nourishments, and history otherwise than by way of the tropes of "the feminine" and "maternity."

In *Totality and Infinity* Levinas discusses the need in childhood for "the protective existence of the parents" (*TI* 278). Instead of leaving the pluralization of parenting alone, Levinas immediately claims that in order to understand the protectiveness of childhood, "the notion of maternity must be introduced" in order to connect the father-son relation to history (278). Levinas is quite content, in his discussions of paternity, to leave things configured as a father-son relation. But he briefly mentions maternity, particularly when he has to invoke a permanent, indubitable "past" to which every child has "recourse." Levinas must be thinking of the sense of "protective existence" that no father can offer a child: a uterus (278, 147).

The emphasis across this section of *Totality and Infinity* is on paternity and filiality, but the feminine "must be introduced" to understand the past. Levinas makes this connection rather clear: "The notion of maternity must be introduced here to account for this recourse. But this recourse to the past, with which the son has nonetheless in his ipseity broken, defines a notion of distinct continuity, a way of resuming the thread of history—concrete in a family and in a nation. The originality of this resumption, distinct from continuity, is attested in the revolt or the permanent revolution that constitutes ipseity" (*TI* 278). The *ipseity*, which here means something like an independent selfhood, is doomed to be undermined, even as it is established in the son. Jeffrey Bloechl points out that the claim to any primacy of the ipseity is "Levinas's real target."[20] The mysterious relationship between the father and the son is his "prototype" for the relationship with the future in *Totality and Infinity* (279). The future is rupture and recourse, the pulsing of breaking away and rejoining, labor and enjoyments. The other-qua-son is alterity and unicity, distance and proximity. This tension actually helps Levinas avoid a transcendence that might seem disembodied and fleshless. Here both father and son are related beyond "causality" (279).

They are torn up from history, in their relation. Levinas wished to underscore the nonhistorical nature of paternity by declaring the father's relation to the son as a "paternal election," similar to creation *ex nihilo*. Like the created world, the child cannot exist "on his own." Levinas writes, "Creation contradicts the freedom of the creature only when creation is confused with causality. Whereas creation as a relation of transcendence, of union and fecundity, conditions the positing of a unique being, and his ipseity qua elected" (*TI* 279). The father's election of the son and the son's establishment in the world relative to that election are not reduced to the causes and effects of being, and they are therefore liberated from the interiority that Levinas wishes to reject.

In *Totality and Infinity*, only a relation that exceeds biology can be an exteriority. Relations of need, necessity, and economy are the elementals of being. Bodies and food, nourishments and rest, labor and enjoyment, these are the materials of being, but they are the accoutrements of essence, of interiority. They are not faces, and they cannot give rise to the ethical. The elemental is *present*, a component of the *instant* that is interior to the economy of labor and dwelling. The elemental, which Levinas casts as feminine in *Totality and Infinity*, is necessary for goodness but not the Good.

The *mother* is invoked as an afterthought in this account, precisely because there has to be *some* matrix in being that becomes the site for this alterity, the transcendence of the son. There has to be a safe venue, a vertex that plays host to this intersection with the infinite. Placing the feminine at this intersection brings Levinas into an intriguing correlation with the introduction of the concept of *khora* in Plato's *Timaeus.* One of the later Platonic writings, *Timaeus* ponders a question that is less obvious in the binary opposition that is outlined in the *Republic.* Timaeus, whose speech dominates this dialogue, grants the basic veracity of Plato's metaphysical theory, including the disembodied *eidos* and the matter in which we detect their imperfect representations. But in what venue do these two categories intersect? Timaeus suggests,

> This new beginning of our discussion of the universe requires a fuller division than the former; for then we made two classes, now a third must be revealed. The two sufficed for the former discussion: one, which we assumed, was a pattern intelligible and always the same; and the second was only the imitation of the pattern, generated and visible. There is also a third kind which we did not distinguish at the time, conceiving that the two would be enough. But now the argument seems to require that we should set forth in words another kind, which is difficult of explanation and dimly seen. What nature are we to attribute to this new kind of being? We reply, that it is the receptacle, and in a manner the nurse, of all generation.... Wherefore, the mother and receptacle of all created and visible and in any way sensible things...is an invisible and formless being which receives all things and in some mysterious way partakes of the intelligible, and is most incomprehensible.[21]

Intelligibility, claims Timaeus, requires a venue on which the forms, which exceed the embodied and temporary history of matter, can make their impressions.[22] Timaeus is struggling to articulate the possibility of origins and turns his attention to the milieu or venue upon which creation comes to be. He invokes *khora,* a rather commonplace Greek term meaning something like the word "receptacle."[23] However, it quickly becomes apparent that khora resembles a clay receptacle in name only, and since khora is the very possibility for words, Timaeus struggles to articulate its nature and character. In this enigmatic section of *Timaeus,* khora is the condition for the intersection between Plato's finite and his infinite.[24] But what is the node, the point of contact, the intersection between his *eidos* and the matter in which it leaves its signs and indicators? The earlier dialogues lack accessible language to describe the milieu within which Plato's metaphysical game may unfold, and Timaeus suggests that it be understood in the manner of a "mother," "a nurse of all generation," "a receptacle."[25] The question is one of venue; Timaeus suggests a matrix, complete with the maternal and uterine metaphors that appear at a crucial juncture in Levinas's philosophy. Thus, Derrida takes interest in khora because he sees it as a subtle indication that philosophy is, in the words of John Caputo, "in a certain amount of trouble here."[26]

The parallels between these appeals to the maternal are stunning, despite broad and obvious differences between Plato's metaphysics and Levinas's philosophy of responsibility. Perhaps we should not be surprised by this correlation, inasmuch as Levinas once said that "all philosophy is Platonic."[27] Levinas credits Plato, in fact, with the very phrase "beyond-being" (*IR* 249). For both philosophers, and despite Levinas's rejection of many aspects of Plato's system, there are two primary terms in play: being and its beyond. Timaeus points to the need for a middle term, a point of contact between two distinct realms. For Levinas, being is utterly incapable of perceiving the Good, which appears through diachrony, through the trace of the saying in the said. For Levinas, being must somehow be capable of being arrested and irrupted by being's otherwise. In both cases, a *matrix* is invoked, a *mother*.

For Levinas and Plato's *Timaeus*, the maternal functions as the "node" for the intersection between being and beyond being (*OB* 76). The appearance of exteriority within the interiority of being is for Levinas "a marvel" (*TI* 292). And since this node opens to radical transcendence, we can no longer address it with "the categories of being" (293). This move, the appearance of transcendence in the face, draws Levinas close to Plato; "We thus encounter, in our own way, the Platonic idea of the Good beyond Being" (293). Indeed, in both the *Timaeus* and Levinas's configurations, the matrix acts as the ontological ground in which metaphysical transcendence can be made known. The ontological realm is a "faceless generous mother, matrix of particular beings" (46).[28]

In *Totality and Infinity,* Levinas uses the feminine to meet this critical need in his philosophical structure. *Something* must bind (masculine) alterity to the neutrality of history and being. Without reducing the transcendence of the son to a relative transcendence, Levinas must find "a way of resuming the thread of history—concrete in the family and in the nation" (*TI* 278). The prototype for this paradoxical relationship is the parents. Parents protect the thread of history. They engender children; they participate in the causal process that presses forward the nation and the family in historical time. The mother is invoked as the recourse to the past, the steady and necessary bind between alterity and proximity. She is the metaphorical connection between humans and history, the connection between totality and the infinite. But in this role, the woman cannot be afforded the alterity of the son and the father; she is what allows the break from "the contingent order of being" to that which is "unrelated to its essential production" (279). Levinas claims that these family relations break free "from their biological limitations" (279; cf. *EI* 70–71). The feminine is only invoked here as maternity to allow the child to have a *historical* past, a body. The time of the feminine is the time of history, the time of bodies and birth and the elemental. The mother makes diachrony possible, but is denied it herself.

In this portion of *Totality and Infinity,* it is unclear how the woman can be thought to exceed the biological function of childbearing, which is the only reason she is invoked. The maternal rescues Levinas from a pure dualism, from a totality that knows nothing of infinity or from an infinity that is gnostic or Manichean.[29] We cannot know, but perhaps this is what Timaeus was attempting to enact on behalf of Platonic metaphysics: to curtail the Neoplatonic developments that forge a unity between being and nonbeing by way of gradient transcendence. For Levinas, transcendence implies a *from,* and this *from* is a feminine dwelling, an original receptacle, indeed, "the nurse of all generation."

Levinas invokes the phrase "uterine existence" in a peculiar, neutralized way in his discussion of the way enjoyment is the throbbing of egoist being. Levinas writes, "Enjoyment is the very production of a being that is *born,* that breaks the tranquil eternity of its seminal or uterine existence to enclose itself in a person, who in living from the world lives at home with itself" (*TI* 147). This "seminal or uterine existence" that is the precondition for birth is also that which makes every birth a living-from. The subject repeatedly, in enjoyment, "frees itself from that past" (147), setting aside prior enjoyments in pursuit of more enjoyment. The throbbing of egoist being repeats the movement from birth to self-enclosure and then back out into another birth. Enjoyment is the moving out, a freedom that is "the possibility of commencement" (148). This movement out is still not exteriority, but a feature of interiority, and the separation involved in enjoyment "is not on the same plane as the movement of transcendence" (148). All that is involved in the pulsation of enjoyment and return is still finite.

This origin of the ego relates directly to time. The relationship with the uterus, birth, enjoyment, internality, and the throbbing of egoist existence points to Levinas's instant, which in *Totality and Infinity* is expanded into the dwelling. The instant, in *Existence and Existents,* is configured according to the absolute interiority of the moment and the isolated subject's inability to rupture the closure of the instant upon itself. In *Totality and Infinity,* we see an expanded treatment of the interiority of the instant, now including the pulsating pattern of separation and return. As in *Existence and Existents,* the ego is locked away from the world, despite the appearance of a direct experience of others. Persons, perceived in the modes of separation and enjoyment, are encountered like objects. The frozen present of Levinas's earlier writings is now understood as the whole economy of being. Levinas will not again claim that "being is evil" (*TO* 51),[30] for being is the condition for irruption. Without enjoyment, without "the primordial relation of man with the material world," there is no site for the irruption of the Good (*TI* 149).

In a very obvious invocation of spatial language, Levinas claims that interiority must have a door to exteriority that is "at the same time open and closed" (*TI* 148). There opens a dimension, within interiority, "through which it will be able to await and welcome the revelation of transcendence" (150). This means the instant is a place for enjoyment and *patience.* The feminine, Levinas says repeatedly, facilitates the gentle intimacy of the home; the woman is the "other who welcomes in intimacy" but is not the "*you* of the face that reveals itself in a dimension of height" (155). This brings us into familiar territory. "Woman" is locked in the interiority of the instant, hostage to the role she plays in making possible an open door from interiority to exteriority. She makes it possible for the man to encounter a face.

What interests me now, however, is the way uterine existence becomes the condition for the opening of being, the node that intersects eternity with history. In both the subtle invocation of the uterus and the more explicit discussion of the mother, Levinas utilizes these maternal images to provide a contact point between bodily history and transcendence. Both are cases where Levinas appeals to motherhood, and both instances share the troubling implications of Levinas's overall invocation of the feminine. The uterine existence is still prior even to the pulsation of enjoyment and representation. The only role played by the uterus in that discussion is a kind of primitive, chaotic matrix from which the enjoyment of birth could be issued (*TI* 147). As in *Timaeus,* she is the "soft material" on which the meaningful imprint of the saying can make itself evident. This too is a deeply problematic invocation of maternity.

There is much to commend in Levinas's later invocation of maternity, which performs a number of reversals and recoveries from the treatment of the feminine in *Totality and Infinity.* But has Levinas fundamentally changed his treatment of the mother? Or does the mother remain, as in *Totality and Infinity,* pinned to the role of matrix, the role of the one who gives life but to whom nothing is owed? In *Totality and Infinity,* the woman as mother is a faceless origin; maternity is the condition for life, enjoyment, and pleasure; she holds the family to being so that the father and son can catch sight of the infinite that transcends her in all her earthiness.

This has not substantially changed in *Otherwise than Being,* where the invocation of maternity allows Levinas to express critical components of substitution and responsibility. Instead, the mother remains the quiet hand that provides the bread that the "responsible one" can give away.[31] Levinas *does* abandon the spatial metaphors of the home, the dwelling, to which the feminine was confined in his earlier work, but there still remains a time-related problem with Levinas's use of maternity in *Otherwise than Being*.

A Matrix for Diachrony

Levinas returns repeatedly to Pascal's accusation that the line "this is my place in the sun" is the beginning of the "usurpation of the whole world" (*OB* vii).[32] Levinas does not ask, however, what prepares a person to lay claim to his or her place in the sun. In *Otherwise than Being*, there is no place in the sun that is *mine;* even my language betrays my inability to lay claim to my own words, which are already heavy with the diachronous proximity of the other. So as the ego of *Otherwise than Being* awakens to discover itself blocking the sun needed to warm the flesh of another, the result is *shame*.[33] This awareness always comes too late, after one has already blocked the sun's rays. The corporeality of this analogy is palpable; I am shameful for the body that is between the other and her need. Drew Dalton summarizes Levinas's use of shame as "the weight of having to be one's self."[34] Perhaps this is similar to the sensation I have, having already squeezed onto the subway car, that *I* make the load one-too-many. What am I to do with my body when the physical nature of my flesh is already an obstacle for the other? This is the pressure of diachrony in which I find myself already in a situation where I am the last one on the scene, where I am late for my responsibility and eating food that is needed by another.

What Levinas does not investigate, at least not with clarity, is *how* I came to be placed here in the sun. It is clear, of course, that I am not my own origin; I am "created" and am here in passivity beyond passivity. But the invocation of creation offers a beginning whose causal eruption is not its primary meaning and significance. For help in this venture, Levinas turns back to Husserl whose originary, primary impression (*Urimpression*) "does not grow up (it has no seed); it is primal creation" (*OB* 33).[35] Elsewhere, in reference to prime matter in Plato and matter as cause in Aristotle, Levinas claims that "philosophers have always wished to think of creation in ontological terms, that is, in a function of preexisting and indestructible matter" (110).

The Good, for Levinas, is not to be found in investigations of being or matter, but in the other, who evades captivity into the themes and sensations of bodies and physical forces. But the said is *necessary,* just as the "plastic form" of the face is necessary to invoke the Good that is its "beyond."[36] So I stand, in the sun, *ex nihilo,* and guilty. But where did I get this body? How did I come to be in this place?

This move Levinas sometimes calls "incarnation," which is "the sensible experience of the body" that is "already and from the start incarnate" (*OB* 76). Levinas associates this initial sensibility with "maternity, vulnerability, apprehension" and marks these as indications that the self is now a character in a *plot* that precedes and exceeds apperception or any coming to consciousness (76). To be incarnate is to become aware of the "maternal relations," along with the elemental enjoyments of food and dwelling, and to know these to be the material from which one lives. Brody summarizes, "The trope of maternity is not one term among others with a detachable univocal significance, but the very reference to the density of sensibility as the one-for-the-other: sensibility as skin, contact, the caress.... The figure of maternity is not to be read as somehow sanitized of all bodily incarnation—on the contrary. It is the very figure of embodiment, flesh and blood, and sensorial contact."[37] The maternal is, therefore, for Levinas, the very conditions for being able to give bread, to cede my place in the sun. I am first in space, in a body, gifted with flesh.[38] The maternal is the elemental *par excellence,* the condition for all incarnation, that which is prior to "flesh and blood," prior to "entrails in a skin," and prior, certainly, to the bread that one can give (77).

Levinas continues to need the metaphor of the maternal, the "matrix" (*OB* 104, 106, 188), to play a neutral and preparatory role in making possible the ethical relation. Levinas writes: "The self involved in maintaining oneself, losing oneself or finding oneself again is not a result, but the very matrix of the relations or events that these pronominal verbs express. The evocation of maternity in this metaphor suggests to us the proper sense of oneself. The oneself cannot form itself; it is already formed with absolute passivity" (104).[39] The maternal is the abundant passivity upon which the struggle to maintain oneself rests. For that matter, this maternity or the matrix continues to be both the preparation and the condition for the experience of alterity and for the birth of consciousness.[40] Certainly Levinas brings maternity into the light, into representation and theme, and, in an obviously positive role, as the exemplar of responsibility: "Is not the restlessness of someone persecuted but a modification of maternity, the groaning of the wounded entrails by those it will bear or has borne?" (75). Or as Alphonso Lingis summarizes, "the figure of maternity is an authentic figure of responsibility" (xix).

As the paradigm of responsibility and persecution for the other, the maternal is *represented*. In this representation, the mother solves the puzzle that troubles *Timaeus:* she provides the khora-womb in which transcendence irrupts. Levinas compares the mystery of this relationship to the elemental the "Gordean knot of the body" (*OB* 77). I cannot say anything *meaningful* about how I found myself here under the sun without falling again into ontology, without turning over the mystery of the beyond being to something identifiable within being. This effort to recover "the origin of the sense ascribed" to the body is a "regressive movement," which Levinas knows to be doomed. One runs into a "non-thematizable alterity" that assails one from an origin older than incarnation, from before the origination of the body (77). This is a knot that cannot be untied; the maternal is a saying that is evident in the said that is my flesh. As I discover that my body blocks the sun, I am already aware that I was *born*. Because someone gave birth to me, I can stand aside and give warmth or bread to the other. And it is in *this* movement, not in my arrival from wherever to be in the sun, that I become a part of the plot that is "larger" than myself (76). For Levinas, diachrony is not a feature of bodily origins, but a feature of the ethical relation. Responsibility is the giving up of myself; but what of the one who gave me the flesh, blood, and bread?

What of the one who situated me here under the sun? Despite the obvious elevation of maternality to the height of responsibility, Levinas nowhere revokes the consignment of childbearing to a function of preethical being. This is evident not just in the event of birth, but also in unspoken roles of preparation and enablement that are being laid at the feet of this metaphorical appropriation of the feminine as mother.

The elevation of the maternal to "bearing par excellence" in *Otherwise than Being* masks an ongoing dependence on a primordial enablement that remains virtually unchanged from *Totality and Infinity*. For Levinas, the mother is needed to keep history tied together, to bring children into the world, to nourish them with milk and then bread, and to place them in the sun, where they can be confronted with the opportunity to become themselves by losing themselves. The problem is not the beauty of this metaphor for bearing others, which indeed invokes a remarkable instance in which every person was always already borne by another. The gifts of flesh, bread, skin, sunlight, and speech are indeed conditioned on the prior reception of the elemental, which moves most initially from uterine existence. But does not the ethical *role* played by the maternal in *Otherwise than Being* still confine the trope of the mother to being? We are in need, I suggest, of a way to think about the grateful receptions of these gifts that is less dependent upon the trope of maternity, as stunning as this metaphor may appear.

The concept of time in Levinas brings these problems into clear focus. By retaining the insinuation from *Totality and Infinity* that the maternal provides the *condition* for ethics, as well as for sensibility and consciousness, Levinas has continued to bind the maternal to the dwelling, to the *instant*. The mother is the condition of the body, which Levinas called "interiority" in *Totality and Infinity* (a term completely abandoned in his later works). In *Otherwise than Being*, interiority is reconfigured as a "psyche"[41] that has already been invaded and pervaded with the diachronic presence of the other. Levinas writes in *Totality and Infinity* that "to be a body is to have time" (*TI* 117). But this phrase appears when Levinas is invoking time as history, not as eschatology. To be a body requires a birth, and birth remains immanence. As the condition for the psyche, already incarnate and late to the scene, the maternal abides on this side of the transcendence of diachrony. The past that is the maternal is not diachronic but immanent, historical, and *essential*. It is a presence.

What is called for here is a reconsideration of Levinas's Gordean knot, which surely will not be untied, but needs to be addressed differently. The invocation of the maternal does not provide a satisfactory site for the mystery of the intersecting node between the saying and the said. There is richness, undoubtedly, in the metaphor of responsibility and diachrony that is available in maternity, but beneath this trope, there lies a persistent problem with Levinas's philosophy of time that remains too formal, that needs more work. Like Chanter, I doubt that this is a trivial problem for Levinas's understanding of time, alterity, and ethics. He has fashioned a philosophy of time that situates the ego outside of its own time and in the time of the other. But what are we to do with the fact that every body has a history and that the history of each body leads back to the maternal? This is the mystery of the intersection of diachronous time with bodies and faces. This is the intentionally ambiguous, unthematizeable relationship between being and otherwise than being, between the saying and the said, between the plasticity of the face and the other who is encountered there as trace. Chanter claims that the feminine is the "privileged unthought of Levinas's philosophy."[42]

Chanter argues that by failing to evaluate the way the feminine relates to time, bodies, and history, Levinas has continued to leave a gaping hole in his philosophy of intersubjectivity: "The feminine as such remains captive to its preparatory role in Levinas's work, and in this way Levinas repeats the all too familiar gesture of confining women to the hearth, the home, the private realm, and excluding them from the public, political domain, which is reserved for the seriousness of masculine affairs."[43] For the most part, and despite an otherwise felicitous reading of Levinas's work, Chanter has discovered that these texts do not liberate *la femme* from the position of enablement. To move past the trope of the maternal is to set aside neither the many other problems that this metaphor presents nor any benefits to our understanding of time and ethics that the maternal might provide. These deliberations must continue. But because it simply does not suffice to rest the node of incarnation on the maternal relation, we can at least wonder if there is another way to configure this fundamental mystery. We cannot expect it to rest on anything representable, which would only repeat the steady reliance on the maternal as preparation and condition for diachrony. Nor can we revoke the creation *ex nihilo* of the subject as responsible, at least not without resuming ontology and embracing as fundamental the ontological relation to being.[44]

Yet before suggesting an alternative, there is another problem that must be addressed within Levinas's use of the maternal, and this too is tightly interwoven with the concept of time. The problems I have already associated with Levinas's invocation of the maternal deal directly with the material history to which the maternal is tied. Yet for Levinas, the feminine may also play an unnoticed role in the future. Levinas hesitates to discuss the future, particularly in his later writings and interviews. The danger, he expresses, is that we presume *anything* about the future and therefore constrain it to the present. Even a *hoped for* future, which is so critical in *Existence and Existents,* is denied to the ego in Levinas's final writings. Diachrony is primarily a function of the past, though he does invoke, briefly, the possibility of a "diachrony of the future" (*AT* 36). Levinas, intriguingly, *does* bind the future to the maternal; death, he states in *God, Death, and Time,* is "a return to a maternal element, to a level situated *beneath* the phenomenological sphere" (*GDT* 86).[45] As Levinas expressed it in *Totality and Infinity,* the feminine footsteps "reverberate the secret depths of being" (*TI* 156). This reverberation is not goodness, or evil, but the condition for the encounter with diachrony, with alterity, with the neighbor. The maternal is earth time; the other is diachrony. This is why the maternal is not a metaphor among many, in *Otherwise than Being,* but the *very possibility of metaphors.*

Bread from Where?

One wonders about the unspoken *hope* for a future that accompanies every sacrifice of bread, flesh, and sunlight. Does not one hope to be restored to another moment, if only to age for-the-other? If I indeed find myself by losing myself, in the responsibility that takes bread from my mouth and hands it to the other, do I not love the other enough to hope that I be fed nonetheless? Only by the banalities of bread, shelter, and rest do I become myself in responsibility again tomorrow. My role in the plot that is beyond me includes responsibility for myself but *precludes* my struggle to maintain my place in the sun over and against the other. This problem is exacerbated by the presence of the third, who always interrupts the encounter with the singular other. In the periphery, I see that feeding this *one* person leaves others hungry. As I give to *this* other of the flesh and bread I have already received as gifts, must I not hope that tomorrow I can give my bread to another person in hunger? One can appreciate Levinas's reluctance to prescript the future of the other, but it hardly seems like cheating the future to hope that I might be around to again take up my place in the story that is not mine. This hope is not a subtle return of virility, or even the presumption that I can do anything helpful for the other tomorrow. The diachrony of the future has no such handles, and the ego must move into this sort of future prepared at any point to surrender its life, its obedient senescence, to the life of the other.

By neither speaking of this future nor the hope of a tomorrow in which I can be-for-the-other, Levinas fails to see that the embodiment of his philosophy depends on just such a hope. Feminine or otherwise, *someone* has to bake bread and make beds and otherwise prepare the world for the bright moment when the ego becomes aware that this is not its story after all. This is a question of hope, of dependence, and of restoration, and it receives little attention anywhere in Levinas's work. In the unrestricted gift of oneself to the other, do I not depend on someone to bandage me, feed me, and give me rest? His reasons for reluctance are obvious; *restoration* is almost always configured as a for-itself. But can restoration be reconfigured as an event both from and for the other? To give my bread and my flesh to another is to *hope* that despite the thoroughness of this gift I might be resurrected, restored to another moment such that I might give again. This is not a redemption of the ego or a restoration of the subject to its own story, but a gratitude for a renewal of diachrony and another moment in which to be for-the-other.

No sooner have I given away my bread than I already feel the rumbling of my stomach. As soon as I absorb the violence intended for another, the flow of blood cries out for bandages. These are not secondary, peripheral, or insignificant components of the way I confront the destabilization of diachrony. I am ripped out of my own time, but not out of my own body, which remains the condition for my responsibility. And these restorations require persons, enfleshed bodies whose hands nourish and redeem me. In *Totality and Infinity*, Levinas at least acknowledges the existence of these hospitable persons, naming them *feminine*. But in *Otherwise than Being*, they are merely presumed, assumed. This is a dangerous assumption, for it overlooks the manner in which I *hold the other hostage*. Does not my blood and my hunger arrest and captivate the other who finds herself or himself diachronically in debt to *me*? Should not my "responsibility for the responsibility of the other" include this phenomenon (*OB* 117)? Levinas insists that despite all of his emphasis on hyperbolic responsibility, these obligations fall uniquely to the self, "for in me alone can innocence be accused without absurdity. To accuse the other, to ask of the other more than he owes, is criminal" (195; cf. *EI* 98–99).

Is it not then criminal to overlook the person whom I presume when I hand over bread and cede my place in the sun? Perhaps this is an unspoken *fourth* figure in Levinas's interpersonal dynamic. The first person is the other, the second person myself, preceded by the first person from time immemorial. The third is the also-needy face that breaks the enclosure with the other and summons forth philosophy and justice. Levinas devotes some attention in his later works to the third person, who complicates, destabilizes, and irrupts within my relation to the other. But is there not yet another person to consider, perhaps unnoticed in this said, who is the representation of the interplay of persons? Quite often, in the sincerity of a sacrificial act, one depends on such a fourth quite unconsciously. One hardly notices the kindness of strangers until disaster befalls, and suddenly hands one has never anticipated are mending and restoring the body that is uniquely mine. Is this not an *other*? Is this other not the diachrony of a *future* beyond the selfless renunciation of *my* hope?

This is obviously not a generous reading of Levinas's updated invocation of gendered bodies in *Otherwise than Being*. Yet it seems necessary to press toward an application of his diachrony that addresses the unseen fourth, the other who restores me despite myself. Levinas's thought presses us beyond the invocation of gender, toward another way of understanding the diachrony of the future. This diachronic future is not a hope for the self, for my ageing, but a hope for the other.

This is not a dismissal of sexual difference, not a reduction to sameness, nor any kind of statement whatsoever about the function of the feminine in philosophy.[46] Neither do I wish to challenge or dismiss the intriguing appropriations of Levinas's thought by gender theorists or feminists who find his invocation of the maternal to be promising or helpful.[47] However, gendered difference cannot and should not be the pivot point between being and being's otherwise, the point-of-contact that allows the infinite to appear in the finite. To move beyond the maternal expressions for this intersection is not a failure to express gratitude for the rich sense in which every life is bound to the maternal. Hannah Arendt called for a natality, a gratitude for being born; Levinas at least demonstrates a rich way to think about such gratitude.[48] But as a metaphorical appeal to the primordial, uterine existence that is the condition first for enjoyment and then for diachrony, the trope of the maternal is problematic. The intersection between embodied history and diachrony must rest otherwise than on gender and childbearing. Others may find ways to breathe new life into Levinas's structural invocation of the maternal or the feminine, but these avenues seem to be fraught with perilous representations and more "consternation and pain." There do appear to be numerous other possibilities for thinking about the node that makes possible the encounter with the other and this otherwise than essence. One possible direction for further diachronic time of responsibility and history might be reconfigured under a more embodied sense of Levinas's *à-Dieu,* the appeal to-God.

God and Philosophy

Merold Westphal claims that Levinas presents "a philosophy of sin without salvation."[49] This is an intriguing claim, with no small degree of accuracy. The "I am guilty" is abundantly evident in Levinas's philosophy, and not just guilty for what I have done, but for all suffering in the world. This divine burden does not offer any evasion or propose any slackening of the guilt that accompanies the gravity of being in the world. If there is salvation to be found, it is not for *me* as a divided self, who rises above the ranks of the unsaved by any belief or activity. I remain, for Levinas, bound to the suffering of the other, and the suffering of *every* other. If there is a hope for salvation, it must indeed invoke hope that is configured as messianic: a hope in a time where there will be no more suffering, or more radically, a time when suffering *will not have been.*

Westphall's claim distorts, however, the sense in which Levinas proposes that responsibility is the "very fact of finding oneself by losing oneself" (*OB* 11). The concept of an individual somehow ascending from guilt and shame to some salvation from these features is not to be found in Levinas, but recall that for Levinas the narrative plot with its desperate *conatus essendi,* which would fight tooth and nail for self-salvation, has been revoked. He believes there is no plot where *I* am justifiably concerned about my salvation. Salvation has always already arrived in the diachrony of the other and lies ahead in a restoration that is only ever a hope for-the-other. Levinas, in fact, levels a number of critiques against mysticism and spirituality that might focus on the salvation of the self. How are these not reproductions of being's economy?[50] For Levinas, the struggle of a being for its own salvation is a struggle against the face and against the God who in the face of the other comes to mind. Levinas's philosophy, if nothing else, eliminates the need to struggle for one's own salvation. The transition from the plot that is natural for the *conatus essendi* is characterized by shock that Levinas construes as traumatic.[51]

Levinas invokes the name of God carefully but regularly in both his philosophical and confessional writings. Christina Gschwandtner points out that for Levinas, God is "the only word that cannot be reduced, that always escapes our grasp."[52] Levinas claims that the word God "neither extinguishes nor smothers nor absorbs its Saying" (*GDT* 204). God, who comes to mind in the face of the other cannot be refused in the same way that I can refuse the human other.[53] The other who stands before me is flesh and blood, participating in the economy of being as incarnate. God, on the other hand, is the debt that I bring with me even as I evade the other. God is the saying in this said that is flesh, the trace that my evasions can ignore but not efface. Because God never falls into a said, this word erases its own utterance, denying the rules of representation. God is the purity of an election to responsibility from time immemorial, from time that is anarchically prior to an establishment of the self. There can be no evasion of God because this diachrony is the very condition of my existence.

Despite this seemingly theological language, Levinas remains adamant that he is doing philosophy and *not* theology (*BPW* 30). Levinas believes that theology is "linked to ontology" (*GDT* 204; cf. *CPP* 153–55). To speak the name of God, Levinas acknowledges, is to run the risk that one might invoke the powerful God who acts as a "protector of egoisms" (*OB* 161). Yet for Levinas the idea of God functions as the permanent summons to justice, forever irrupting closed societies for the sake of a universal justice. God is therefore a "third person," unsettling any reversion to egoism and any idolatry that would ossify God into representation (*GDT* 203). Levinas appeals to the concept of God to address the need for a common life together that is bound by a logic, a "reason" other than the reason that synchronizes being (*OB* 160). Levinas points to this concept of God without theological force or a clear inclination of how one might utilize these invocations of God to face the daunting relationship between the ethical and the political. It is politics, Levinas teaches us, that drives us back to being. The exigency of politics arises from the haunting sense that even as one helps a neighbor, a third person stands hungry in the periphery. To attend to the suffering of one other is to be aware of a fraternity of others that her face already announces. This is not a new concern for Levinas, who worried about this problem of a "closed society" even in the 1950s (*CPP* 31). As Cohen summarizes in the preface to *Otherwise than Being*, "to give all to one is to leave others destitute" (xvi). The face-to-face encounter is troubled by the absent face.

The important idea of God, which plays a key role in Levinas's final philosophical moves, is bound up in diachrony. The to-God, or *à-Dieu*, of Levinas's philosophy presents us with an intriguing puzzle. According to Derrida, Levinas shared that his deepest interest was not ethics but *holiness*. And at the point where we find Levinas closest to discussing the question of politics and the "third person," he invokes God. It is only, writes Levinas, "with the help of God" that we can hope for justice (*OB* 160). In the face of the difficult question of politics, Levinas appeals to the God who is the "original locus of justice, a terrain common to me and the others where I am counted among them" (161). Such an idea, if it is truly the idea of God, is for Levinas immediately opposed to any egoism.[54] Any configuration of politics that does not arise "with the help of God" is doomed to the return of egoism, violence, and the synchrony of the same. What Levinas hopes for here is not the synchronization of history beneath the time of the ego, but a peace that irrupts diachronically for all humans.

It is in the concept of God, which can be read alongside Levinas's earlier invocations of the infinite, that he can introduce, at last, something that the subject, the other, and the third person have in *common:* diachrony. When Levinas invokes the idea of God, he does it with an awareness of the synchronization that is involved in this claim. The pursuit of justice is therefore the pursuit of a sociality where egoism has been abandoned. Only, finally, in this situation will Levinas claim that "my lot is important" (*OB* 161; cf. 158). The one on whom such a synchronization rests is God, for the concept of God defies any synchrony within the horizon of the ego. This justice for which one can work and hope is therefore riveted to the saying in its unapproachable purity. One cannot thematize the God who is what all of being has in common; the God who comes to mind is "always subject to repudiation" (161).

This repudiation is the evasion of God from the time of history, from time that would be characterized as anticipation, and from every hope that might be configured according to the perspective of the self. This makes God the permanent unsettling of the structures of the said, the laws and governments of history. God is therefore the permanent trace of the "to come" (*avenir*) in the "now" (*GCM* 95). Increasingly reluctant to speak of the future, Levinas feared that the future would forever retain the hope of consolation for the self, the "future according to the manner proper to me" (95–96). As such, Levinas says we should direct our attention toward the future in "passivity or patience without assumption," as an "awaiting without an awaited" (95). To wait for the future is to wait for God, to wait for the purity of a saying that is already too congealed in any approximation or anticipation we might direct toward it. Cohen considers Levinas's final move in his development of time to be the *à-Dieu.* He claims that Levinas's idea of God "provides the ultimate support for the dimensions of time."[55]

This approach to the future, to the messianic, plays a quiet role in Levinas's late philosophy; he admits that it is obscured by his interests in the irruption of responsibility from the diachronic past (*GCM* 95–96). But might we hold to Levinas's considerations about diachrony and wonder if there is another way to think about both the diachronic past and the diachronic future that takes into account the other person I have called the "fourth"? Ongoing work in this direction needs to address, better than Levinas, the node of intersection between holiness and being. Such developments must also seek a more vigilant awareness of the fourth, often unseen other into whose hands I trust myself when I act in responsibility. Both of these developments relate to Levinas's problematic invocation of the feminine, and both relate directly to Levinas's understanding of time.

Levinas's ethical philosophy rests precariously on his understanding of maternity and the feminine. Vasey outlines this ongoing dependence: "For Levinas *woman* is the always already forgotten other, the taken-for-granted, hence always overlooked other; without her, there is no being-at-home, no dwelling, no enjoyment of the elements, no separation, no consciousness of, no encounter with the serious face of the other (who is, by implication, essentially masculine)."[56] What makes possible the encounter with the other, the face, the diachrony of a past and future that are different from the natural history of the ego? What arrests the *conatus essendi* and reverses its concern for its own survival? Levinas approaches this question in a number of different ways. From his phenomenological perspective, it is most compelling to simply analyze that such an irruption of holiness takes place in the flesh. In *Totality and Infinity,* Levinas utilizes his analysis of the face in this way, though he is compelled to attempt an explanation of *how* the ego might become capable of such an encounter. In "The Trace of the Other" Levinas offers another way to demonstrate this point of contact; in the present, in my present, the other has left a trace of "an utterly bygone past" (*BPW* 60). Then in *Otherwise than Being*, Levinas builds on these insights and points to evidence for the intersection of being and its "otherwise" in the diachronic structure of language (*OB* 6). The traces of the saying in the said mark the site where being gives way to the Good. Levinas is clear in all three of these expressions that the flesh is the site for this irruption. And as we have seen, he leans unrelentingly on maternity to bring into the world persons who are capable of encountering this diachrony. This distorts and limits the feminine, thematizing what is among the most astounding and mysterious events of all: that women give birth to ethical beings.

However, there is no *reason* at all for this intersection, and Levinas's attempts to establish a reasonable sight for the irruption of holiness in the world only succeed in obfuscating the mystery of the trace, the mystery of diachrony. Perhaps for fear of dogmatism or leaving behind philosophy, Levinas persistently seeks a way to express the philosophical condition for the encounter with the other, just as Timaeus went in search of a similar medium.

Chanter claims that "to thematize the role of the feminine in Levinas's philosophy is to produce a philosophy of mediation and to undercut the radicality of God's alterity."[57] The feminine must remain unthematized, but not by way of introducing another mediation that replaces the role of the feminine as the grounded possibility of the irruption of diachrony. The problems inherent in utilizing the feminine as a trope do not call for a new representation, nor the tweaking of old metaphors to better suit this function. Both Timaeus and Levinas work too hard to make the possibility of transcendence understandable, and in so doing consign the feminine to connect being with its otherwise.

Rather, there is no rational *matrix* that is the condition for the irruption of the face within being. There is no philosophical explanation for the appearance of the saying as trace in the said. This intersection, the node, like the saying itself, evades representation and does not belong properly to essence or meaning. Levinas utilizes the feminine to explain the appearance of this possibility in being, but perhaps being simply does not produce this site, through the feminine or anywhere else. The node itself would then be received as pure gift. This would mean the intersection between holiness and history is made possible by being's otherwise, God. Other terms are possible, and perhaps preferable. What is critical about this suggestion is that it implies an abandonment of the pursuit of any logic that might make sense out of the *possibility* of diachrony. The node of intersection is itself already diachronic, arriving *ex nihilo.* Perhaps future research could focus on the *bare fact* of the irruptive event within being without needing to tie this site back into gender and maternity.

This bare fact is the condition for the Good and the appearance of diachrony. It would be more faithful to Levinas's project to suggest that there is no *how* that explains this node of contact between the saying and the said; it is beyond the reach of philosophical investigation. Such claims, and the utilization of religious language, need not indicate any God who operates "behind the scenes" (*OB* 8). But the uncreated capacity of being to open to its other is a good thing and is best named as such. This node makes the impossible possible. Despite its predisposition to violence and economy, fleshed bodies can nevertheless be sites for the irruption of the holy.

Philosophy and Lamentation

The stark and harsh themes of *Otherwise than Being* have frequently and justifiably absorbed much of the attention directed toward this book and Levinas's later writing in general. There may be, behind the glaring concepts of hostage and persecution, a subtle theme that illuminates both this text and what it means to work in the wake of Levinas. *Otherwise than Being* opens with a dedication to the "memory of those who were closest among the six million assassinated by the National Socialists, and the millions on millions of all confessions and all nations, victims of the same hatred of the other man, the same anti-semitism" (*OB* v). This dedication is more than just a reverent gesture that stands in distinction from the text that follows. The issue of memory is frequently invoked in *Otherwise than Being*, often juxtaposed with the loaded term "history." Memory relates directly to the concept of time, and in this case serves as an antecedent to the whole book. Furthermore, Levinas speaks not of memory in abstraction, but the very intimate memory of *those who were closest.* This memory is *his* memory, a secret he otherwise guards well. Yet in light of the dedication of *Otherwise than Being,* perhaps there is a global reading of this book that can be helpful in thinking both about and beyond Levinas's latest writings.

The dedication of a book about absurd, hyperbolic responsibility is offered to the victims of atrocious violence and genocide. The act of dedication is therefore itself a declaration of lament. The hope of this book is undeniably a future that is justice and peace. The purpose of the book, and of Levinas's career for that matter, is the prevention of such atrocities in the future. Yet the tenor and focus of *Otherwise than Being,* and many of Levinas's later writings, is undeniably on the past. The sentiment, emphasized by the vernacular of diachrony and its inflection on the anarchic and irrecoverable past, is somber. Some misunderstandings may develop from this fixation. If, as Levinas repeatedly emphasizes, one is caught up in the drama of the other person, then shouldn't the result be both joy and suffering? Am I not hostage to the other person's happiness as well as his or her pain?

This theme is not absent in *Otherwise than Being,* but neither is it emphasized. It is the other's "destitution" that receives nearly all attention. The abiding fixation on the suffering may be related, in part, to the register of *lament* to which this book is attuned from the outset. The guilt of the survivor, a burden Levinas undoubtedly carried, is unrelenting. Beneath the personal nature of this dedication there may be a way of reading *Otherwise than Being* that better disposes readers to its content. What if the entirety of this enigmatic text is an act of lament?[58] And what if the proper response to the impossibility of being responsible is best approached in the posture of lament?

The act of lament, of mourning, and of testifying to my impossible responsibilities disposes me in a new way to the other person, encountered in any and every setting. Levinas provides here something noticeably absent in *Totality and Infinity:* a way to operate in the economy of being despite my impossible disadvantage, guilt, and obligation. I can mourn. I must mourn. I can encounter the doctor, the banker, the pilot, the baker and the plumber within the obvious economies of these encounters. These encounters are overwhelmingly frequent. They fill my world full of faces that press on me from time immemorial, bottomless stories that are both *prior* to me and also my immediate *priority*. I must sigh and let the mail carrier walk away without even asking about the pain written on her face, perhaps evident in her brief words. My baby cries from the other room. The phone rings. A pipe leaks. A student emails. My stomach grumbles. There is not enough time. I am guilty, obliged, caught, and captive to the suffering of the mail carrier, but I must choose and attend to some of my responsibilities even as I mourn the responsibilities to which I cannot attend. The choice may be arbitrary, or it may be clear. But for Levinas, even when I cannot do anything to help the other because I am caught up attending to other responsibilities, I remain nonetheless hostage to the suffering of the other. I remain responsible. Lament both encapsulates the remorse for these insurmountable responsibilities and the ongoing need for vigilance and even hope.

Otherwise than Being reminds us that all encounters with other people are diachronous. Every encounter is tardy, not just the ones we are careless about, or the ones that obviously conform to the economy of being with its synchronous clocks and calendars. Something also needs to be said about the *mood* in which I encounter the other, for in *Otherwise than Being*, Levinas meets the suffering of the other in a particular state-of-mind, that of mourning. This mood is not an internal orientation generated and sustained by the self, but is itself a gift from the other. Levinas's work on time and diachrony help to illuminate the role of lamentation as the attunement of *Otherwise than Being*.

The concept of "mood" received some close attention from Heidegger, both in *Being and Time* and afterward (*BT* 172–77).[59] For Heidegger, the *Stimmung* (mood) is a way of talking about Dasein's state of mind (*Befindlichkeit,* literally "state in which one may be found") (172n2). For Heidegger, a mood is something one discovers as already present in the world of Dasein. Dasein is always in a mood, and that mood is always Dasein's own, even as it arises as seemingly foreign or exterior. Mood carries being to Dasein, and it is in the disclosure and mastery of moods that Dasein understands itself (172–77). "A mood assails us," writes Heidegger, "It comes from neither from 'inside' or 'outside' but arises from being-in-the-world, as a way of such being" (176). Levinas, of course, contests Heidegger's appropriation of *Stimmung* from as early as 1935 in *On Escape* (*OE* 66–67). From the outset, the hermeneutical circle proposed by Heidegger, where Dasein comes to know the meaning of being by way of its own mood, troubled Levinas. It is from this perspective, which leaves one "enclosed in a tight circle that smothers," that one dreams of escape (66). The *Stimmung* that Levinas appropriates there is *nausea,* as the vertiginous situation where one's only frame of reference is the self. Levinas does not deny that the self is always found to be in some *Stimmung* or another. Rather, in *On Escape,* he denies the circularity between moods and the self. This denial will escalate to radical proportions by the time he writes *Otherwise than Being* almost four decades later.

For Levinas, one discovers in the analysis of that which ossifies into a said vestiges and traces of a saying that has already past and now remains in the mode of an echo. Mood itself is a said, at least inasmuch as it is analyzed and identified. There are intriguing correlations, in fact, between Heidegger's discussion of mood and Levinas's discourses on the saying and the said. For Heidegger, the awareness of a mood is not its mastery, nor does Dasein ever circumvent a mood to fully understand what it discloses. But the disclosure of moods for Dasein is always self-disclosure. For Levinas, one discovers quite the opposite, in naming a mood or in uttering words we declare a said that echoes and reverberates with a saying of the other. For Levinas, my mood is already itself something given; it is not my own, even if I deem myself to be the author.

Levinas writes very little about moods, particularly in his later work. If there were a Levinasian philosophy of mood it would surely focus on the attunement or disposition of the self toward the other that is utter passivity. My mood has already been given to me in a passivity that precedes passivity. I have a mood that is *older* than me, a disposition toward the other that is already a gift from the other. Heidegger suggests that we fight moods with "counter-moods," and perhaps this is the way we silence the saying in a said, muting the reverberation of the gift that is mood (*BT* 175). The mood of *Otherwise than Being* is undeniably somber and reverent, but is not this intonation of the text itself a testament to the others for whom the book is dedicated?

Otherwise than Being—and Levinas's latest work only continues this trend—does not allow Levinas to become a historical footnote on the Holocaust. It attempts to describe the situation of philosophy as itself a kind of mourning over a past that philosophy cannot console. Levinas points out that consolation is not philosophy's role; philosophy's focus remains on the "immemorial past," which is a punishing and rigorous fixation that never relents (*GCM* 96). He leaves consolation, and the hope of the future, mostly to religion. Philosophy's mood, therefore, is determined not by its own said but by the saying to which philosophy must be attuned, constantly gathering remnants of the passing of the other for which philosophy is already impossibly late. The tardiness of philosophy is not, for Levinas, a technical or academic reality but a sorrowful state. Philosophy is a kind of faithful dirge, a persistent mourning for my lamentably belated arrival in the drama of the other person. The mood of lament is attentive and vigilant, aware of the incapacity of the self without caving to the urge to despair of responsibility.

Diachrony and Restoration

Lament can also be stirred by hope. Levinas's use of maternal themes, combined with his nearly exclusive emphasis on the anarchic past in his later writings, have conspired to obscure the other person I have named the "fourth." This fourth person is taken for granted, overlooked, and often burdened without warning or request. Chanter points out that the feminine "plays a silent role, baking bread, and providing a home for those who give bread 'taken from one's own mouth,' even in those texts of Levinas that do not name women as such."[60] Levinas suggests I cede my place in the sun and the bread from my mouth, but he does not identify the bodies, feminine or otherwise, who through labor or sacrifice gave these gifts to me. This problem I consider similar to the other whose future escapes all anticipation but who may catch and restore me, even from my sacrificial movements of responsibility. The idea of the fourth person can only be sketched here, though it presents one potential for future development in Levinas's understanding of time as diachrony.

The fourth person appears in the often unidentified hope that new bread will find its way into my hand, as restoration for a senescence that is for-the-other. For the person who turns over bread and blood in the hemorrhaging of one's life for the other, hope is hope for the impossible restoration of blood and food and life. These restorations are not egoisms, a hope for oneself, but a hope for the other. With each gift of bread and warmth, I give myself over to death. I do not rush to death, for my death is not my own. To live again despite this forfeiture is a "duration despite oneself" (*OB* 52). To stave off death, by eating and sleeping and working and breathing, is to live in *obedience.*

Suicide, on the other hand, abandons the possibility of helping another. That I might be helpful to the other is not to be presumed; this too must be given over to hope. But suicide, conceived through diachrony, puts to death the other that is always primordially present in the I. The "sin" of suicide is not the violation of a divine decree but the abandonment of the other. So I live, I breathe, and I depend. I hope in a future that is *peace,* and I hope that I might be a site where history is opened by the diachrony of the holy. I cannot see any way to act in hyperbolic responsibility without risking that some other will be burdened with my blood or my hunger.

To rely on the fourth person is to repeatedly concede life and yet hope that one's life might be restored. This hope certainly relates to the banal way in which my belly rumbles and desires food. But because my life takes place in the time of the other, in the diachrony of my relationship to the other who anarchically precedes me, my restoration to life is a restoration to the other. I give to the other and hope that my life will be restored by the gift of another moment, another day, another loaf, but all for the sake of the other. This is a hope, perhaps the "birthplace of hope" (*GDT* 96), that begins and remains always a forfeiture of my history. This is the hope that I will breathe again tomorrow in order to be for the other under yet another sun, and even the hope that my senescence is what the other needs. This is hope as diachrony, which is always hope *for the other,* even as this hope for the other is established within the life that appears to be mine.

To hope in this way as a community again raises the concept of justice, which Levinas rests precariously on a communal embrace of the evacuation of the ego (*OB* 160). Traces of *politics* abide here, in the question of how one might be thought to live in the world of laws and judges and justice. Such a pursuit follows Levinas in routinely playing at the unstable borders between ethics, religion, and politics.[61]

At the conclusion of this analysis of the progression of the concept of time in Levinas's work, it is apparent that Levinas arrives somewhere very different from his beginnings. The theme of time consumes his career at many stages and ends up being critical to his philosophical argument for radical responsibility. Diachrony is the rupture of time that makes possible the encounter with the other; it is the mysterious irruption of the other that is already at work before I am even aware. Levinas's philosophy of time rests on the bodies who find themselves irrupted and interrupted by a time that is not their own. The interruption of the present, of the self, is an event so disconnected from the horizon of selfhood that it calls for the "idea of God" to be inserted again into the vocabulary of philosophy. Such an event is no proof of God's existence, but a trace of the "fall of God into meaning" (*EN* 173). The *à-Dieu* is not just the mystery of transcendence, but also the mystery of the flesh. The way of devotion, "responsibility for the other [person]" to the point of death, becomes for Levinas "the *à-Dieu* of theology!" (173).

From the vantage point of this conclusion, I find a striking resonance with one of Levinas's very first independent essays. In his prescient reflection on Hitlerism, Levinas invoked the term *grace* to describe the miracle that allows for forgiveness. It is the genius of Christianity and Judaism, Levinas wrote, that forgiveness breaks the chains of time and defeats the clock time that would otherwise bind sinners to their fate (*UH* 14). The problem, Levinas contended even in 1934, is that Western philosophy and society *forgot* that it was relying on grace and usurped the concept, presuming the freedom and liberation that forgiveness allows. This forgetfulness is the *presumption* of grace, the forgetting of one's dependence on God to rupture time. In its forgetfulness, Western philosophy unwittingly presumed grace as a property and turned toward the world as though we owed nothing and were free to make the world into whatever we wanted. This is the philosophy of Hitler.

Much work remains to be done in this direction, exploring the marvel of grace and restoration, the undetected fourth person, and the mysterious node that makes possible the irruption of the Good within the fleshed bodies of history. At the heart of Levinas's final reflections on time is the complex concept of diachrony. The other who supports, who forgives, who restores, and who catches me, is owed *diachronically*, even before I am restored. Diachrony neither neglects nor loathes the flesh, which is the site of the appearance of forgiveness. Debt as diachrony is finding oneself torn up from the history to which the ego naturally clings; the subject exists between times, suspended between the past for which I am responsible to the other and the future that I hope to give to the other. Debt carried by diachrony is not just debt for a past too old for my remembrance, but for a future restoration beyond anticipation.

Notes

Notes to Introduction

1. Chanter, *Time Death, and the Feminine,* 1.

2. Indeed, chronological studies on Levinas's work are also strangely sparse, despite a growing interest in his work, particularly in the Anglo-American context. As Samuel Moyn notices, "Surprisingly, the origins of Levinas's thinking have never... been studied carefully by intellectual historians, with a method that calls for illuminating a body of philosophy by reading it chronologically and understanding it contextually" (Moyn, *Origins of the Other,* 4). Moyn attempts a chronological study not of Levinas and time, but of the relationship between revelation and ethics in Levinas. This book investigates the chronological development of Levinas's thinking about time, beginning with his first independent essays and concluding with some of his final essays.

3. Cohen, *Elevations,* 133–61; Wygoda, "Phenomenology of Time," 283–301; Tauber, "Outside the Subject," 439–59; Bernet, "L'autre du temps," in Levinas, *Positivité et transcendence,* 143–63; Robert Legros. "L'expérience originaire du temps, 77–97. Paul Olivier wrote an article in 1983 proposing to "compare" the theme of time to the main themes of Levinas's thought (Oliver, "L'être et le temps chez Emmanuel Lévinas," 337–80). Though Oliver's study traces many of the themes expanded upon in this volume, I am staking the claim that time *is* a "main theme" in Levinas's oeuvre.

4. Wygoda, "Phenomenology of Time," 283.

5. Tauber, "Outside the Subject," 439.

6. Chanter makes this case not only throughout *Time, Death, and the Feminine* but also in her essays "Hands that Give and Hands that Take: The Politics of the Feminine in Levinas" and "Conditions: The Politics of Ontology and the Temporality of the Feminine."

7. Rose, *Broken Middle,* 252.

8. Llewelyn, *Emmanuel Levinas,* 5.

9. Jeffrey Bloechl points out that it is a "dangerously misleading premise" to suppose that Levinas foresees in the beginning what he will reveal in detail in the end (Bloechl, *Liturgy of the Neighbor,* 12).

10. Lingis, translator's introduction, *EE* xxii. It is significant that Lingis makes this note in introducing *Existence and Existents,* as it is Levinas's first book that is a development of his own philosophical project.

11. Bloechl, *Liturgy of the Neighbor,* 5.

12. There is little agreement regarding the translation of Levinas's French terms *autre, Autre, autrui,* and *Autrui.* Levinas vacillates in his capitalization of these terms, often without rhyme or reason, and this has perplexed translators and confused readers. I have followed the later strategy of Alphonso Lingis who,

in translating Levinas's *Otherwise than Being,* renders all four terms as simply "other." There is some risk of losing intonations that Levinas intended with the capitalized terms, but this risk is preferable to the danger of adding mystical intonations where they were not intended.

13. Levinas has essentially left us three levels of primary sources. First, we have the philosophical texts that Levinas published with careful deliberation. Second, we have religious and confessional writings that Levinas wrote for a mostly Jewish audience. While my focus is on the former, the confessional writings turn out to be critical in investigating Levinas's understanding of time. Levinas considered these to be a separate genre, yet they appear to gestate ideas and terms that Levinas later introduces to his philosophical writings. I introduce them carefully and avoid building philosophical arguments from the claims he makes in his confessional texts. Third, there is a significant collection of interviews, anecdotal stories, and private notes that express his philosophical positions in less polished forms. Levinas demonstrated a deep willingness to participate in conversations and to have his interviews published, despite the obvious risks involved when a person speaks without preparation. There are multiple volumes of Levinas's interviews. Some of these are helpful, and some leave a troubling trail of idiosyncratic opinions. These were risks Levinas undertook willingly. The notes and interviews are often lucid and candid, a quality sometimes lacking in the more polished philosophical writings. Their clarity can confirm the genesis of certain elements as they appear in the published philosophical texts.

14. The publication of *Carnets de captivité et autre inedits,* volume 1 of Levinas's *Oeuvres,* allows for fresh insight into Levinas's thoughts between 1937 and 1950. I have included references to this volume only as support to the themes from his published writings, and I translated them into English myself, with the kind assistance of Kurtis Jardim and John Fraley.

Notes to Chapter 1

1. "You have kindly agreed to take charge of this task of explication, and you have accomplished it in a most remarkable fashion. You have included my modest essay—in which the consciousness of having no way out was tied to the determined anticipation of impossible new thoughts—in the context of great contemporary ideas. You have made major voices resound in counterpoint to my lines, as you transformed them into echoes of great human whispering. Your generous attention has succeeded in extracting from my words—already then growing silent—the forebodings they still harbored." Levinas's correspondence to Jacques Rolland, December 1981, titled "Letter from Emmanuel Levinas," in *OE* 1–2.

2. Kleinberg, *Generation Existential,* 19.

3. Levinas, "Meaning of Religious Practice," 3. The material in this essay was drawn from Levinas's interview on the French radio program *Voix d'Israël* on April 9, 1937. The interview was originally published in French in *L'Univers*

israélite 37 (May 1937) and can also be found in *Les Cahiers du Judaïsme* 6 (Winter 1999–2000): 74–75. Here, Levinas also discusses a good number of religious principles that he will later abandon or critique.

4. William Richardson, "Irresponsible Subject," 127.

5. Cohen, *Ethics, Exegesis and Philosophy,* 27. Levinas comes to regard Bergson as anticipating the work on time performed by Heidegger. Howard Caygill points out that Levinas "also acknowledged [Bergson] as providing the means to go beyond Heidegger" (Caygill, *Levinas and the Political,* 12).

6. Tina Chanter notes, "Heidegger reiterates a number of times that no one, including Bergson, departs in any significant way from Aristotle's conception of time. . . . In his earlier lectures, Heidegger also maintains that Aristotle's concept of time is retained throughout the tradition, including Bergson" (Chanter, *Time, Death, and the Feminine,* 266n35).

7. In the interviews Levinas gave later in life, he gave much credit to Bergson for his helpfulness in rethinking the concept of time. In a 1986 interview with François Poirié, Levinas credits Bergson with thinking "a little beyond being and otherwise than being, all the marvel of diachrony" (*IR* 31).

8. Husserl, *Logical Investigations,* 1:263.

9. Dermot Moran summarizes Husserl's attitude toward presuppositions as follows: "Already in the First Edition of *Logical Investigations,* Husserl presented phenomenology as pure, *presuppositionless* science of consciousness. The claim, as we have seen, means first of all that phenomenology cannot assume or utilize the results of any other science in its investigations. . . . Husserl made more and more radical claims about the nature of this freedom from presuppositions" (Moran, *Introduction to Phenomenology,* 126).

10. Sokolowski, *Husserlian Meditations,* 138.

11. Malka, *Emmanuel Levinas,* 37.

12. Levinas will later say of his debt to Heidegger, "This is the case, whenever the debt of every contemporary thinker might be to Heidegger, a debt that he often owes to his regret" (*GDT* 8). Elsewhere, Levinas names *Being and Time* "one of the finest books in the history of philosophy" (*EI* 37).

13. Chanter, it bears repeating, sees Heidegger working on time despite the appearance that he moves on from this project (Chanter, *Time, Death, and the Feminine,* 264).

14. Collins, *Introducing Heidegger,* 96.

15. Malka, *Emmanuel Levinas,* 35.

16. Levinas writes in the essay "Freiburg, Husserl, and Phenomenology" (1929), "To be sure of having a seat at his five o'clock lecture in one of the largest halls at the university, I had to retain it by ten o'clock in the morning at the latest" (*UH* 64).

17. Chanter, *Time, Death, and the Feminine,* 1.

18. John Caputo summarizes Heidegger's assessment of Plato and Aristotle as such: "The temporality of Dasein—its experience of what is now—operates behind the back of ancient ontology." Heidegger's investigation seeks to interrogate this

omission. Caputo continues: "It is in Plato and Aristotle that we see where the move to think Being as presence was first made, to which everybody else thereafter just consented without question" (Caputo, *Demythologizing Heidegger,* 13).

19. Heidegger is quoting from book 4 of Aristotle's *Physics.* For context see Aristotle, *Physics,* 220a24.

20. Heidegger's critique of the traditional concept of time is nicely summarized by Chanter in *Time, Death, and the Feminine,* 25–27.

21. "How is 'time' in its course to be touched even the least bit when a man who has been present-at-hand 'in time' no longer exists? Time goes on, just as indeed it already 'was' when a man 'came to life'" (*BT* 477).

22. Aristotle, *Physics,* 219a1–5.

23. See ibid., book VI, part 7. "Movements that have simultaneous limits have the same time, yet the one may in fact be fast and the other not, and one may be locomotion and the other alteration; still the time of the two changes is the same if their number also is equal and simultaneous; and for this reason, while the movements are different and separate, the time is everywhere the same, because the number of equal and simultaneous movements is everywhere one and the same" (223b6–12).

24. Ibid., 223b5.

25. Heidegger, *Metaphysical Foundations of Logic,* 204.

26. Ibid., 204.

27. Ibid., 204.

28. Hodge, "Ethics and Time," 127–28.

29. For a strong and lucid critique of Heidegger's thought and writing, see Walter Kaufmann, *Nietzsche, Heidegger, and Buber.*

30. Caputo, *Demythologizing Heidegger,* 17.

31. Ibid., 4.

32. For a discussion of Heidegger's turn away from the promised second half of *Being and Time,* see Crowell, "Metaphysics, Metontology, and the End of *Being and Time,*" 307–31. Crowell carefully outlines Heidegger's turn to "metontology" and the consequences this turn has for the original implications of *Being and Time.*

33. Spiegelberg, *Phenomenological Movement,* 360. See also, Kisiel, *Genesis of Heidegger's* Being and Time, 1. Kisiel quotes Spiegelberg approvingly for identifying the "absence of the projected second half" of *Being and Time.*

34. Heidegger writes, "Time must be brought to light—and genuinely conceived—as the horizon for all understanding of Being and for any way of interpreting it" (*BT* 39). In light of Levinas's eventual attack on "light," Heidegger's unflinching use of vision and light to refer to time is striking.

35. The two translations of Levinas's essay cited here are by Nidra Poller ("Some Thoughts on the Philosophy of Hitlerism," in *UH*) and by Seán Hand ("Reflections on the Philosophy of Hitlerism," abbreviated as RPH). The sometimes stark differences in translation are evident in the different English titles the translators used. Both translations are reasonably good; I have moved between

the two translations in this discussion according to the renderings that seemed most straightforward. The original article first appeared as "Quelques réflexions sur la philosophie de l'hitlérisme" in 1934.

36. As Homer quotes Hecuba, mother of the slain Hector, "this is the doom that strong Fate spun out, our son's life line drawn with his first breath—the moment I gave him birth—to glut the wild dogs, cut off from his parents, crushed by the stronger man" (Homer, *Iliad,* 595).

37. Derrida will make much of the way violence is at the heart of all truth and meaning, particularly in the 1960s and in his first essay on Levinas's work, "Violence and Metaphysics" (1964). Derrida especially accuses Levinas of failing to identify the roots of violence in his own doctrine of primordial passivity and gentleness (VM 117).

38. Levinas uses this phrase in a preface he wrote for the republication of this essay in 1990.

39. Tina Chanter summarizes this tendency: "In both Christianity and modern liberalism Levinas identifies a tendency which he will often cite as a decisive hallmark of the Western tradition, from which he will rigorously distinguish his own philosophy: the primacy of freedom. This primary freedom carries with it as an inherent possibility the freedom of never being definitively committed to any ideal" (Chanter, *Time, Death, and the Feminine,* 173).

40. Manning, "Serious Ideas Rooted in Blood," 133.

41. Davidson, "Introduction to Musil and Levinas," 42–43. Davidson is quoting passages found in Hitler's, *Mein Kampf,* 407, 403.

42. "For Levinas, the return from captivity also meant the discovery of horror. His whole family in Lithuania had been murdered. His father, his mother, his two brothers. All of them were executed by machine-gun fire in Kaunas" (Malka, *Emmanuel Levinas,* 80).

43. Hand, *Emmanuel Levinas,* 29.

44. Lingis, "Sensuality and the Sensitivity," 229.

45. Allen, "Loving the Good Beyond Being," 89.

46. It is Jacques Rolland who suggests that this statement in the early pages of *Existence and Existents* echoes a latent and unspoken sentiment in *On Escape.* See Rolland, "Getting Out of Being by a New Path," in *OE* 9.

47. Peperzak, preface to *BPW* viii.

48. It has been suggested by Walter Kaufmann that the appeal of Heidegger's thesis about time and mortality originates in its novel presentation of "a secularized Christian preaching about guilt, dread and death." See Dutton, "Kaufmann, Heidegger, and Nazism," 325–36. At worst, which is how Kaufmann seems to read him, Heidegger stumbled on a truly important question for philosophy but shrouded his incomplete answers in his formulaic and spurious writing style.

49. Gary Mole discusses the subtle critique of Heidegger here as follows: "Levinas devotes the final section of the essay to a brief consideration of the problems he sees in Western philosophy, where ontologism, whether in realism of idealism, has prevented it from going beyond being.... That Levinas writes

these words in the shadow of events about to befall Europe and in particular the Jews indicates an urgency and topicality to Levinas's essay not immediately apparent and an inherent critique of Heidegger" (Mole, *Lévinas, Blanchot, Jabès,* 30).

50. Mole, *Lévinas, Blanchot, Jabès,* 30.

51. Rolland utilizes one of Levinas's closing lines: "It is a matter of getting out of being by a new path."

52. Richard Cohen appears to concur: "Jacques Rolland is no doubt on target in discovering the birth of Levinas's own thought, his intersubjective ethics...in a 1935 article entitled 'De l'évasion'" (Cohen, foreword to *TIH* xxv).

53. This is Richard Cohen's phrase, and it is also the title for part two of Levinas's essay collection, *Discovering Existence with Husserl.*

54. Peperzak, *Beyond,* 41.

55. Ibid., 42.

56. I am thinking here particularly of the way the *il y a* functions in *Existence and Existents,* as well as the notions of "insomnia" and the very idea of an "existence without existence." These ideas rely on the bracketing of other fields of experience (daylight, things, nourishments) to expose elements that are normally obscured or repressed. This methodology is seldom traced to Husserl, but it seems like a helpful way to think about Levinas's early writings and argumentation.

57. Rodemeyer, "Developments in the Theory of Time-Consciousness, 136.

58. Rajan, *Deconstruction and the Remainders of Phenomenology,* 15.

59. Drabinski, *Sensibility and Singularity,* 2.

60. Ibid., 2.

61. Peperzak, *Beyond,* 44. Drabinski may be pushing back against the standard interpretation summarized by Peperzak. See Drabinski, *Sensibility and Singularity,* 2.

62. Drabinski, *Sensibility and Singularity,* 3.

63. Ibid.

64. Ibid.

65. Levinas, "Questions et reponses," 72. Steven G. Smith quotes this article, and translates it into the English line quoted here in his "Reason as One for Another," 69–70. Cf. *OB* 183; *TI* 28, where Levinas also vocalizes a commitment to Husserlian phenomenology.

66. Lotz, *From Affectivity to Subjectivity,* 74.

67. Biographical details of Levinas's life, including his years in Stalag XIB, are meticulously outlined by Malka's book, *Emmanuel Levinas.*

Notes to Chapter 2

1. "These studies begun before the war were continued and written down for the most part in captivity" (*EE* xxvii).

2. Malka, *Emmanuel Levinas,* 80.

3. Moyn, *Origins of the Other,* 22.

4. Ibid., 22.

5. See, for instance, Kant, *Groundwork for the Metaphysics of Morals.*

6. Moyn, *Origins of the Other*, 22.

7. Ibid., 22.

8. In 1984, Levinas entitles a key essay "Ethics as First Philosophy" (*LR* 75–87).

9. Moyn, *Origins of the Other*, 22.

10. Malka, *Emmanuel Levinas,* 71.

11. As Adriaan Peperzak summarizes, "During the war, however, Levinas began to develop a philosophy of his own" (Peperzak, *Beyond,* 49).

12. Levinas, *Carnets de captivité,* 217–19. The reflections on Bergson, penned immediately after Levinas's release, will be a critical component of the discussion later in the chapter.

13. Levinas, *Carnets de captivté,* 207.

14. Moran, *Introduction to Phenomenology,* 353. Moran concludes that it is "entirely unclear how this phenomenology of alterity can be a phenomenology at all. Because of its dense style and apparent abandonment of rational argument and justification in favour of repetitive, dogmatic assertions which have the character of prophetic incantations and quasi-religious absolutist pronouncements, Levinas's work is largely ignored among analytic philosophers" (352). He provides his unsympathetic reading of *Existence and Existence* on pages 333–38.

15. Levinas repeatedly employs phrases such as "anonymity of the night," "horror of the night," "darkness," and "nocturnal space" (*EE* 55, 58, 53).

16. "Let us take some time to look at the example of food; it is significant for us because of the place it occupies in everyday life, but especially because of the relationship between desire and its satisfaction which it represents, and which constitutes what is typical of life in the world. What characterizes this relationship is a complete correspondence between desire and satisfaction. Desire knows perfectly well what it wants. And food makes possible the full realization of its intention" (*EE* 34–35).

17. "The 'I' . . . when purified of all that is not authentically human in it—is given to peace with itself, completes itself, closes on and rests upon itself" (*OE* 49).

18. Levinas uses the term *need* rather than *desire* in *On Escape* to designate this drive: "The being that has not satisfied its needs dies. But this indisputable statement has an extrinsic origin. In itself, need does not foreshadow the end. It clings fiercely to the present, which then appears at the threshold of a possible future. One heartrending need is the despair over a death that does not come. Moreover, the satisfaction of a need does not destroy it. Not only are needs reborn, but disappointment also follows their satisfaction" (*OE* 59).

19. Levinas quotes Theophile Gautier's line, "I am one of those for whom the external world exists" (*EE* 27).

20. "When random memories, fantasies and associations run out, and with the gradual suffocation of a will even to sleep, the night itself draws near: I neither

invite it nor resist it, but protest without the possibility of words or deeds. This is horror. I simply *am*" (Bloechl, *Liturgy of the Neighbor,* 138).

21. Ibid., 138.

22. "In Heidegger sociality is completely found in the solitary subject. The analysis of Dasein, in its authentic form, is carried out in terms of solitude" (*EE* 98).

23. Chanter, *Time, Death, and the Feminine,* 152.

24. "This structure of Being, which belongs to the essential 'is an issue,' we shall denote as Dasein's 'Being-ahead-of-itself' " (*BT* 236).

25. "When one has an understanding of Being-towards-death—toward death as one's *ownmost* possibility—one's potentiality-for-Being becomes authentic and wholly transparent" (*BT* 354).

26. In a footnote to Levinas's use of the term *il y a* in *Time and the Other,* translator Richard Cohen explains the importance of this term across Levinas's career: "The *there is* again appears in *Totality and Infinity,* where it is also called 'the elemental.' It is a notion of continued significance for all of Levinas's subsequent thought, and is always assumed when it is not explicitly invoked" (*TO* 46n15).

27. "An interpretation of the meaning of the antecedence cannot overlook Levinas's claims concerning the paradox of being and the problem of origins both in *On Escape* and *Existence and Existents*" (Thomas, *Emmanuel Levinas,* 50).

28. Levinas, *Carnets de captivité,* 218.

29. Ibid., 218.

30. "In the end, after all of their many elaborate descriptions and forceful conclusions, Husserl, Heidegger, and now Sartre, will have simply reified what no one can deny is at least the appearance that all experience refers properly to oneself." (Bloechl, *Liturgy of the Neighbor,* 142).

31. "Even the term *experience* is problematic" (ibid., 139).

32. In fact, as Levinas is exploring the concept of the *il y a* he utilizes Bergson positively. Immediately after he turns critically against Heidegger, whose anxiety is insufficient and still allows "escape" in death, he speaks of Bergson's "critique of nothingness" as a helpful way to think about negation with a "positive meaning" (*EE* 58–59). This may be an example of the way that Levinas utilizes Bergson to move past Heidegger, even as Heidegger repeatedly insists that he has left Bergson far behind.

33. Levinas, *Carnets de captivité,* 217.

34. There is an uncanny, and perhaps not accidental, parallel between Bergson's choice to redefine the "real" according to the existential experience of real life (*CE* 317) and Heidegger's redefinition of the "The Real." Heidegger claims that "the Real is essentially accessible only as entities within-the-world" (*BT* 246).

35. "What we actually obtain in this way is an artificial imitation of the internal life, a static equivalent which will lend itself better to the requirements of logic and language, just because we have eliminated from it the element of real time" (*CE* 4).

36. Bergson scholar F. C. T. Moore argues that the English term *duration* may not be the best way to render the French *durée,* so important to Bergson's thought. For Bergson, argues Moore, the *durée* refers to "*the fact or property of going through time.*" (Moore, *Bergson Thinking Backwards,* 58).

37. Levinas uses the melody example to introduce Bergson's concept of the *durée* (*EE* 21–23). This is a favorite example of Bergson's. See for instance, *CE* 73–74 and Bergson, *Matter and Memory,* 148–49.

38. Levinas, *Carnets de captivité,* 218.

39. "Intellect shall never pass to intuition" (*CE* 268).

40. "We shall thereby restore to the ordinary conception the autonomy which is its rightful due, as against Bergson's thesis that the time one has in mind in this conception is space" (*BT* 39). Then, "Every subsequent account of time, including Bergson's, has been essentially determined by [Aristotle's essay on time]" (49).

41. This opinion on Heidegger's relationship to Bergson is carefully outlined, and convincingly supported, by Richard A. Cohen in his book *Ethics, Exegesis, and Philosophy: Interpretation after Levinas.* He writes, "There are thinkers who followed Bergson—most obviously Heidegger—who though often credited with a revolution in thought are but continuing a revolution that has already taken place, beneficiaries of it" (Cohen, *Ethics, Exegesis, and Philosophy,* 27).

42. See, for instance, Moulard-Leonard's recent discussion of Bergson's originality in *Bergson-Deleuze Encounters,* especially 1–32.

43. Heidegger defines care (*Sorge*) as "Dasein's primordial state of Being" (*BT* 273).

44. "The more it advances in this work, the more will it perceive that intuition is mind itself, and, in a certain sense, life itself: the intellect has been cut out of it by a process resembling that which has generated matter. Thus is revealed the unity of the spiritual life.... Philosophy introduces us thus into the spiritual life" (*CE* 268).

45. Bergson devotes an entire book—*Time and Free Will: An Essay on the Immediate Data of Consciousness*—to developing this relationship between time and free will.

46. Thomas, *Emmanuel Levinas,* 38.

47. Levinas will also later reflect on the ways that Bergson provides the resources to move *past* Heidegger. Howard Caygill makes this point and explores some of the implications in *Levinas and the Political,* 12–13.

48. "The authentic coming-towards-itself of anticipatory resoluteness is at the same time a coming-back to one's ownmost Self, which has been thrown into its individuation. This ecstasis makes it possible for Dasein to be able to take over resolutely that entity which it already is. In anticipating, Dasein *brings* itself *again forth* into its ownmost potentiality-for-being" (*BT* 388).

49. "As an authentic Present, or waiting-towards, the moment of vision permits us *to encounter for the first time* what can be 'in a time' as ready-to-hand or present-at-hand" (*BT* 368). Heidegger goes on to explain that the "moment of vision" is temporalized as "the authentic future" of Dasein (388).

50. Bergson makes reference to this doctrine: "In short, *the world the mathematician deals with is a world that dies and is reborn at every instant—the world which Descartes was thinking of when he spoke of continued creation*" (*CE* 22).

51. Pyle, *Malebranche,* 112–13.

52. "And as for motion, we shall best understand it if we think only of local motion and do not enquire into the force by which it is produced" (Descartes, *Principles of Philosophy,* 30).

53. Ibid., 30.

54. Descartes, Descartes to Henry More, 200–01. The issue of "force of motion" in Descartes is nicely summarized in Hankins, *D'Alembert: Science and the Enlightenment,* 154.

55. "Such being the case, is there not much to be said for the hypothesis of a conscious force of free will, which, subject to the action of time and storing up duration, may thereby escape the law of conservation of energy?" (Bergson, *Time and Free Will,* 154).

56. One should bear in mind that what Levinas means by a resurrection is not some afterlife after physical death but a "resurrection in the son in whom the rupture of death is embodied" (*TI* 56–57).

57. Franck, "Body of Difference," 25.

58. Ibid., 24.

59. John Caputo outlines the ease with which some of Levinas's earlier texts can be read to insinuate that "being is evil" (Caputo, *Weakness of God,* 332). This phrase appears in Levinas, *Time and the Other,* 51, but is often read out of context. Caputo appears to be aware of Levinas's extensive discussions of evil and being elsewhere that determine otherwise. Caputo points to Blond's critique, which strikes Caputo as "overly enthusiastic" (See Blond, "Emmanuel Levinas," 195–228). Caputo realizes, to extend this claim and to extract it from *Time and the Other* without context, is to misunderstand Levinas's attitude toward both being and evil. Indeed, Richard J. Bernstein writes: "We must be careful not to misinterpret what Levinas is saying here. There is nothing evil about the law of being in itself. . . . Furthermore, as human beings, we do and must act to preserve our own being. But if we act as if we were beings exclusively concerned with our own *conatus essendi,* if we fail to respond to the demands, needs and suffering of the other, then we are succumbing to the 'law of evil" (Bernstein, "Evil and the Temptation of Theodicy," 265).

60. Bloechl, *Liturgy of the Neighbor,* 142.

61. Cohen, *Elevations,* 138.

62. Defoe, *Robinson Crusoe,* 121.

63. "We will leave the theories on the side. We want to know nothing of the image but what the reflection can teach us. . . . For now I want only to attempt a 'phenomenology' of the image. The method is simple: produce images in ourselves, reflect on these images, describe them, which is to say, try to determine and classify their distinctive characteristics" (Sartre, *The Imaginary,* 5).

64. Webber, introduction to *The Imaginary,* xiii.

65. Sartre, *The Imaginary,* 80.

66. "The act of imagination, as we have just seen, is a magical act. It is an incantation destined to make the object of one's thought, the thing one desires, appear in such a way that one can take possession of it" (ibid., 125).

67. "Comprehension of a word therefore is given as the sudden appearance of an object. So that the spatial determinations are not signs or images of the structural relations that constitute the thing: they are apprehended as those very relations. They are the relations constituted by a piece of knowledge that is incorporated in a series of movements" (ibid., 106).

68. "I have tried, on the contrary, to show that there is an internal relation between the horse and its image, which I have called a relation of possession: through the analog on it is the horse itself that appears to consciousness" (ibid., 83).

69. "[Thought] tries to make the object appear before it, to *see* it, or better still to *possess* it" (ibid., 122).

70. Bloechl, *Liturgy of the Neighbor,* 166.

71. Levinas ends this phrase with a question mark.

72. Bloechl, *Liturgy of the Neighbor,* 171.

73. Robbins, *Altered Reading,* 75.

74. Levinas points to the way poetic activity "enraptures and transports the interlocutor" (*TI* 203). And yet few would deny that *Totality and Infinity,* and other works by Levinas, have a lyrical and even poetic rhythm.

75. Riera, " 'Possibility of the Poetic Said,' " 14.

76. Heidegger writes of how "Dasein understands itself in its own *superior power,* the power of its finite freedom" (*BT* 436).

Notes to Chapter 3

1. Levinas, *Carnets de captivité,* 74. This collection includes several notebooks kept by Levinas from the period between 1937 and 1950. The notebook with this "to do" list is labeled "1942."

2. Parmenides lived in the early fifth century BCE and was one of the Presocratic philosophers who laid the groundwork for Platonic philosophy; all that remains of his thought is some of the heritage marked in the works of later philosophers and a fragmentary poem. See Kirk, Raven, and Schofield, *Presocratic Philosophers,* 239–62. The poem describes reality as unified, denies the possibility of change, and emphasizes the timelessness of existences. For Rosenzweig, and then for Levinas, Parmenides became an important emblem for the philosophical desire to unify all knowledge and being beneath the towering, logical unity of the One. "Parmenides' arguments and his paradoxical conclusions had an enormous influence on later Greek philosophy; his method and his impact alike have rightly been compared to those of Descartes' *cogito*" (ibid., 241).

3. "Only one story of a path is now left; namely, that it is. On this path there are very many signs, showing that what is, is uncreated and indestructible, whole, unique, unmoved and perfect" (Parmenides, *Fragments of Parmenides,* 88).

4. "[Being] is all together now, one and coherent" (ibid., 88).

5. This is an excerpt of a footnote Cohen provides discussing separation, Rosenzweig, and Parmenides.

6. Flew, *A Dictionary of Philosophy,* s.v. "Hypostasis."

7. "In 1947, in reaction to what he sees as a lack of concretion in Heidegger's Dasein, Levinas offers an account of the instant that emphasizes its substantiality. *Hypostasis* is the name Levinas gives to the event by which the subject (or existent) takes up subjectivity (or existence). With the notion of hypostasis, Levinas reworks Heidegger's ontological difference, the difference between Being and beings. In this early period of his work, Levinas describes the present as prior to time, and the I as separate from the Other" (Chanter, *Time, Death, and the Feminine,* 32).

8. Bloechl, *Liturgy of the Neighbor,* 128.

9. "Nor is [Being] divided, since it is all alike, and [Being] is not any more or any less in any way, so as to prevent itself from being coherent, but it is all full of that which is" (Parmenides, *Fragments of Parmenides,* 88).

10. Levinas returns to the themes of insomnia, among other places. See *TO* 48; *TI* 258; *OB* 30, 64, 68, 87, 93, 163, 192.

11. Despite the intensity of Levinas's reading of Heidegger, there is considerable consensus, even among Levinas devotees, that Levinas is not always accurate in his summaries and invectives against Heidegger's work. This is not to say that he is a *careless* reader of Heidegger; his appreciation for many of the subtleties of Heidegger's greatest work is noteworthy. After writing *Being and Time,* Heidegger undergoes a series of transitions, though Levinas does not appear to track the later work of Heidegger very carefully. He remains fixed on the innovations and problems of *Being and Time.* For Heideggerians in particular, this constitutes a failure in Levinas to keep his relationship with Heidegger's work relevant. For this reason, I will pay close attention to Heidegger's work as we explore Levinas's claims about *Being and Time* and "What Is Metaphysics?" which are directly and indirectly referenced here. "What Is Metaphysics?" is an article published by Heidegger in 1929 that contains the phrase "Nothingness nothings" that Levinas refers to (*TO* 49). See Heidegger, "What Is Metaphysics," 353–93.

12. "Nothing is neither an object nor anything that 'is' at all. Nothing occurs neither by itself nor 'apart from' what-is, as a sort of adjunct. Nothing is that which makes the revelation of what-is as such possible for our human existence. Nothing not merely provides the conceptual opposite of what-is but is also an original part of essence (*Wesen*). It is in the Being (*Sein*) of what-is that the nihilation of Nothing *(das Nichten des Nichts)* occurs" (Heidegger, "What Is Metaphysics?," 370).

13. F. Ruth Irwin writes, "Heidegger is wary of nothingness. He understands nothingness as nihilism which is a road leading to loss, annihilation, or as he puts it, the forgetting of Being" (Irwin, "Heidegger and Nietzsche," 194).

14. "The 'nothing' with which anxiety brings us face to face, unveils the nullity by which Dasein, in its very *basis,* is defined; and this basis itself is as thrownness into death" (*BT* 356).

15. "The forgetting which is constitutive for fear, bewilders Dasein and lets it drift back and forth between 'worldly' possibilities which it has not seized upon. In contrast to this making-present which is not held on to, the Present of anxiety is *held on to* when one brings oneself back to one's ownmost thrownness. The existential meaning of anxiety is such that it cannot lose itself in something with which it might be concerned. If anything like this happens in a similar state-of-mind, this is fear, which the everyday understanding confuses with anxiety" (*BT* 394).

16. "Original dread can be awakened in Da-sein at any time. It need not be awakened by any unusual occurrence. Its action corresponds in depth to the shallowness of its possible cause. It is always on the brink, yet only seldom does it take the leap and drag us with it into the state of suspense" (Heidegger, "What Is Metaphysics?," 374).

17. "The nullity by which Dasein's Being is dominated primordially through and through, is revealed to Dasein itself in authentic Being-towards-death" (*BT* 354).

18. *TO* 49. Quoted from "What Is Metaphysics?," 369.

19. "The utter insignificance which makes itself known in the 'nothing and nowhere,' does not signify that the world is absent, but tells us that entities within-the-world are of so little importance in themselves that on the basis of this *insignificance* of what is within-the-world, the world in its worldhood is all that still obtrudes itself" (*BT* 231).

20. "In the structure of thrownness, as in that of projection, there lies essentially a nullity. This nullity is the basis for the possibility of *in*authentic Dasein in its falling; and as falling, every authentic Dasein factically is. *Care itself, in its very essence, is permeated with nullity through and through*" (*BT* 331).

21. "Selfhood is to be discerned existentially only in one's authentic potentiality-for-Being-one's-Self—that is to say, in the authenticity of Dasein's Being *as care*" (*BT* 369).

22. Levinas supports this with an intriguing, but underdeveloped discussion of suicide, which has the effect of confusing the reader and making the *il y a* seem theoretical and abstract. He provides a clearer discussion in one of the later lectures in this volume, to which I will turn shortly (*TO* 71–73).

23. The concept of ontology is a complicated one in *Being and Time,* and it represents an instance where Levinas's reading of Heidegger lacks nuance. He opens the first lecture in *Time and the Other* questioning Heidegger's "ontological structure of Dasein" and questioning whether his conception of solitude is not "ontologically obscure" (*TO* 40). What Heidegger really offers here is not ontology but a critique of the timeless, theoretical ontology that has been delivered to us by Plato and Aristotle. Heidegger therefore appeals to the "pre-ontological"

in order to recover what has been lost and reposition philosophy accordingly (*BT* 242–43). It is not necessarily the case that Levinas is *wrong* in claiming that Heidegger is offering an ontological structure, but this is an oversimplification that Levinas makes and, for the most part, retains throughout his career.

24. "The relationship with the Other is indeed posed by Heidegger as an ontological structure of *Dasein,* but practically it plays no role in the drama of being or in the existential analytic" (*TO* 40).

25. "We may not summarize our characterization of authentic Being-towards-death as we have projected it existentially: anticipation reveals to Dasein its lostness in the they-self, and brings it face to face with the possibility of being itself, primarily unsupported by concernful solicitude, but of being itself, rather, in an impassioned **freedom toward death**—a freedom which has been released from the Illusions of the 'they,' and which is factical, certain of itself, and anxious" (*BT* 311).

26. "The 'they' provides a *constant tranquillization about death.* At bottom, however, this is a tranquillization not only for him who is 'dying' but just as much for those who 'console' him. . . . *The 'they' does not permit us the courage for anxiety in the face of death*" (*BT* 298).

27. A summary of Heidegger's low opinion of "idle talk" appears in *BT* 211–24, but there are more than fifteen other extended references to this phenomenon in *Being and Time* alone.

28. For a thorough analysis of this accusation and possible oversights in the generalization that Heidegger failed to address ethics, see Hodge, *Heidegger and Ethics* and to a lesser degree, Olafson, *Heidegger and the Ground of Ethics.* For reflections on this theme in relationship to Levinas, see Morgan, *Discovering Levinas.* Morgan addresses this question in his first chapter, "Auschwitz, Politics, and the Twentieth Century" (1–38), and revisits the question later in his book as well (230).

29. See, for instance, Cohen, *Elevations.*

30. Levinas, "Entre deux mondes (Biographie spirituelle de Franz Rosenzweig)," 121–37 ; translated as " 'Between Two Worlds': The Way of Franz Rosenzweig" (*DF* 181–201).

31. Some scholars have pointed out that Levinas's appropriation of Franz Rosenzweig as an ethical thinker does not represent an accurate reading of Rosenzweig's *The Star of Redemption.* See, for instance, Gordon, *Rosenzweig and Heidegger,* 9–12.

32. See, for instance, where "separation" and "holiness" are listed alongside one another (*OB* 59), but also where holiness is a "distance from a theme" and a "reclusion" (162).

33. One of Levinas's most important late essays is "Diachrony and Representation." Reflecting on his early deliberations on time, Levinas credits "Bergson, Rosenzweig, and Heidegger, each in his own way," for their parts in helping him think about the deformalization of time (*EN* 176).

34. "And even in Kant's case the concept of the All again carried off the victory over the individual through his formulation of the law of morality as the universally valid act" (*SR* 10).

35. Cohen, *Elevations,* xiv.

36. Levinas writes, "The petrification of the instant in the heart of duration—Niobe's punishment—the insecurity of a being which has a presentiment of fate, is the great obsession of the artist's world, the pagan world" (*CPP* 11).

37. Cohen boldly presents Rosenzweig as an *ethical* thinker and connects Levinas's insistence on the transcendence of the other person to Rosenzweig's discussions of the transcendence of God (Cohen, *Elevations,* 62). Cohen emphasizes transcendence and love, which appear to establish Rosenzweig as a distinctly ethical thinker, and align him more clearly with Levinas. Cohen has also shown, with precision and careful textual alignments, that Levinas appears to have gleaned his famed concept of the "face," which is so crucial to *Totality and Infinity,* from *The Star of Redemption* (Cohen, "The Face of Truth in Rosenzweig, Levinas, and Jewish Mysticism," 175–204). Additionally, Robert Gibbs has demonstrated a number of points at which Levinas depends on Rosenzweig in his study *Correlations in Rosenzweig and Levinas.* Peter Eli Gordon writes critically of the efforts of Cohen and Gibbs, claiming that the alignment of these two thinkers obscures their marked differences. He finds the attempt to correlate Rosenzweig and Levinas at the point of ethics to be troubled by Rosenzweig's repeated insistence that ethics must *not* be established by relations in the world (Gordon, *Rosenzweig and Heidegger,* 9–12). Gordon's interest is in demonstrating that the alignment with Levinas obscures commonalities between Heidegger and Rosenzweig. Gordon claims, in what is a direct opposition to Cohen's interpretation, that "Rosenzweig was a holist; that is, he was committed to the doctrine that meaning depends upon a coherent existential horizon, a bounded and self-sustaining sphere of common practices, shared language and experience" (ibid., 11–12.

38. Gordon, *Rosenzweig and Heidegger,* 11–12.

39. Epicurus to Menoeceus, 31.

40. Levinas is quoting Shakespeare. See *Macbeth,* 86.

41. Shakespeare, *Macbeth,* 86.

42. The exact origin of this phrase is uncertain, but its origins are often associated with Cicero for its appearance in his *Letters to Atticus:* "As a sick man is said to have hope for as long as he has breath, so I did not cease to hope so long as Pompey was in Italy." Cicero, *Letters to Atticus,* 229. Levinas refers to this saying in *TO* 73.

43. Levinas points out that "we recognize the other as resembling us," though this resemblance gives way to mystery on closer examination (*TO* 75).

44. "The Other as Other is not only an alter ego: the Other is what I myself am not. The Other is this, not because of the Other's character, or physiognomy, or psychology, but because of the Other's very alterity. The Other is, for

example, the weak, the poor, 'the widow and the orphan'" (*TO* 83). I will make much of the alignment of biblical themes in later chapters of this study, but it is worth noting that Levinas has directly quoted the Hebrew Bible here in order to emphasize its concern for "orphans and widows." This phrase occurs regularly in the Hebrew Scriptures (cf. Exod. 22:21; Deut. 10:18, 24:17, 19, 20, 21, 26:12, 27:19; Isa. 1:17, 9:16, 10:2; Jere. 7:6, 22:3; Ezek. 22:7; Zech. 7:10; Mal. 3:5; Psalm 68:6, 109:9, 146:9; Lam. 5:3).

45. Caygill, *Levinas and the Political,* 23.

Notes to Chapter 4

1. In introducing "Is Ontology Fundamental?," Simon Critchley writes, "Levinas is engaged here in a questioning of Heidegger's project of fundamental ontology, that is to say, his attempt in *Being and Time,* to raise anew the question of the meaning of Being through an analysis of that being for whom Being is a question: *Dasein.* In Heidegger's earlier work, ontology—science of Being in the Aristotelian sense—is fundamental, and Dasein is the fundament or condition of the possibility for an ontology, a being whose 'a priori' structure must first be clarified in an existential analysis" (*BPW* 1).

2. "Do we in our time have an answer to the question of what we really mean by the word 'being'? Not at all. So it is fitting that we should raise anew *the question of the meaning of Being.* But are we nowadays even perplexed at our inability to understand the expression 'Being'? Not at all. So first of all we must reawaken an understanding for the meaning of this question. Our aim in the following treatise is to work out the question of the meaning of *Being* and to do so concretely. Our provisional aim is the Interpretation of *time* as the possible horizon for an understanding whatsoever of Being" (*BT* 19).

3. "The ontological structure of that entity which, in each case, I *myself* am, centers in the Self-subsistence of existence. . . . The proposition 'Dasein is historical,' is confirmed as a fundamental existential ontological assertion" (*BT* 381).

4. Heidegger demonstrates the subordination of phenomena to ontology in his discussion of death, among other things, which must be "sketched out by the ontology of Dasein. Within the ontology of Dasein, which is *superordinate* to an ontology of life, the existential analysis of death is, in turn, *subordinate* to a characterization of Dasein's basic state" (*BT* 291).

5. Tina Chanter summarizes, "Dasein remains the center of Heidegger's analysis" (Chanter, *Time, Death, and the Feminine,* 182). Chanter's entire fifth chapter, "Giving Time and Death" (170–88), provides a meticulous summary of the relationship between Levinas and Heidegger on time and death.

6. "Nothing theological, nothing mystical, lies hidden behind the analysis that we have just given of the encounter with the other (*autrui*)" (*BPW* 8).

7. "Is reason domination by which the resistance of being as such is surmounted, not in an appeal to this very resistance but as a ruse of the hunter who ensnares all that such a being contains of strength and irreducibility on the basis

of its weaknesses, the abdication of its particularity, its place in the horizon of universal being" (*BPW* 8).

8. Levinas adopts the word "horizon" from *Being and Time,* where for Heidegger it refers to the temporal field "for any understanding whatsoever of Being" (*BT* 19).

9. Levinas does not mention Heidegger here, but he is clearly referring to Dasein's fixation on a being-toward-death that is finally "personal" and "private" and "returns incessantly to itself, even when it seems to flee itself."

10. "Here the formula 'before being in relation with a being, I must first have comprehended it as being' loses its strict application" (*BPW* 7).

11. "The person with whom I am in relation I call being, but in so calling him, I call to him" (*BPW* 7).

12. "He is my *partner* in the heart of a relation which ought only to have made him present to me" (*BPW* 7).

13. Levinas does not address the fact that Heidegger agreed that the Dasein of others does not function like other things encountered in the world (*BT* 153–63).

14. Levinas opens the preface of *Totality and Infinity* by invoking the relationship between philosophy and violence: "Everyone will readily agree that it is of the highest importance to know whether we are not duped by morality. Does not lucidity, the mind's openness upon the true, consist in catching sight of the permanent possibility of war? The state of war suspends morality; it divests the eternal institutions and obligations of their eternity and rescinds ad interim the unconditional imperatives" (*TI* 21).

15. Levinas writes in "Philosophy and the Idea of Infinity," "The other's face is the revelation not of the arbitrariness of the will, but its injustice. Consciousness of my injustice is produced when I incline myself not before facts, but before the other. In his face the other appears to me not as obstacle, not as a menace I evaluate, but as what measures me" (*CPP* 57–58).

16. The alternative, Levinas explains, is "avoiding the gaze."

17. Malka, *Emmanuel Levinas,* 65.

18. "To love is to exist as though the lover and the beloved were alone in the world. The intersubjective relationship of love is not the beginning, but the negation of society.... Love is the ego satisfied by the you, apprehending the other in the justification of its being. The presence of the other exhausts the content of such a society.... The love of the neighbor depends on chance proximity; it is hence love of one being to the detriment of another, always privilege even if it is not preference.... Love makes blind the respect which is impossible without blindness toward the third person and is only pious intention oblivious of the real evil" (*CPP* 31).

19. "Limited to the intimate society, faced with the only freedom which the act concerned, I could, in dialogue, receive absolution for it. The ego, in dialogue, would thus recover, be it only after the fact, through pardon, its solitary sovereignty. The ego, capable of forgetting its past and renewing itself, but which

by its actions creates the irreparable, would through the pardon be liberated from this last shackle to freedom, since the only victim of its act would or could consent to forget it" (*CPP* 30–31).

20. "One could legitimately accept pardon only if the other is God or a saint" (*CPP* 21).

21. Summarizing his article in the final paragraph, Levinas asks a series of questions that are clearly directed at Heidegger: "Can the self present itself to itself with so much natural complacency? Can it appear, shamelessly, in its own eyes? Is narcissism possible?" (*CPP* 59). On the same page, Levinas points to three essays from earlier in the decade when he had already taken up these (apparently Heideggerian) themes in various ways. He writes, "We have dealt with the different themes relevant to this matter in three articles published in the *Revue de Métaphysique et de Morale*." The three articles Levinas lists are "Is Ontology Fundamental?," "Freedom and Command," and "The Ego and the Totality."

22. Levinas was always reluctant to participate in the alignment of Heidegger's thought with his disastrous alignment with Nazism. He does, however, make a comment of this sort in a 1990 letter that is the preface to Seán Hand's translation of "Reflections on the Philosophy of Hitlerism." In that short note, Levinas associates Heidegger's ontology with the possibility of "elemental evil" that he found embodied in National Socialism (RPH 63).

23. Editor Adriaan Peperzak adds in a footnote that Levinas's use of "the set of concepts Same and Other is taken from Plato's *Sophist* (254b–256b), where they figure as the highest of the categories of being. Cf. also *Timaeus* 35ab and *Theaetetus* 185cd" (*CPP* 48n3).

24. Cf. Plato, *Phaedo,* 77a–c.

25. Taminiaux provides a helpful summary of the relationship between Heidegger's ontology and the history of metaphysics, claiming that Heidegger turns the metaphysical thoughts from the history of philosophy into "timber for its own fire" (Taminiaux, *Heidegger and the Project of Fundamental Ontology,* vii).

26. Levinas refers to this commandment from Exodus 20:3 in his discussion of *creatio ex nihilo* (*CPP* 58).

27. "Distance alone does not suffice to distinguish transcendence from exteriority. Truth, the daughter of experience, has very lofty pretentions; it opens upon the very dimension of the ideal. In this way philosophy means metaphysics, and metaphysics inquires about the divine" (*CPP* 47).

28. Lévy-Bruhl, *Notebooks on Primitive Mentality,* 49.

29. It is puzzling that Levinas does not invoke or critique Sartre in "The Ruin of Representation" (1959), for instance. Ten years earlier, as we previously observed, Levinas was rather furiously attacking Sartre's understanding of representation and images, especially in "Reality and Its Shadow" (1948). Traces of Levinas's engagement with Sartre remain, however, as when he appears to follow Sartre by invoking Aristotle's "Third Man" to discuss the importance of the opening of ethical relation to the third person. See Sartre, *Being and Nothingness,* 259; Aristotle, *Metaphysics,* book 1; *CPP* 29. The emphasis on Husserl in the "Ruin of Representation" makes one think that Levinas's opposition to repre-

sentation is more directly a quibble with Husserl and less an issue with Sartre, despite the critique of Sartre on this theme in "Reality and Its Shadow."

30. The concept of the "hostage," a controversial but central theme in *Otherwise than Being,* appears many years earlier in one of Levinas's lectures on the Talmud (*OB* 184–85; cf. *NT* 170–71). In chapter 6, I will explore the relationship between his Jewish writings and his escalating emphasis on time.

31. "Ontology, allegedly authentic, coincides with the facticity of temporal existence" (*BPW* 3).

32. Levinas, *Carnets de captivité,* 134.

33. Levinas eventually admits some of these trends in his thought. See, for instance, the semi-autobiographical essay, "Signature," where he writes, "The ontological language which *Totality and Infinity* still uses in order to exclude the purely psychological significance of the proposed analysis is henceforth avoided" (*DF* 295).

Notes to Chapter 5

1. Morgan, *Discovering Levinas,* 214.

2. *Totality and Infinity* was the main thesis for Levinas's *Doctorat d'État,* required for serious academic posts in the *l'Académie Française.* After publishing *Totality and Infinity,* Levinas was appointed Professor of Philosophy at the University of Poitiers (Moran, *Introduction to Phenomenology,* 326).

3. "We recognize the other as resembling us, but exterior to us; the relationship with the other is a relationship with a Mystery. The other's entire being is constituted by its exteriority, or rather its alterity, for exteriority is a property of space and leads the subject back to itself through light" (*TO* 75–76).

4. "The relationship with Being that is enacted as ontology consists in neutralizing the existent in order to comprehend or grasp it" (*TI* 45–46). Levinas also writes, "For Heidegger intersubjectivity is a coexistence, a *we* prior to the I and the other, a neutral intersubjectivity."

5. "Infinity is characteristic of a transcendent being as transcendent; the infinite is the absolutely other. The transcendent is the sole *ideatum* of which there can be only an idea in us; it is infinitely removed from its idea, that is, exterior, because it is infinite" (*TI* 49).

6. "History is worked over by the ruptures of history, in which a judgment is borne upon it. When man truly approaches the Other he is uprooted from history" (*TI* 52).

7. Frederick Beiser calls Hegel's system a "panlogicism" in which "the philosopher sees things from the perspective of the whole" and should therefore "know that everything happens of necessity" (Beiser, *Hegel,* 78). This panlogicism is often called the "view from nowhere." See, for instance, Pinkard, *Hegel's Phenomenology,* 123.

8. "War does not manifest exteriority and the other as other; it destroys the identity of the same. The visage of being that shows itself in war is fixed in the concept of totality, which dominates Western philosophy" (*TI* 21).

9. Heraclitus, *Fragments of Heraclitus,* 48, 49.

10. "Everything comes to be according to this *logos*" (ibid., 39).

11. "What is, is uncreated and indestructible, whole, unique, unmoved and perfect.... It *is* all together now, one and coherent" (Parmenides, *Fragments of Parmenides,* 88). "Having heard not me, but the *logos,* it is wise to concur that all things are one" (Heraclitus, *Fragments of Heraclitus,* 39).

12. "But I do not expect any popular approval or indeed any great crowd of readers. On the contrary I would not urge anyone to read this book except those who are able and willing to meditate seriously with me, and to withdraw their minds from the sense and from all preconceived opinions" (Descartes, *Meditations on First Philosophy,* 8).

13. Heidegger names Descartes as the classic example of an atemporal thinker who posits a spatial atomism that utterly divides the ego from both the world and its fundamental character as temporal. Levinas employs Descartes fully aware that he is turning to a deeply anti-Heideggerian resource.

14. "Anxiety discloses an insignificance of the world; and this insignificance reveals the nullity of that with which one can concern oneself" (*BT* 393).

15. Unsurprisingly, this is exactly the methodology that Derrida will soon use against Levinas himself.

16. For Heidegger, there is no abstract principle of free will that provides the foundation for human freedom. Instead, Levinas thinks that Heidegger's ontology identifies a freedom "that comes from obedience to being: it is not man who possesses freedom; it is freedom that possesses man." Despite this, Heidegger still reconciles freedom and obedience through a dialectic that "presupposes the primacy of the same" (*TI* 45). This dialectic is resolved, in Heidegger, by situating the pursuit of truth within the relationship between Dasein and being. Heidegger still locates the problematic and resolution of ontology within the field of the same; philosophy's great mystery remains a function of reconciling being to the self. The other, and therefore, justice, plays a secondary and subordinate role.

17. This is conveyed succinctly in Levinas, "Diachrony and Representation" (*EN* 159–78). This theme also pervades Levinas's second major work, *Otherwise than Being.*

18. "The last question, which indeed could be Levinas's question to Husserl, would be to demonstrate as soon as *he speaks* against Hegel, Levinas can only confirm Hegel, has confirmed him already" (VM 120).

19. "The presence of the Other is equivalent to this calling into question of my joyous possession of the world" (*TI* 75–76).

20. "In separation—which is produced in the psychism of enjoyment, in egoism, in happiness, where the I identifies itself—the I is ignorant of the other" (*TI* 62).

21. "He prefers the certainty of tomorrow to today's enjoyments. He demands guarantees in the present against the future, which introduces unknowns into those solved problems from which he lives. What he possesses becomes capital,

carrying interest or insurance against risks, and his future, thus tamed, is integrated in this way with his past" (*OE* 50). The discussion of labor and dwelling in *Totality and Infinity* follows remarkably similar lines. For all the transitions and alterations in his philosophical development, Levinas cannot be accused of losing sight of the themes that absorbed his earliest reflections.

22. As in Heidegger, the experience of time introduces a steady erosion of confidence in the connection between the ego and its egoistic enjoyments. In the face of the "nothingness of the future," the ego anticipates pain and suffering and braces against these eventualities: "man holds in his hands the remedy for his ills, and the remedies preexist the ills" (*TI* 146).

23. "Interiority will appear as a presence at home with oneself, which means inhabitation and economy" (*TI* 110).

24. "The psychic life, which makes birth and death possible, is a dimension in being, a dimension of non-essence, beyond the possible and the impossible. It does not exhibit itself in comprehension of being in which interiority is sacrificed. The present work proposes another option. The real must not only be determined in its historical objectivity, but also from interior intentions, from the *secrecy* that interrupts the continuity of historical time. Only on the basis of this secrecy is the pluralism of society possible" (*TI* 58).

25. Hegel, *Phenomenology of Mind*, 95.

26. Ibid., 96.

27. Interestingly, Levinas here gestures not only toward the *face* as this rupture but also *speech and language*. We will find Levinas, in his later writings, focusing increasingly on language as it reveals the rupture of history.

28. It is interesting that Levinas both selects this example and excludes from his retelling the presence of the prospector's friend, "Big Jim," who helps the prospector stabilize his thoughts and survive the danger by escaping the house first.

29. For Levinas, the product of labor is "fixed between the four walls of the home, is calmed in possession. It appears there as a thing, which can, perhaps, be defined by tranquility—as in a 'still life.' The grasp operated on the elemental is labor" (*TI* 158).

30. Though he does not mention Heidegger here, Levinas opens the section with the obvious reference to Heideggerian images like the hammer and the pen (*TI* 152). By invoking this terminology, Levinas brings to mind the notion of *zuhanden* (ready-to-hand), which for Heidegger characterizes tools in their prereflective and functional manifestation. The dwelling is like a tool that functions to nourish and sustain life. This particular tool has a "privileged place" among other tools, for it is the very "condition" of human activity, the source and commencement of all knowing and acting. Heidegger knew that tools examined in abstraction (*vorhanden:* being-at-hand) are not the same as tools in their natural habitat (*zuhanden*). Both the scientific method and philosophy after Socrates have forgotten that something escapes the analysis of *vorhanden*. The reduction of tools to objects for analysis is exemplified by Heidegger in the awareness that

one has of a hammer when it becomes something *present* and not something being used instrumentally, hammering nails. A broken hammer, for instance, becomes an object for analysis and no longer *zuhanden*. In his sweeping critique of Western thought since Socrates, Heidegger claims that analyses of things as *vorhanden* levels them down, reducing them to a metaphysics of presence (*BT* 102).

31. "Hence the subject contemplating a world presupposes the event of dwelling" (*TI* 153).

32. "The extraterritoriality has a positive side. It is produced in the gentleness or the warmth of intimacy" (*TI* 150).

33. "The infinity, stronger than murder, already resists us in a face, in his face, is the primordial expression, is the first word: 'you shall not commit murder'" (*TI* 199).

34. John Wild writes in the introduction to *Totality and Infinity*, "History itself is not the final judge of history" (*TI* 19).

35. Diane Perpich's theory about the transition between Levinas's two major works will be discussed extensively in the final two chapters of this study. She describes the transition in terms of a shift away from the ego as protagonist, or away from the narrative of self-meets-world that abides in *Totality and Infinity*. "*Totality and Infinity* engaged in an extended narrative that purported to show how a separated and atheist ego could nonetheless come to be commanded by and responsible for the other" (Perpich, *The Ethics of Emmanuel Levinas*, 118).

36. For a review of these interpretations and a wide range of opinions on status of the feminine in Levinas, see Chanter, ed., *Feminist Interpretations of Emmanuel Levinas*.

37. Richard Cohen provides a prolonged discussion of this controversy in *Elevations*, 195–204. Cohen's work on this topic is sharpest in his exegesis of Levinas and the gendered themes of Levinas's writings. Cohen does not demonstrate similar exegetical care, as Claire Elise Katz points out, in his representation of the feminist reaction to Levinas (Katz, *Levinas, Judaism, and the Feminine*, 164).

38. Peperzak, *To the Other*, 129.

39. "Ancient Greek philosophy showed that alterity, otherness, is the same thing as negation, therefore Evil. To pose the Other is to define a Manichaeism. That is why religions and codes of law treat woman with such hostility as they do. By the time humankind reached the stage of written mythology and law, the patriachate was definitively established: the males were to write the codes" (Beauvoir, *The Second Sex*, 79).

40. Ibid., 79.

41. Sandford, "Levinas, Feminism and the Feminine," 139.

42. Beauvoir, *The Second Sex*, xxxv.

43. Ibid., xxii.

44. Ainley, "Feminine, Otherness, Dwelling," 8.

45. Chanter discusses the way Levinas plays with and against masculine "virility" in her introduction to *Feminist Interpretations of Levinas*, 1–25.

46. Bloechl, *Liturgy of the Neighbor,* 199, 320–21n88. Cf. Aristotle, *The Nicomachean Ethics,* book 8, para. 10.

47. Bloechl, *Liturgy of the Neighbor,* 199, 320–21n88.

48. Katz, *Levinas, Judaism, and the Feminine,* 59.

49. Peperzak, *To the Other,* 195.

50. Sandford, *Metaphysics of Love,* 47.

51. Sonia Sikka, "Delightful Other," 107.

52. Ibid., 107.

53. "Hence does Levinas's theory of time return, now plainly underwritten by a relation in which the other has touched me even before I begin the struggle that is temporalization" (Bloechl, *Liturgy of the Neighbor,* 199).

54. The title of this subsection is a phrase that appears in *TI* 155.

55. Chanter, *Time, Death, and the Feminine,* 56.

56. Chanter, *Time, Death, and the Feminine,* 56.

57. Ibid., 74.

58. Katz, *Levinas, Judaism, and the Feminine,* x.

59. Bloechl *Liturgy of the Neighbor,* 199. In this passage Bloechl is summarizing Levinas's approach to gender in *Totality and Infinity.*

60. Ibid., 199.

61. Chanter comments that this relegates women to "the private, corporeal, domestic realm, to watch over children, to take care of men's needs, to provide solace and love and sustenance, to give a break, to interrupt monotony, create a delightful lapse in being" (Chanter, *Time, Death, and the Feminine,* 57).

62. Chanter, *Time, Death, and the Feminine,* 57.

63. Richard Cohen has commented on this particular trajectory in Levinas's development across his career (Cohen, *Elevations,* 144–45).

64. Ibid., 133–61.

65. "The relation with the Other alone introduces a dimension of transcendence, and leads us to a relation totally different from experience in the sensible sense of the term, relative and egoist" (*TI* 193).

66. Levinas uses this phrase to talk about the erotic encounter: "The face fades, and its impersonal and inexpressive neutrality is prolonged, in ambiguity, into animality. The [erotic] relations with the Other are enacted in play; one plays with the other as with a young animal" (*TI* 263).

67. Desmond, "Marcel, Jaspers, Levinas," 166.

68. John Caputo discusses Heidegger's eschatology his book *Radical Hermeneutics,* 187–208.

Notes to Chapter 6

1. A "discussion and correspondence" followed Levinas's presentation of "Transcendence and Height" to the members of the *Société Française de Philosophie* on January 27, 1962 (*BPW* 11). The editors of *Basic Philosophical Writings* included a published transcript of that discussion at the end of "Transcendence and Height."

2. This phrase is a section heading in Levinas's 1989 essay "Philosophy and Transcendence" (*AT* 29–37).

3. Levinas concludes an extended reflection on proximity (*OB* 81–97) by comparing this concept to the insufficient preparation within the self for the "plot that forms in the face of another, trace of an immemorial past, arousing a responsibility that comes from before and goes beyond what abides in the suspense of an epoque" (97).

4. While the essay was published as part of "Meaning and Sense" in 1964, it was published independently in 1963. See Levinas, "La trace de l'autre," 605–23.

5. "Both the fact that the totality overflows the sensible given and the fact that vision is incarnated would belong to the essence of sight. Its original and ultimate function would not consist in reflecting being as in a mirror. The receptivity of vision should not be interpreted as an aptitude to receive impressions. A philosophy such as that of Merleau-Ponty, who guides the present analysis, was able to be astonished by the marvel of a sight essentially attached to an eye" (*BPW* 39).

6. In his introduction, Peperzak points out that "Meaning and Sense" was written under the influence of Merleau-Ponty's emphasis on the many ways that "the cultural horizons of phenomena are emphasized and thematized" (*BPW* 34).

7. Husserl's moves in this direction are insufficient for Levinas. In one sense, Husserl "marks the end of this notion of meaning" that Levinas calls "sensualist empiricism." But Husserl nevertheless returns to "intellectualism" in the way he "accounts for meanings by a return to the given" (*BPW* 35). Merleau-Ponty is helpful by pressing Husserlian phenomenology on exactly this point. Merleau-Ponty "refused to resolve" the prereflective realm "by simple recourse to the finitude of the subject, incapable of a total reflection" (56).

8. By the late 1960s, Levinas's writings avoid these spatial themes, and when Levinas does employ concepts like "exteriority," he will only do so with careful disclaimers and caveats. Below I will evaluate the theory that Derrida's 1964 critique of Levinas played a key role in this terminological transition.

9. "Therefore fundamental ontology, from which alone all other ontologies can take their rise, must be sought in the existential analytic of Dasein" (*BT* 34). Heidegger frequently refers to his project in *Being and Time* as fundamental ontology. This phrase is complicated in Heideggerian studies by Heidegger's eventual abandonment of this project, though Levinas paid little attention to this move. Derrida, who followed the later progressions of Heidegger more closely, summarizes, "After desiring to restore the proper ontological intention dormant within metaphysics, after having reawakened the 'fundamental ontology' beneath 'metaphysical ontology,' Heidegger, faced by the tenacity of traditional ambiguity, finally proposes to abandon the terms 'ontology' and 'ontological' (*Introduction to Metaphysics*)" (VM 311n3).

10. One of the unanswered questions left open in *Totality and Infinity* is frequently revisited throughout the rest of Levinas's career: what is the relationship between God as *other* and the human other? This question remains critical in the early 1960s, and it is not a question easily answered. Peperzak takes it that Levinas knows there is work to be done on this front and reads "Meaning and Sense" as critical in this endeavor (*BPW* 35). I will address this question in the context of Levinas's engagement with Judaism and the Talmud below.

11. Levinas asks, "How is such a production possible?" (*BPW* 53).

12. There is a puzzling error in an introductory footnote to this essay in *Collected Philosophical Papers*—entitled in that volume as "Phenomenon and Enigma." Peperzak notes there that the essay was published in French in 1957 in the *Revue de Métaphysique et de Morale* 62 (*CPP* 61n). The essay Levinas published in that issue was actually "La philosophie et l'idée de l'infini," while "Phenomena and Enigma" first appeared in 1965 as "Énigme et phénomène" in *Esprit.* This error, which places the writing of "Phenomenon and Enigma" before *Totality and Infinity,* might be insignificant in other investigations of Levinas's work, but it is of utmost importance within the context of this study, since I am tracing Levinas's use of the term *diachrony,* which appears in this essay. Thus, it is critical to understand that contrary to Peperzak's footnote, "Phenomenon and Enigma" was first published *between* the publication of *Totality and Infinity* and *Otherwise than Being.*

13. Ciocan and Hansel, *Levinas Concordance,* s.v. "diachronie." The claim that this is the first appearance of "diachrony" in Levinas's oeuvre is based on my study of the *Levinas Concordance* as well as various writings not indexed in the concordance.

14. Ibid., s.v. "diachronie."

15. Saussure, *Course in General Linguistics,* 94, 97

16. There are surprising differences between Saussure's presentation of his own work and the articulation of linguistics as reproduced by his students. Since Levinas was working with the term in the *Course in General Linguistics,* these developments are not significant for this study.

17. Saussure, *Course in General Linguistics,* 67.

18. Ibid., 76.

19. Fabian, "Rule and Process," 105.

20. Ibid., 105.

21. Levinas discusses "diachronic transcendence" as a "disturbance" of rational speech, and then he asks, "How can such a disturbance occur?" (*BPW* 66–67, 68). He responds by outlining the "enigma" of the other person, who can "appear without renouncing his radical alterity" (68–69).

22. "But the other distinguishes himself absolutely, by absolving himself, moving off, passing, passing beyond being, to yield his place to being" (*BPW* 74).

23. Zahavi, *Husserl's Phenomenology,* 83.

24. This relationship to phenomenology is nicely summarized in Bettina Bergo's essay "Levinas's Weak Messianism in Time and Flesh," 225–48.

25. "Light makes objects into a world, that is, makes them belong to us.... The apprehension which is at the bottom of all our sensations is the origin of property in the world" (*EE* 40). "Thus a radical *passivity,* a radical exposure, prior to all the syntheses which have hitherto defined time, subjectivity, being, and truth for philosophy, and a radical *alterity,* again beyond all the syntheses which have hitherto defined time, subjectivity, being, and truth for philosophy, are related by means of ethics" (*TO* 17).

26. Bergo, "Levinas's Weak Messianism," 228–29.

27. Cohen, *Ethics, Exegesis and Philosophy,* 101.

28. Levinas writes, "Time accomplishes this freedom; it does not exist prior to the mind, does not engage it in a history in which is could be overwhelmed. Historical time is constituted. History is explained by thought" (*DEH* 78).

29. As Levinas describes Husserl's *Logical Investigations,* "which defines phenomenology so badly, but proves it so well, for it does so as one proves movement—by walking" (*DEH* 113).

30. Alweiss, *World Unclaimed,* 44.

31. In this critique, Levinas follows Immanuel Kant against Husserl. For Kant, the experience of alterity resides not in the realm of theory but in the moral realm; alterity is the result of a moral exigency, a superiority of the experience of the absolute. Alterity therefore requires not an intensification of the theoretical appropriation of objects as they are encountered by consciousness, but what Richard Cohen names a specifically "moral reading." For Kant this moral reading requires a pursuit of a common, universal, moral law that functions as a common legislation for the self and for every other person. But Levinas suspects in Kant a reduction and totalization, and therefore wishes to push further than both Husserl and Kant (Cohen, introduction to *DEH* xvii).

32. Levy, "Emmanuel Levinas on Secularization," 29.

33. This is argued vigorously by Tauber, "Outside the Subject," 439–59.

34. Levinas invokes the story of Jonah to underscore our inability to evade "God" (*BPW* 29).

35. Bloechl, *Face of the Other,* xiv.

36. "There is even in *Totality and Infinity,* the evocation of the *tzimtzum* [the idea in kabbalistic writings of the self-contraction of God in order to create the void in which creation can take place], but I won't venture into that" (Levinas, "Interview with Emmanuel Levinas," 107). Bracketed explanation of *tzimtzum* added by Wyschogrod.

37. Chalier, "Levinas and the Talmud," 114.

38. Thales of Miletos (ca. 624 BCE–ca. 546 BCE) is credited by Aristotle with being the "first philosopher" (Aristotle, *Metaphysics,* 983b20). There may be some historical validity to the legend that he inaugurates philosophy by outthinking the powerful mythology surrounding the solar eclipse. The legendary beginning of philosophy occurred when Thales anticipated an eclipse that

occurred during a battle between the Lydians and the Medes. When his predication was verified by the darkening of the sun, the combatants are said to have immediately made peace (Gottlieb, *Dream of Reason*, 5).

39. Tauber, "Outside the Subject," 452.

40. This claim is set up in opposition to Richard Cohen, who normally can be found supporting the religious components of Levinas's thought, but nevertheless claimed that Levinas's philosophy "stands or falls independent of its relation to Judaism or Jewish thought" (Cohen, *Elevations*, 127).

41. Levinas claims that there is "nothing more hypocritical than the messianic prophetism of the comfortable bourgeois" (*DF* 96).

42. "Here it becomes clear that the production of time in teaching, procreating, or loving is not only a way to ensure that time will still exist at the moment when the Messiah decides to come. Rather, the production of time in touch, in caress, is already a production of messianic triumph in the here and now" (Kavka, *Jewish Messianism and the History of Philosophy*, 177).

43. Wyschogrod, *Emmanuel Levinas*, 178.

44. Ibid., 179.

45. Burggraeve, "Bible Gives to Thought," 155–57.

46. Herbert Spiegelberg writes, "In the Paris of the years immediately before and after World War II Levinas came to know the crisis of modern civilization but also the crisis of traditional religiosity from close up. It too assumed existential significance for him. Is the God of Abraham, Isaac, and Jacob dead? What can religion still mean for a Jew after Auschwitz? It can be assumed with certainty that Levinas has struggled very personally with the problem of atheism. The fact that eventually he found in Talmudic piety . . . a source of inspiration cannot by any means be considered an automatic return to the traditions which had accompanied his childhood" (Spiegelberg, *Phenomenological Movement*, 613).

47. Levinas's interpretation: "In my own reading of this verse, *lemor* means 'in order to speak': 'Speak to the children of Israel in order that they might speak'; teach them profoundly enough so that they begin to speak, so that they hear to the point where they start speaking" (*BV* 80).

48. Levinas's encounter with Chouchani on this passage is narrated for the sake of its exegetical importance in Ouaknin, *Burnt Book*, 16–17.

49. Levinas repeats phrases like this to introduce most of his talmudic readings. He takes pains to point out that this humility is sincere and not posed or formal.

50. "Levinas does not bow to the superiority of the text, its power to teach and judge, *after* the fact of reading but *before* the fact of reading. There is a certain prerequisite attitude in which the text must be interpreted" (Aronowicz, translator's introduction, *NT* xxvi).

51. Cohen, *Ethics, Exegesis and Philosophy*, 259.

52. Ibid., 259.

53. Catherine Chalier points out that Levinas finds in the talmudic tractates "the extraordinary trace that Revelation leaves in a thought that, beyond the

vision of being, hears the word of God" (Chalier, "Levinas and the Talmud," 100–01).

54. Chalier, "Levinas and the Talmud," 114.

55. Claire Elise Katz summarizes this sentiment: "Some have suggested that [*Otherwise than Being*] is Levinas's response to Derrida's criticisms of *Totality and Infinity,* expressed in his essay 'Violence and Metaphysics' (1968). *Otherwise than Being* is often thought to respond to the charge that *Totality and Infinity* remained trapped in ontological language, despite Levinas's insistence that he provided a description of an ethical relation that is pre-reflective." (Katz, *Levinas, Phenomenology and his Critics,* 5).

56. Malka, *Emmanuel Levinas,* 172.

57. Bernasconi, "Deconstruction and the Possibility of Ethics," 129.

58. Ibid., 129.

59. Bernasconi, "Trace of Levinas in Derrida," 19. This entire essay is an extensive and helpful reading of "Violence and Metaphysics."

60. Alan Bass, translator's introduction to *Writing and Difference,* xiii.

61. Bernasconi, "Deconstruction and the Possibility of Ethics," 129.

62. Ibid., 129.

63. Malka, *Emmanuel Levinas,* 176.

64. Derrida wonders in parentheses if this is really as spatial as it seems.

65. Derrida compares this return to interiority and exteriority to the way one can write about something while also crossing out what one has written or the way one might rub a "rusty and devalued old coin."

66. Bernasconi, "Trace of Levinas in Derrida," 24.

67. Ibid., 14.

68. Ibid., 15. Bernasconi carefully outlines the differences between the two editions of Derrida's essay.

69. Ibid., 15.

70. These movements with and against Levinas are summarized in Geoffrey Bennington and Jacques Derrida's book *Jacques Derrida,* 302–12.

71. Derrida, *Margins of Philosophy,* 21.

72. "During the same period, Levinas had expelled the concept of *exteriority*" (VM 166).

73. Levinas never relies heavily on the language of symmetry and asymmetry, using these words just a handful of times across his philosophical works, and never in instrumental ways.

74. Bennington, "Derrida and Politics," 13.

75. Vries, *Religion and Violence,* 280.

76. Bennington, "Derrida and Politics," 13.

77. Derrida contests that Levinas has wrongly defined history, supposing that it is the "totality transcended by eschatology, metaphysics or speech" (VM 117).

78. He claims that Derrida writes with a "marvelous rigor, learned at the school of phenomenology, by devoting extreme attention to Husserl's discrete

moves and Heidegger's more sweeping ones, but applied with consistency and consummate skill" (*PN* 56).

79. "The exteriority has to be emphasized. It is not objective or spatial, recuperable in immanence and thus falling under the orders of—and in the order of consciousness; it is obsessional, non-thematizable and, in the sense we have just defined, anarchic" (*OB* 102). The only time Levinas mentions Derrida in *Otherwise than Being* is in a footnote that references Derrida's claim about "the exteriority of language" and the "allegedly inward aspect of meaning" (189n23).

80. Bernasconi, "Trace of Levinas in Derrida," 14.

81. Wyschogrod, "Derrida, Levinas, and Violence," 182.

82. Paola Marrati, "Derrida and Levinas," 70.

83. Ibid., 70.

84. Bernasconi, "Trace of Levinas in Derrida," 18–21.

85. For instance, see the essay "Substitution," which was first printed in 1968 in *La revue philosophique de Louvain* and later was modified as chapter 4 of *Otherwise than Being*. The English translation is found in Levinas, *Basic Philosophical Writings*, 79–95.

86. In his introduction to "Substition," Bernasconi claims that these three terms were first used in Levinas's confessional writings (*BPW* 79).

Notes to Chapter 7

1. "Let us look into this more closely. The response of the responsible one does not thematize the diachronical as though it were retained, remembered or historically reconstructed" (*OB* 11).

2. Cf. Pascal, *Pensées*, 25; *AT* 164; *OB* vii.

3. Levinas describes this past as "more profound than all I can reassemble by memory, by historiography, all that I can dominate a priori—in a time before the beginning" (*OB* 88).

4. "Proximity thus signifies a reason before the thematization of signification by a thinking subject, before the assembling of terms in a present, a pre-original reason that does not proceed from any initiative of the subject, an anarchic reason" (*OB* 166).

5. Later Levinas writes, "Chosen without a choice! If this passivity is not reducible to the passivity of an effect in a causal relation, if it can be conceived to be on the hither side of freedom and non-freedom, it must have the meaning of a *'goodness despite itself,'* a goodness always older than the choice.... Goodness is always older than the choice; the Good has always already chosen and required the unique one" (*OB* 56–57).

6. Levinas entitles a late collection of essays, *De Dieu qui vient à l'idée* (1982), which appears in English as *Of God Who Comes to Mind*.

7. This is the subtitle to Bloechl's *Liturgy of the Neighbor*.

8. "The Biblical notion of the Kingdom of God—kingdom of a non-thematizable God, a non-contemporaneous, that is non-present, God—must be not be

conceived as an ontic image of a certain 'époque' of the 'history of Being,' as a modality of essence" (*OB* 52).

9. Cohen, *Elevations,* 156.

10. Perpich, *Ethics of Emmanuel Levinas,* 118. Perpich credits this insight to Ricoeur, *Autrement,* 3.

11. Perpich, *Ethics of Emmanuel Levinas,* 118.

12. Levinas uses the phrase "hither side" dozens of times (see *OB* 43–49).

13. "Temporalization as lapse, the loss of time, is neither an initiative of an ego, nor a movement toward some telos of action. The loss of time is not the work of a subject" (*OB* 51).

14. Bernasconi, "What Is the Question," 250.

15. Smith, "Reason as One for Another," 68–69.

16. Levinas can now write of consciousness and the hypostasis: "Prior to the return to itself proper to the consciousness, this hypostasis, when it shows itself, does so under the borrowed mask of being. The event in which this unity or uniqueness of the hypostasis is brought out is not the grasping of self in consciousness. It is an assignation to answer without evasions, which assigns the self to be a self" (*OB* 106).

17. Perpich's bibliography indicates a familiarity with Ciaramelli's work but she does not cite his theory in developing her own, which appears to have been developed mostly independently.

18. Ciaramelli, *Transcendance et éthique,* 92–95.

19. Bettina Bergo, in a helpful summary of Ciaramelli's thesis, writes, "Whereas *TI* explored the emergence of the 'I' through its sensuous enjoyment of the elemental and labored creation of the habitation, the work never paused to inquire specifically into the ethical meaning of this genesis" (Bergo, *Levinas Between Ethics and Politics,* 141).

20. Bergo, "What Is Levinas doing?," 130.

21. Ciaramelli, *Transcendance et éthique,* 95–96.

22. "Henri Bergson, who, for the first time in the history of ideas, attempts to conceive of time outside of [the] failure of eternity, has characterized the destiny of that notion in philosophy as that of a becoming that passes for a privation of eternity" (*AT* 13).

23. "Husserl—who brings ultimate intelligibility back to temporality—brings the ideas in all their eternity (eternity understood here as omni-temporality) back to temporality" (*GDT* 107).

24. "That which Husserl still calls time-consciousness, i.e., consciousness of time, is precisely time, itself, in the primordial sense" (Heidegger, *Metaphysical Foundations of Logic,* 204).

25. "Or as thought it were formulated before every possible present, in a past that shows itself in the present of obedience without being recalled, without coming from memory, being formulated by him who obeys in his very obedience. But this is still perhaps a quite narrative, epic way of speaking" (*OB* 13).

26. "Proximity is no longer in knowing in which these relations with the neighbor show themselves, but do so already in narration, in the said, as an epos and a teleology.... The 'three unities' are not exclusively a matter of theatrical action; they command every exposition, assemble into a history, a narration, a tale, the bifid or bifocal relationship with the neighbor" (*OB* 83).

27. "It is the impossibility of the dispersion of time to assemble itself in the present, the insurmountable diachrony of time, a beyond the Said. It is diachrony that determines the immemorial; a weakness of memory does not constitute diachrony" (*OB* 38).

28. Perpich, *Ethics of Emmanuel Levinas,* 79.

29. "The *otherwise than being* is stated in a Saying that must also be unsaid in order to thus extract the *otherwise than being* from the Said in which it already comes to signify but a *being otherwise*.... Can this Saying and this *being unsaid* be assembled, can they be at the same time? In fact to require this simultaneity is already to reduce being's other to *being* and *not being*" (*OB* 7).

30. "In an approach I am first a servant of a neighbor, already late and guilty for being late" (*OB* 87).

31. "The subjectivity of a subject is vulnerability, exposure to affection, sensibility, a passivity more passive still than any passivity, an irrecuperable time, an unassemblable diachrony of patience, an exposedness always to be exposed the more, an exposure to expressing, and thus to saying, thus to giving" (*OB* 50).

32. "The fecundity of subjectivity, by which the I survives itself, is a condition required for the truth of subjectivity, the clandestine dimension of the judgment of God" (*TI* 247).

33. Levinas's language indicates the secondary nature of the mother: "The notion of maternity must be introduced here to account for this recourse" (*TI* 278).

34. The phrase "trope of the feminine" is borrowed from Stella Sandford. Responding to Adriaan Peperzak's defense of Levinas's use of "*le femme*" in *Totality and Infinity* (see Perperzak, *To the Other,* 129), Sandford writes, "Presumably he would not be foolhardy enough to argue that, a metaphor, the trope of the feminine has no connection whatsoever, no linguistic or cultural reference at all, to empirically existing women" (Sandford, *The Metaphysics of Love,* 47).

35. "The subject is inseparable from this appeal or this election, which cannot be declined" (*OB* 53; cf. 15, 57, 122, 127).

36. "The problem of transcendence and of God and the problem of subjectivity irreducible to essence, irreducible to essential immanence, go together" (*OB* 17).

37. The phrase "individual singularity" is my own translation of Levinas's French word "*semelfacticité*" (*AE* 72). The word is virtually untranslated by Lingis, as "semelfacticity," which is hardly helpful for English readers. The term creates something of a puzzle in *Otherwise than Being* that warrants some

consideration. The word "*semelfacticité*" provides echoes of philosopher Vladimir Jankélévitch, a Russian-born, French philosopher who was a contemporary of Levinas. Levinas was familiar with Jankélévitch, citing him occasionally across his writings. Jankélévitch's use of "*semelfacticité*" is summarized by Colin Smith: "our life seen as dissimilar in its course from every other ... universally unique" (Smith, "Philosophy of Vladimir Jankélévitch," 317). The word is a derivative of the term *semelfactif,* which Levinas uses at least four times in his work according to the *Levinas Concordance,* s.v. *semelfactif.* Michael B. Smith and Barbara Harshav translate *semelfactif* as "unique singularities" in "The Other, Utopia, and Justice" (*EN* 229).

38. Perpich, *Ethics of Emmanuel Levinas,* 79, 118. Perpich also calls the narrative of *Totality and Infinity* a "tale," picking up on Levinas's own imagery as he moves to critique the ego-centered "tale" of the "unique exemplar" (79).

39. "In this plot I am bound to others before being tied to my body" (*OB* 76).

40. Bergo, *Levinas between Ethics and Politics,* 142. Bergo is here discussing Ciaramelli's theory of the transition between *Totality and Infinity* and *Otherwise than Being.*

41. For a thoughtful analysis of how this theme is reflected in both *Existence and Existence* and *Totality and Infinity,* see Kavka, *Jewish Messianism and the History of Philosophy,* 181–82.

42. Reinhard, "Kant with Sade, Lacan with Levinas," 793.

43. Ciaramelli, *Transcendence et éthique,* 95–96. These phrases are translated from Ciaramelli's French by Bettina Bergo in *Levinas Between Ethics and Politics,* 141–43.

44. Levinas refers to creation *ex nihilo* in *Otherwise than Being,* but it appears several times in his work (*OB* 113; *TI* 63, 89, 104).

45. Ricoeur, *Oneself as Another,* 340.

46. Ibid., 340.

47. Richard Cohen claims, against Ricoeur, that "there is no surreptitious reversal of a reversal in Levinas's account, but the intensification of an original reversal" (Cohen, "Moral Selfhood," 146). Cohen methodically examines each of Ricoeur's critiques of Levinas in this text, as well as in Cohen, *Ethics, Exegesis and Philosophy,* 283–325.

48. Chalier, "Levinas and the Feminine," 126.

49. Kagan, *The Limits of Morality,* 3–4. Perpich points to this text in Perpich, *The Ethics of Emmanuel Levinas,* 82.

50. On the topic of Levinas and theodicy, see Bernstein's excellent treatment in "Evil and the Temptation of Theodicy," 252–67.

51. Dostoyevsky, *Brothers Karamazov,* 347.

52. Portions of the discussion here of Dostoevsky are published in my introduction to the collection: Severson, ed., *I More than Others,* 1–4. The story of Markel can be found in Dostoyevsky, *Brothers Karamazov,* 344–49.

53. Dostoyevsky, *Brothers Karamazov,* 345.

54. Ibid., 346.

55. This line has been translated several different ways, and this rendering amalgamates several translations. Andrew MacAndrew translates this as, "we are all guilty toward others and I am the guiltiest of all" (ibid., 347).

56. Tina Chanter points out that understanding the role of time in the differentiation between the saying and the said is "of central importance to everything Levinas has to say about ethics, to his rethinking of subjectivity, to his demand that we acknowledge the founding role of the Other for the enterprise of philosophy" (Chanter, *Time, Death, and the Feminine,* 145).

57. Levinas initially outlines this distinction in *OB* 5–9, but he returns to these concepts throughout *Otherwise than Being.*

58. "By comparison being appears like a game. Being is play or detente, without responsibility, where everything possible is permitted" (*OB* 6). Levinas claims that the play between words and meanings can mimic the "fallacious frivolity of play" (6).

59. In *Totality and Infinity,* Levinas ascribed to the face the very "first word," the phrase "you shall not murder" that is spoken from every face before any sound (*TI* 199).

60. Waldenfels, "Levinas on the Saying and the Said," 94.

61. This utterance, Levinas tells us, is from the face, from the eyes. My speech is already a response to this; I never speak the first word: "The face, preeminently expression, formulates the first word: the signifier arising at the thrust of his sign, as eyes that look at you" (*TI* 178).

62. "It is the coming of the order to which I am subjected before hearing it, or which I hear in my own saying. It is an august command, but one that does not constrain or dominate and leaves me outside of any correlation with its source. No *structure* is set up with a correlate. Thus the saying that comes to me is my own word. Authority is not somewhere, where a look could go seek it, like an idol, or assume it like a logos" (*OB* 150).

63. Peperzak, *To the Other,* 221.

64. "Saying prolongs this extreme passivity, despite its apparent activity" (*OB* 153).

65. "Here the identity of the subject comes from the impossibility of escaping responsibility, from the taking charge of the other. Signification, saying—my expressivity, my own signifyingness qua sign, my own verbality qua verb—cannot be understood as a modality of being" (*OB* 14).

66. "The responsibility for the other cannot have begun in my commitment, in my decision. The unlimited responsibility in which I find myself comes from the hither side of my freedom" (*OB* 10).

67. Chanter, *Time, Death, and the Feminine,* 145–47.

68. "But in reducing the said to the saying, philosophical language reduces the said to breathing opening to the other and signifying to the other its very signifyingness. This reduction is then an incessant unsaying of the said, a reduction of the saying always betrayed by the said, whose words are defined by

non-defined words; it is a movement going from said to unsaid in which the meaning shows itself, eclipses and shows itself" (*OB* 181).

69. "The for-the-other of responsibility for the other does not proceed from any free commitment, any present, in which its origins would germinate, or in which an identity identifying itself would catch its breath. . . . Responsibility without a prior commitment, without a present, without an origin, anarchic, is thus an infinite responsibility of the one for the other who is abandoned to me without anyone being able to take my place as the one responsible for him" (*OB* 153).

70. In the next sentence, Levinas suggests that "this divergency perhaps deserves the name glory."

71. "The accusation is in this sense persecuting; the persecuted one can no longer answer it. More exactly, it is accusation which I cannot answer, but for which I cannot decline responsibility. Already the position of the subject is a deposition, not a *conatus essendi.* It is from the first a substitution by a hostage expiating for the violence of the persecution itself" (*OB* 127).

72. Hayat, preface to *AT* xix.

73. Plotinus, *Enneads* 356.

74. Ibid., 356.

75. Ibid., 424.

76. Plato, *Symposium,* 211d–e.

77. Naas, "Lending Assistance Always to Itself,", 99n5. Naas provides a nuanced look at the way Levinas relates to Plato differently at different points in his career. Across the 1940s and 1950s Plato remains for Levinas the "essential ontological thinker," though the later Levinas is increasingly inclined to point to the places where Plato's work undermines ontology and points to a beyond.

78. "Questions and Answers" is the transcript of a dialogue session in which Levinas participated at the University of Leyden on May 20, 1975 (see *GCM* 79).

79. Sounding very much like what he would write in his later works, Levinas wrote in *Totality and Infinity,* "Though of myself I am not exterior to history, I do find in the other a point that is absolute with regard to history—not by amalgamating with the other, but in speaking with him. History is worked out over the ruptures of history, in which judgment is borne upon it. When man truly approaches the other he is uprooted from history" (*TI* 52).

Notes to Chapter 8

1. Peperzak, *To the Other,* 7.

2. Diane Perpich claims that "commentators have generally understood deformalization to involve the provision of a concrete experience in which the formal structure is realized" (Perpich, *The Ethics of Emmanuel Levinas,* 60). She points to Peperzak, who demonstrates Levinas's methodological push to find lived experiences in which the formal expression of his philosophy "can realize itself" (see Peperzak, *To the Other,* 61). This is indeed a critical aspect of Levinas's philosophy. Levinas was bothered since the 1920s by the formalities and theo-

reticism that he felt inhabited too much philosophical discourse, particularly in the tradition of Husserl. This is an effect of Heidegger's turn against Husserl. Philosophy becomes too intellectual when it over-intellectualizes the encounter with human beings in the world. When theory absorbs the attention of the philosophy, key components of an experience are potentially occluded. When Levinas questions the intentionality of consciousness, he is in fact attempting to be more Husserlian than Husserl. I suspect this is why he claims to the end that he is a phenomenologist; his moves away from Husserl are attempts to deformalize the structure of philosophy, to draw it closer to the real world of lived experience. His pursuit of a less *formal* understanding of time has to be read in the context of his response to the formalizations he sees in Husserl, and still in Heidegger.

3. Ricoeur discusses his problems with Levinas's work in the seventh and tenth studies of *Oneself as Another*. Ricoeur dismisses Levinas's language in *Otherwise than Being* as hyperbolic "to the point of paroxysm" (Ricoeur, *Oneself as Another,* 238). But he never mentions diachrony and never directly investigates the distinction between the saying and the said.

4. Fruitful projects are possible in this regard, aligning Levinas's diachrony with understandings of time in Kant, Kierkegaard, Heidegger, and many others.

5. Derrida, *Adieu to Emmanuel Levinas,* 4.

6. Kigel, translator's postscript in Nemo, *Job and the Excess of Evil,* 236.

7. Chanter's book-length treatment of gender and time in Levinas and Heidegger (*Time, Death, and the Feminine*) has been instrumental in preceding chapters. In this section, I will explore the implications of that text alongside two of her later articles.

8. Chanter, "Conditions," 311–12.

9. Ibid., 310.

10. Brody, "Levinas's Maternal Method," 53–77.

11. See also Chanter, introduction to *Feminist Interpretations of Emmanuel Levinas,* 25.

12. Vasey, "Faceless Women and Serious Others," 329, 317. These quotes appear in Sonia Sikka's article "The Delightful Other," 106–07.

13. "There is not this measure of agreement among feminists about what the 'feminine' consists in, and attempts to define the feminine that claim universality for themselves . . . cannot help but fall into essentialism" (Sikka, "The Delightful Other," 107).

14. Ibid., 107.

15. Perpich, *Ethics of Emmanuel Levinas,* 192.

16. Grace Jantzen writes that "feminists have been anxious to resist romanticizing motherhood, either the labour of giving birth or the labour of bringing up children: still less is it a feminist view that womanhood should be equated with motherhood" (Jantzen, *Becoming Divine,* 143). For a full treatment of these themes relative to Levinas's work, see the essay collection: Chanter, ed., *Feminist Interpretations of Emmanuel Levinas.*

17. Perpich, *Ethics of Emmanuel Levinas,* 193.

18. "Those silent comings and goings of the feminine being whose footsteps reverberate the secret depths of being are not the turbid mystery of the animal and feline presence whose strange ambiguity Baudelaire likes to evoke" (*TI* 156).

19. Brody, "Levinas's Maternal Method," 74.

20. Bloechl says of the concept of "ipseity" in Levinas, "Levinas's real target is therefore any approach to human existence and experience that supposes or defends the primacy of what has been called *ipseity*" (Bloechl, *Liturgy of the Neighbor,* 87).

21. Plato, *Timaeus,* 48–53.

22. John Sallis offers an exceedingly helpful reading of this theme in his book *Chorology.*

23. Sallis, *Chorology,* 98.

24. We must be cautious not to presume, as Derrida seems to, that Plato is at work deconstructing his own system (Derrida, "*Khōra,*" 89–130). This idea of the khora receives only passing treatment in the *Timaeus* dialogue. Plato's use of ventriloquism defies any easy association between *Timaeus's* suggestions and Plato's overall system.

25. Plato, *Timaeus,* 48–53.

26. Caputo, ed., *Deconstruction in a Nutshell,* 99.

27. In a 1985 interview, Françoise Armengaud asked Levinas, "Do you not also say that 'all philosophy is Platonic'?" Levinas responded, "That is a quotation from an earlier text, I believe. I do not reject it, to the extent that the link between philosophy and transcendent alterity is affirmed in the Platonic theory of ideas, where the problem or the anxiety of that radical alterity—even though there is an attempt to reduce it—seem to me to authenticate philosophy.... I do not reject my attachment to Platonism, because to owe the daring formulation *beyond-being* to Plato is good luck" (*IR* 249). And on another occasion he describes his philosophy as a "return to Platonism in a new way" (*CPP* 101).

28. Levinas uses these phrases to indicate the structure of Heideggerian ontology.

29. Philip Harold writes, "The acceptance of diachrony and of skepticism is the overcoming of gnosticism. To which there is nothing more foreign than suffering for what cannot be known" (Harold, *Prophetic Politics,* 151).

30. This early statement is part of Levinas's description of the *il y a* and its irremissible character (*TO* 51). The claim here, that being is the condition for the appearance of diachrony, is not so far from the sense Levinas advocates in his earlier work, where being's terror is related to the inability of the subject to evade.

31. Chanter, "Hands that Give," 48–62.

32. Pascal, *Pensées,* 25. Cf. *AT* 164.

33. Drew Dalton defines Levinasian shame as "the very way in which one is inexorably bound to oneself. Shame thus functions as a revelation of the terrible weight of having to be one's self.... Whereas one can escape the more pedestrian

shame of having somehow failed by seeking forgiveness or by attempting to rectify the wrong, the experience of shame before the other shakes the subject to its core, for it is the result of an accusation of the very way in which one is—one cannot simply escape it then by merely plucking out the proverbial eye that offends" Dalton, *Longing for the Other*, 128–29). Dalton provides an extensive discussion of shame in this volume (114–37).

34. Dalton, *Longing for the Other*, 129.

35. Cf. Husserl, *Phenomenology of Internal Time Consciousness*, 131.

36. "The face which breaks through its own plastic image, but must be revealed, simultaneously with this presence, in its withdrawal and in its absence" (*TI* 155).

37. Brody, "Levinas's Maternal Method," 73.

38. "The subject called incarnate does not result from a materialization, an entry into space and into relations of contract and money which would have been realized by a consciousness, that is a self-consciousness, forewarned against every attack and first non-spatial" (*OB* 77).

39. "The sensible—maternity, vulnerability, apprehension—binds the node of incarnation into a plot larger than the apperception of the self. In this plot I am bound to others before being tied to my body" (*OB* 76).

40. Chanter, "Conditions," 328.

41. "Indeed in the transcendence of intentionality diachrony is reflected, that is, the psyche itself, in which the inspiration of the same by the other is articulated as a responsibility for another, in proximity" (*OB* 67). Cf. Stähler, "Getting under the Skin," 68–69.

42. Chanter, *Time, Death, and the Feminine*, 74.

43. Chanter, "Conditions," 333.

44. For a discussion of how "ontological difference" figures in Levinas's resistance to Heidegger's ontology, see Tina Chanter's chapter "Ontological Difference, Sexual Difference, and Time," in *Time, Death, and the Feminine*, 37–74.

45. Chanter references this passage in "Hands that Give," 57.

46. Luce Irigaray has pointed out that one of the more dangerous generalities is the eradication of the "*difference between the sexes*," which simply defaults back to the basically masculine subject. This simply serves to "*reduce all others to the economy of the Same*" (Irigaray, *Sex Which is Not One*, 74).

47. Lisa Guenther's recent book *The Gift of the Other: Levinas and the Politics of Reproduction* is noteworthy in this regard. Guenther acknowledges that "Levinas himself was not a feminist in any recognizable sense of the word" (Guenther, *Gift of the Other*, 6). Yet she sees promise in the analogy between maternity and ethics, such that "maternity would not refer to a biological or social imperative for women to reproduce, but rather an ethical imperative for each of us to bear the stranger *as if* she were already under my own skin, gestating in my own flesh" (7). Guenther's political applications for these thoughts take her a considerable distance from Levinas.

48. Guenther, *Gift of the Other,* 31. Guenther quotes Hannah Arendt: "The lifespan of man running toward death would inevitably carry everything human to ruin and destruction if it were not for the faculty of interrupting it and beginning something new, a faculty which is inherent in action like an ever-present reminder that men, although they must die, are not born in order to die but in order to begin" (Arendt, *Human Condition,* 246).

49. Westphal, "Transparent Shadow," 278.

50. "But does this gap [between the subject and reality] come from the subject? Does it come from a *being* concerned about its being and persevering in being, from an interiority clothed in an essence of a personage, from a singularity taking pleasure in its exception, concerned with its happiness—or with its salvation—with its private intentions in the midst of the universality of the true?" (*GCM* 5).

51. Michael Newman connects the concept of trauma with the confrontation in Levinas's understanding of time: "The trace is thus absolutely—and paradoxically—primary, and memory becomes the response to our receiving of an immemorial affection which as having-never-been-present-as-such is no different from the future to come. The affection of time is thus not auto-affection but a hetero-affection with a structure that is at once traumatic and messianic whereby the immanence of the present is broken open by the absolute Other.... In *Otherwise than Being,* the traumatic-messianic account of time will be extended into a description of responsible subjectivity" (Newman, "Sensibility, Trauma, and the Trace," 110).

52. Gschwandtner, "Neighbor and the Infinite," 246.

53. Ibid., 246.

54. "But justice can be established only if I, always evaded from the concept of the ego, always desituated and divested of being, always in non-reciprocal relationship with the other, always for the other, can become an other like the others" (*OB* 160).

55. Cohen, *Elevations,* 158.

56. Vasey, "Faceless Women and Serious Others," 327.

57. Chanter, "Hands That Give," 59.

58. This approach to *Otherwise than Being* was suggested to me by Jeffrey Bloechl.

59. Cf. Heidegger, *Fundamental Concepts of Metaphysics,* 182–84.

60. Chanter, "Hands That Give," 59.

61. John D. Caputo attempts to expose the political edge of Levinas's philosophy in his essay "*Adieu—sans Dieu:* Derrida and Levinas." Here, Caputo claims that "This ethics of hospitality must certainly translate into politics, must have a bearing, must be borne across the borders of ethics and politics" (290).

Bibliography

Ainley, Alison. "The Feminine, Otherness, Dwelling: Feminist Perspectives on Levinas." In *Facing the Other: The Ethics of Emmanuel Levinas,* edited by Seán Hand, 7–20. London: Curzon, 1996.

Allen, Sarah. "Loving the Good Beyond Being: The Paradoxical Sense of Levinas's 'Return' to Platonism." In *Emmanuel Levinas 100: Proceedings of the Centenary Conference, Bucharest, 4–6 September 2006,* edited by Cristian Ciocan, 75–107. Bucharest: Zeta Books, 2008.

Alweiss, Lilian. *The World Unclaimed: A Challenge to Heidegger's Critique of Husserl.* Athens, OH: Ohio University Press, 2003.

Arendt, Hannah. *The Human Condition.* Chicago: University of Chicago Press, 1958.

Aristotle. *The Metaphysics.* Translated by Hugh Lawson-Tancred. London: Penguin Books, 1998.

———. *The Nicomachean Ethics.* Translated by J. A. K. Thompson. London: Penguin Books, 2003.

———. *Physics.* Translated by Robin Waterfield. Oxford: Oxford University Press, 1996.

Beauvoir, Simone de. *The Second Sex.* Translated by H. M. Parshley. New York: Everyman's Library, 1993.

Beiser, Frederick. *Hegel.* New York: Routledge, 2005.

Bennington, Geoffrey. "Derrida and Politics." In *Deconstruction: Critical Concepts in Literary and Cultural Studies,* vol. 4, edited by Jonathan Culler, 3–23. New York: Routledge, 2003.

Bennington, Geoffrey, and Jacques Derrida. *Jacques Derrida.* Translated by Geoffrey Bennington. Chicago: University of Chicago Press, 1999.

Bergo, Bettina. *Levinas between Ethics and Politics: For the Beauty that Adorns the Earth.* Pittsburgh: Duquesne University Press, 2003.

———. "Levinas's Weak Messianism in Time and Flesh, or The Insistence of Messiah Ben David." *Journal for Cultural Research* 13, no. 3/4 (July 2009): 225–48.

———. "What Is Levinas Doing?: Phenomenology and the Rhetoric of an Ethical Un-Conscious." *Philosophy and Rhetoric* 38, no. 2 (2005): 122–44.

Bergson, Henri. *Creative Evolution*. Translated by Arthur Mitchell. New York: H. Holt, 1913.

———. *Matter and Memory*. Translated by Nancy Margaret Paul and W. Scott Palmer. Mineola, NY: Dover Publications, 2004.

———. *Time and Free Will: An Essay on the Immediate Data of Consciousness*. Translated by Frank Lubecki Pogson. New York: Macmillan, 1921.

Bernasconi, Robert. "Deconstruction and the Possibility of Ethics." In *Deconstruction and Philosophy: The Texts of Jacques Derrida*, edited by John Sallis, 122–39. Chicago: University of Chicago Press, 1987.

———. "The Trace of Levinas in Derrida." In *Derrida and Différance*, edited by Robert Bernasconi and David C. Wood, 13–29. Evanston, IL: Northwestern University Press, 1988.

———. "What Is the Question to Which 'Substitution' Is the Answer?" In *The Cambridge Companion to Levinas*, edited by Simon Critchley and Robert Bernasconi, 234–51. Cambridge: Cambridge University Press, 2002.

Bernstein, Richard J. "Evil and the Temptation of Theodicy." In *The Cambridge Companion to Levinas*, edited by Simon Critchley and Robert Bernasconi, 252–67. Cambridge: Cambridge University Press, 2002.

Bloechl, Jeffrey, ed. *The Face of the Other and the Trace of God: Essays on the Philosophy of Emmanuel Levinas*. New York: Fordham University Press, 2000.

———. *Liturgy of the Neighbor: Emmanuel Levinas and the Religion of Responsibility*. Pittsburgh: Duquesne University Press, 2000.

Blond, Phillip. "Emmanuel Levinas: God and Phenomenology." In *Post-Secular Philosophy: Between Philosophy and Theology*, edited by Phillip Blond, 195–228. London: Routledge, 1998.

Brody, Donna. "Levinas's Maternal Method from 'Time and the Other' Through *Otherwise than Being*." In *Feminist Interpretations of Emmanuel Levinas*, edited by Tina Chanter, 53–77. University Park: Pennsylvania State University Press, 2001.

Burggraeve, Roger. "The Bible Gives to Thought: Levinas on the Possibility and Proper Nature of Biblical Thinking." In *The Face of the Other and the Trace of God: Essays on the Philosophy of Emmanuel Levinas*, edited by Jeffrey Bloechl, 165–83. New York: Fordham University Press, 2000.

———. *The Wisdom of Love in the Service of Love: Emmanuel Levinas on Justice, Peace, and Human Rights.* Translated by Jeffrey Bloechl. Milwaukee, WI: Marquette University Press, 2002.

Caputo, John D. "*Adieu—sans Dieu:* Derrida and Levinas." In *The Face of the Other and the Trace of God: Essays on the Philosophy of Emmanuel Levinas,* edited by Jeffrey Bloechl, 276–311. New York: Fordham University Press, 2000.

Caputo, John D., ed. *Deconstruction in a Nutshell: A Conversation with Jacques Derrida.* New York: Fordham, 1997.

———. *Demythologizing Heidegger.* Bloomington: Indiana University Press, 1993.

———. *Radical Hermeneutics: Repetition, Deconstruction, and the Hermeneutic Project.* Bloomington: Indiana University Press, 1987.

———. *The Weakness of God: A Theology of the Event.* Bloomington: Indiana University Press, 2006.

Caygill, Howard. *Levinas and the Political.* London: Routledge, 2002.

Chalier, Catherine. "Levinas and the Feminine." In *Re-Reading Levinas,* edited by Robert Bernasconi and Simon Critchley, 119–29. Bloomington: Indiana University Press, 1991.

———. "Levinas and the Talmud." In *The Cambridge Companion to Levinas,* edited by Simon Critchley and Robert Bernasconi, 100–18. Cambridge: Cambridge University Press, 2002.

Chanter, Tina. "Conditions: The Politics of Ontology and the Temporality of the Feminine." In *Addressing Levinas,* edited by Eric Sean Nelson, Antje Kapust, and Kent Still, 310–37. Evanston, IL: Northwestern University Press, 2005.

———. "Hands that Give and Hands that Take: The Politics of the Feminine in Levinas." In *Difficult Justice: Commentaries on Levinas and Politics,* edited by Asher Horowitz and Gad Horowitz, 48–62. Toronto: University of Toronto Press, 2006.

Chanter, Tina, ed. *Feminist Interpretations of Emmanuel Levinas.* University Park, PA: Pennsylvania State University Press, 2001.

———. *Time, Death, and the Feminine: Levinas with Heidegger.* Stanford: Stanford University Press, 2001.

Ciaramelli, Fabio. *Transcendance et éthique: Essai sur Lévinas.* Brussels: Ouisia, 1989.

Cicero. *Letters to Atticus.* Translated by E. O. Winstedt. New York: Macmillan, 1921.

Ciocan, Cristian, and Georges Hansel. *Levinas Concordance.* Dordrecht: Springer, 2005.

Cohen, Richard A. *Elevations: The Height of the Good in Rosenzweig and Levinas.* Chicago: University of Chicago Press, 1994.

———. *Ethics, Exegesis and Philosophy: Interpretation after Levinas.* Cambridge: Cambridge University Press, 2001.

———. "The Face of Truth in Rosenzweig, Levinas, and Jewish Mysticism." In *Phenomenology of the Truth Proper to Religion,* edited by Daniel Guerrière, 175–204. Albany: State University of New York Press, 1990.

———. "Moral Selfhood: A Levinasian Response to Ricoeur on Levinas." In *Ricoeur as Another: The Ethics of Subjectivity,* edited by Richard A. Cohen and James L. Marsh, 127–60. Albany: State University of New York Press, 2002.

Collins, Jeff. *Introducing Heidegger.* 3rd ed. Cambridge: Icon Books, 2006.

Critchley, Simon, and Robert Bernasconi, eds. *The Cambridge Companion to Levinas.* Cambridge: Cambridge University Press, 2002.

Crowell, Steven Galt. "Metaphysics, Metontology, and the End of *Being and Time.*" *Philosophy and Phenomenological Research* 60, no. 2 (2000): 307–31.

Culler, Jonathan, ed. *Deconstruction: Critical Concepts in Literary and Cultural Studies.* Vol. 4. New York: Routledge, 2003.

Dalton, Drew M. *Longing for the Other: Levinas and Metaphysical Desire.* Pittsburgh: Duquesne University Press, 2009.

Davidson, Arnold. "Introduction to Musil and Levinas." *Critical Inquiry* 17, no. 1 (1990): 35–45.

Defoe, Daniel. *Robinson Crusoe.* New York: Barnes and Noble, 2003.

Derrida, Jacques. *Adieu to Emmanuel Levinas.* Translated by Pascale-Anne Brault and Michael Naas. Stanford: Stanford University Press, 1999.

———. "*Khōra.*" In *On the Name,* edited by Thomas Dutoit, translated by David Wood, John P. Leavey, Jr., and Ian McLeod, 89–130. Stanford: Stanford University Press, 1995.

———. *Margins of Philosophy.* Translated by Alan Bass. Chicago: University of Chicago Press, 1982.

———. "Violence and Metaphysics." In *Writing and Difference,* translated by Alan Bass, 79–153. Chicago: University of Chicago Press, 1978.

———. *Writing and Difference.* Translated by Alan Bass. Chicago: University of Chicago Press, 1978.

Descartes, René. *Meditations on First Philosophy.* Translated by John Cottingham. Cambridge: Cambridge University Press, 1996.

———. *Principles of Philosophy.* Translated by Valentine Rodger Miller and Reese P. Miller. Dordrecht: Kluwer Academic, 1991.

———. René Descartes to Henry More, August 1649. In *New Studies in the Philosophy of Descartes,* by Norman Kemp Smith, 200–01. London: Macmillan, 1966.

Desmond, William. "Philosophies of Religion: Marcel, Jaspers, Levinas." In *Continental Philosophy in the 20th Century,* edited by Richard Kearney, 131–74. New York: Routledge, 2003.

Dostoyevsky, Fyodor. *The Brothers Karamazov.* Translated by Andrew H. MacAndrew. New York: Bantam Books, 1981.

Drabinski, John. *Sensibility and Singularity: The Problem of Phenomenology in Levinas.* Albany: State University of New York Press, 2001.

Dutton, Denis. "Kaufmann, Heidegger and Nazism." *Philosophy and Literature* 12 (1988): 325–36.

Epicurus. Epicurus to Menoeceus. In *The Stoic and Epicurean Philosophers: The Complete Writings of Epicurus, Epictetus, Lucretius, and Marcus Aurelius,* edited by Whitney J. Oates, 30–33. New York: Random House, 1940.

Fabian, Johannes. "Rule and Process." In *Time and the Work of Anthropology: Critical Essays, 1971–1991,* 87–109. Amsterdam: Harwood Academic, 1991.

Flew, Anthony. *A Dictionary of Philosophy.* Rev. 2nd ed. New York: St. Martin's Press, 1984.

Franck, Didier. "The Body of Difference." In *The Face of the Other and the Trace of God: Essays on the Philosophy of Emmanuel Levinas,* edited by Jeffrey Bloechl, 3–29. New York: Fordham University Press, 2000.

Gibbs, Robert. *Correlations in Rosenzweig and Levinas.* Princeton, NJ: Princeton University Press, 1994.

Gordon, Peter Eli. *Rosenzweig and Heidegger: Between Judaism and German Philosophy.* Berkeley: University of California Press, 2003.

Gottlieb, Anthony. *The Dream of Reason: A History of Philosophy from the Greeks to the Renaissance.* New York: W. W. Norton, 2000.

Gschwandtner, Christina M. "The Neighbor and the Infinite: Marion and Levinas on the encounter Between Self, Human Other, and God." *Continental Philosophy Review* 40 (2007): 231–49.

Guenther, Lisa. *The Gift of the Other: Levinas and the Politics of Reproduction.* Albany: State University of New York Press, 2006.

Hankins, Thomas. *D'Alembert: Science and the Enlightenment.* New York: Gordon and Breach, 1990.

Harold, Philip J. *Prophetic Politics: Emmanuel Levinas and the Sanctification of Suffering.* Albany: State University of New York Press, 2009.

Hegel, G. W. F. *The Phenomenology of Mind.* 2nd ed. Translated by J. B. Baillie. Mineola, NY: Dover, 2003.

Heidegger, Martin. *Being and Time.* Translated by John Macquarrie and Edward Robinson. New York: Harper and Row, 1962.

______. *The Fundamental Concepts of Metaphysics: World, Finitude, Solitude.* Translated by William McNeill and Nicholas Walker. Bloomington: Indiana University Press, 2001.

———. *The Metaphysical Foundations of Logic.* Translated by Michael Heim. Bloomington: Indiana University Press, 1984.

———. "What Is Metaphysics?" Translated by R. F. C. Hull and Alan Crick, in *Existence and Being,* edited by Werner Brock, 353–93. Chicago: Henry Regnery, 1949.

Heraclitus. *Fragments of Heraclitus.* In *The Presocratics,* by Edward Hussey, 32–59. New York: Charles Scribner's Sons, 1973.

Hitler, Adolf. *Mein Kampf.* Translated by Ralph Manheim. Boston: Houghton Mifflin Harcourt, 1943.

Hodge, Joanna. "Ethics and Time: Levinas between Kant and Husserl." *Diacritics* 32, no. 3/4 (2002): 107–34.

———. *Heidegger and Ethics.* London: Routledge, 1995.

Homer. *The Iliad.* Translated by Robert Fagels. New York: Penguin Books, 1990.

Husserl, Edmund. *Logical Investigations.* Vol. 1. Translated by J. N. Findlay. London: Routledge and Kegan Paul, 1970.

———. *The Phenomenology of Internal Time-Consciousness.* Translated by James S. Churchill. Bloomington: Indiana University Press, 1964.

Hussey, Edward. *The Presocratics.* New York: Charles Scribner's Sons, 1973.

Irigaray, Luce. *This Sex Which Is Not One.* Translated by Catherine Porter. Ithaca, NY: Cornell University Press, 1985.

Irwin, F. Ruth. "Heidegger and Nietzsche: Nihilism and the Question of Value in Relation to Education." In *Heidegger, Education, and Modernity,* edited by Michael A. Peters, 191–210. Lanham, MD: Rowman and Littlefield, 2002.

Jantzen, Grace. *Becoming Divine: Towards a Feminist Philosophy of Religion.* Manchester: Manchester University Press, 1998.

Kagan, Shelly. *The Limits of Morality.* Oxford: Oxford University Press, 1989.

Kant, Immanuel. *Groundwork for the Metaphysics of Morals.* Translated by Mary J. Gregor. Cambridge: Cambridge University Press, 1998.

Katz, Claire Elise. *Levinas, Judaism, and the Feminine: The Silent Footsteps of Rebecca.* Bloomington: Indiana University Press, 2003.

Katz, Claire Elise, ed. *Levinas, Phenomenology and his Critics.* New York: Routledge, 2005.

Kaufmann, Walter. *Nietzsche, Heidegger, and Buber.* Vol. 2 of *Discovering the Mind.* New Brunswick, NJ: Transaction, 1992.

Kavka, Martin. *Jewish Messianism and the History of Philosophy.* Cambridge: Cambridge University Press, 2004.

Kirk, Geoffrey Stephen, John Earle Raven, and Malcom Schofield. *The Presocratic Philosophers: A Critical History with a Selection of Texts.* 2nd ed. Cambridge: Cambridge University Press, 1983.

Kisiel, Theodore. *The Genesis of Heidegger's* Being and Time. Berkeley: University of California Press, 1993.

Kleinberg, Ethan. *Generation Existential: Heidegger's Philosophy in France, 1927–1961.* Ithaca, NY: Cornell University Press, 2005.

Legros, Robert. "L'expérience originaire du temps: Lévinas et Husserl." *Cahiers d'études Lévinassiennes* 1 (2002): 77–97.

Levinas, Emmanuel. "'As Old as the World?'" 1966. In *Nine Talmudic Readings,* translated by Annette Aronowicz, 70–88. Bloomington: Indiana University Press, 1990.

———. *Autrement qu'être ou au-delà de l'essence.* The Hague: Martinus Nijhoff, 1974.

———. *Basic Philosophical Writings.* Edited by Adriaan T. Peperzak, Simon Critchley, and Robert Bernasconi. Bloomington: Indiana University Press, 1996.

———. "'Between Two Worlds': The Way of Franz Rosenzweig." 1963. In *Difficult Freedom: Essays on Judaism,* translated by Seán Hand, 181–201. Baltimore: Johns Hopkins University Press, 1990.

———. *Beyond the Verse: Talmudic Readings and Lectures.* Translated by Gary D. Mole. Bloomington: Indiana University Press, 1994.

———. *Carnets de captivité et autres inédits.* Vol. 1 of *Oeuvres,* edited by Rodolphe Calin and Catherine Chalier. Paris: Grasset/Imec, 2009.

———. *Collected Philosophical Papers.* Translated by Alphonso Lingis. Pittsburgh: Duquesne University, 1987.

———. "Damages Due to Fire." In *Nine Talmudic Readings,* translated by Annette Aronowicz, 178–97. Bloomington: Indiana University Press, 1990.

———. "Diachrony and Representation." In *Entre Nous: Thinking of the Other,* translated by Michael B. Smith and Barbara Harshav, 159–77. New York: Columbia University Press, 1998.

———. "Dialogue with Emmanuel Levinas." Interview translated and conducted by Richard Kearney, in *Face to Face with Levinas,* edited by Richard A. Cohen, 13–33. Albany: State University of New York Press, 1986.

———. *Difficult Freedom: Essays on Judaism.* Translated by Seán Hand. Baltimore: Johns Hopkins University Press, 1990.

———. *Discovering Existence with Husserl.* Translated by Richard A. Cohen and Michael B. Smith. Evanston, IL: Northwestern University Press, 1998.

———. "The Ego and the Totality." In *Collected Philosophical Papers,* translated by Alphonso Lingis, 25–45. Pittsburgh: Duquesne University, 1987.

———. "Enigma and Phenomenon." In *Basic Philosophical Writings,* edited by Adriaan T. Peperzak, Simon Critchley, and Robert Bernasconi, 65–77. Bloomington: Indiana University Press, 1996.

———. "Énigme et phénomène." *Esprit* 33 (June 1965): 1128–42.

———. "Entre deux mondes: Biographie spirituelle de Franz Rosenzweig." In *La conscience juive: Données et débats,* edited by E. Amado Lévy-Valensi and J. Halpérin, 121–37. Paris: Presses Universitaires de France, 1963.

———. *Entre Nous: Thinking of the Other*. Translated by Michael B. Smith and Barbara Harshav. New York: Columbia University Press, 1998.

———. "Ethics as First Philosophy." In *The Levinas Reader*, edited by Seán Hand, 75–87. Oxford: Blackwell, 1989.

———. *Existence and Existents*. Translated by Alphonso Lingis. Pittsburgh: Duquesne University Press, 2001.

———. "Freedom and Command." In *Collected Philosophical Papers*, translated by Alphonso Lingis, 15–23. Pittsburgh: Duquesne University, 1987.

———. "Freiburg, Husserl, and Phenomenology." In *Unforeseen History*, translated by Nidra Poller, 57–66. Chicago: University of Illinois Press, 2004.

———. *God, Death, and Time*. Translated by Bettina Bergo. Stanford: Stanford University Press, 2000.

———. "Humanism and An-archy." In *Collected Philosophical Papers*, translated by Alphonso Lingis, 127–39. Pittsburgh: Duquesne University Press, 1987.

———. "Ideology and Idealism." In *Of God Who Comes to Mind*, translated by Bettina Bergo, 3–14. Stanford: Stanford University Press, 1998.

———. "Intentionality and Sensation." In *Discovering Existence with Husserl*, translated by Richard A. Cohen and Michael B. Smith, 135–50. Evanston, IL: Northwestern University Press, 1998.

———. "Interview with Emmanuel Levinas: December 31, 1982." Conducted and translated by Edith Wyschogrod, in *Philosophy and Theology* 4, no. 2 (Winter 1989): 105–18.

———. "Interview with François Poirié." In *Is it Righteous to Be? Interviews with Emmanuel Levinas*, edited by Jill Robbins, translated by Jill Robbins and Marcus Coelen, 23–83. Stanford: Stanford University Press, 2001.

———. *In the Time of Nations*. Translated by Michael B. Smith. London: Athlone Press, 1994.

———. *Is It Righteous to Be? Interviews with Emmanuel Levinas*. Edited by Jill Robbins. Stanford: Stanford University Press, 2001.

———. "Is Ontology Fundamental?" In *Basic Philosophical Writings*, edited by Adriaan T. Peperzak, Simon Critchley, and Robert Bernasconi, 1–10. Bloomington: Indiana University Press, 1996.

———. "Jacques Derrida: Wholly Otherwise." In *Proper Names*, translated by Michael B. Smith, 55–62. Stanford: Stanford University Press, 1996.

———. "Judaism and the Present." In *Difficult Freedom: Essays on Judaism,* translated by Seán Hand, 208–15. Baltimore: Johns Hopkins University Press, 1990.

———. "Letter from Emmanuel Levinas." In *On Escape,* translated by Bettina Bergo, 1–2. Stanford: Stanford University Press, 2003.

———. "Lévy-Bruhl and Contemporary Philosophy." In *Entre Nous: Thinking of the Other,* translated by Michael B. Smith and Barbara Harshav, 39–51. New York: Columbia University Press, 1998.

———. "Meaning and Sense." In *Basic Philosophical Writings,* edited by Adriaan T. Peperzak, Simon Critchley, and Robert Bernasconi, 33–64. Bloomington: Indiana University Press, 1996.

———. "The Meaning of Religious Practice." Translated by Peter Atterton, Matthew Calarco, and Joëlle Hansel. *Levinas Studies* 5 (2010): 1–4.

———. "Messianic Texts." In *Difficult Freedom: Essays on Judaism,* translated by Seán Hand, 59–96. Baltimore: Johns Hopkins University Press, 1990.

———. "The Name of God according to a Few Talmudic Texts." In *Beyond the Verse: Talmudic Readings and Lectures,* translated by Gary D. Mole, 116–28. Bloomington: Indiana University Press, 1994.

———. *Nine Talmudic Readings.* Translated by Annette Aronowicz. Bloomington: Indiana University Press, 1990.

———. *Of God Who Comes to Mind.* Translated by Bettina Bergo. Stanford: Stanford University Press, 1998.

———. *On Escape.* Translated by Bettina Bergo. Stanford: Stanford University Press, 2003.

———. "On Jewish Philosophy." In *Is it Righteous to Be? Interviews with Emmanuel Levinas,* edited by Jill Robbins, translated by Michael B. Smith, 239–54. Stanford: Stanford University Press, 2001.

———. "The Other, Utopia, and Justice." In *Entre Nous: Thinking of the Other,* translated by Michael B. Smith and Barbara Harshav, 223–33. New York: Columbia University Press, 1998.

———. *Otherwise than Being: Or Beyond Essence.* Translated by Alphonso Lingis. Pittsburgh: Duquesne University Press, 1998.

———. *Outside the Subject.* Translated by Michael B. Smith. Stanford: Stanford University Press, 1994.

———. "The Pact." In *Beyond the Verse: Talmudic Readings and Lectures,* translated by Gary D. Mole, 68–85. Bloomington: Indiana University Press, 1994.

———. "The Philosopher and Death." In *Alterity and Transcendence,* translated by Michael B. Smith, 153–68. New York: Columbia University Press, 1999.

———. "Philosophy and the Idea of Infinity." In *Collected Philosophical Papers,* translated by Alphonso Lingis, 47–59. Pittsburgh: Duquesne University, 1987.

———. "Philosophy and Transcendence." In *Alterity and Transcendence,* translated by Michael B. Smith, 3–37. New York: Columbia University Press, 1999.

———. "Philosophy, Justice, and Love." In *Entre Nous: Thinking of the Other,* translated by Michael B. Smith and Barbara Harshav, 103–21. New York: Columbia University Press, 1998.

———. *Positivité et transcendence: Suivi de Lévinas et la phenomenology.* Edited by Jean-Luc Marion. Paris: Presses Universitaires de France, 2000.

———. "Prohibition against Representation and 'The Rights of Man.'" In *Alterity and Transcendence,* translated by Michael B. Smith, 121–30. New York: Columbia University Press, 1999.

———. *Proper Names.* Translated by Michael B. Smith. Stanford: Stanford University Press, 1996.

———. "Quelques réflexions sur la philosophie de l'Hitlérisme." *Esprit,* no. 26 (Nov. 1934): 199–208.

———. "Questions and Answers." In *Of God Who Comes to Mind,* translated by Bettina Bergo, 79–99. Stanford: Stanford University Press, 1998.

———. "Questions et reponses." *Le nouveau commerce* 36/37 (Spring 1977): 61–86.

———. "Reality and Its Shadow." In *Collected Philosophical Papers,* translated by Alphonso Lingis, 1–13. Pittsburgh: Duquesne University, 1987.

———. "Reflections on the Philosophy of Hitlerism." Translated by Seán Hand. *Critical Inquiry* 17, no. 1 (Autumn 1990): 63–71.

———. "The Ruin of Representation." In *Discovering Existence with Husserl,* translated by Richard A. Cohen and Michael B. Smith, 111–21. Evanston, IL: Northwestern University Press, 1998.

———. "Signature." In *Difficult Freedom: Essays on Judaism,* translated by Seán Hand, 291–95. Baltimore: Johns Hopkins University Press, 1990.

———. "Simone Weil against the Bible." In *Difficult Freedom: Essays on Judaism,* translated by Seán Hand, 133–41. Baltimore: Johns Hopkins University Press, 1990.

———. "Some Thoughts on the Philosophy of Hitlerism." In *Unforeseen History,* translated by Nidra Poller, 13–21. Chicago: University of Illinois Press, 2004.

———. "The Spinoza Case." In *Difficult Freedom: Essays on Judaism,* translated by Seán Hand, 106–10. Baltimore: Johns Hopkins University Press, 1990.

———. "Substitution." In *Basic Philosophical Writings,* edited by Adriaan T. Peperzak, Simon Critchley, and Robert Bernasconi, 79–95. Bloomington: Indiana University Press, 1996.

———. "The Temptation of Temptation." In *Nine Talmudic Readings,* translated by Annette Aronowicz, 30–50. Bloomington: Indiana University Press, 1990.

———. *The Theory of Intuition in Husserl's Phenomenology.* 2nd ed. Translated by André Orianne. Evanston, IL: Northwestern University Press, 1995.

———. "The Thinking of Being and the Question of the Other." In *Of God Who Comes to Mind,* translated by Bettina Bergo, 111–21. Stanford: Stanford University Press, 1998.

———. *Time and the Other: And Additional Essays.* Translated by Richard A. Cohen. Pittsburgh: Duquesne University Press, 1987.

———. *Totality and Infinity: An Essay on Exteriority.* Pittsburgh: Duquesne University Press, 1969.

———. "Toward the Other." In *Nine Talmudic Readings,* translated by Annette Aronowicz, 12–29. Bloomington: Indiana University Press, 1990.

______. "La trace de l'autre." *Tijdschrift voor filosofie* 24, no. 3 (1963): 605–23.

———. "Transcendence and Height." In *Basic Philosophical Writings,* edited by Adriaan T. Peperzak, Simon Critchley, and Robert Bernasconi, 11–31. Bloomington: Indiana University Press, 1996.

———. "The Transcendence of Words: On Michel Leiris's *Biffures.*" In *Outside the Subject,* translated by Michael B. Smith, 144–50. Stanford: Stanford University Press, 1994.

———. "The Translation of the Scripture." In *In the Time of Nations*, translated by Michael B. Smith, 33–54. London: Athlone Press, 1994.

———. "Truth of Disclosure and Truth of Testimony." In *Basic Philosophical Writings*, edited by Adriaan T. Peperzak, Simon Critchley, and Robert Bernasconi, 97–107. Bloomington: Indiana University Press, 1996.

———. *Unforeseen History*. Translated by Nidra Poller. Chicago: University of Illinois Press, 2004.

———. "Useless Suffering." In *Entre Nous: Thinking of the Other*, translated by Michael B. Smith and Barbara Harshav, 91–101. New York: Columbia University Press, 1998.

———. "The Work of Edmund Husserl." In *Discovering Existence with Husserl*, translated by Richard A. Cohen and Michael B. Smith, 47–87. Evanston, IL: Northwestern University Press, 1998.

Levy, Ze'ev. "Emmanuel Levinas on Secularization in Modern Society." *Levinas Studies* 1 (2007): 19–35.

Lévy-Bruhl, Lucien. *How Natives Think*. Translated by Lilian A. Clare. Princeton, NJ: Princeton University Press, 1985.

———. *The Notebooks on Primitive Mentality*. Translated by Peter Riviere. Oxford: Basil Blackwell, 1975.

Lingis, Alphonso. "The Sensuality and the Sensitivity." In *Face to Face with Levinas*, edited by Richard A. Cohen, 219–30. Albany: State University of New York Press, 1986.

Llewelyn, John. *Emmanuel Levinas: The Genealogy of Ethics*. London: Routledge, 1995.

Lotz, Christian. *From Affectivity to Subjectivity: Husserl's Phenomenology Revisited*. New York: Palgrave Macmillan, 2007.

Maldonado-Torres, Nelson. *Against War: Views from the Underside of Modernity*. Durham, NC: Duke University Press, 2008.

Malka, Salomon. *Emmanuel Levinas: His Life and Legacy*. Translated by Michael Kigel and Sonja M. Embree. Pittsburgh: Duquesne University Press, 2006.

Manning, Robert John Sheffler. "Serious Ideas Rooted in Blood." In *Postmodernism and the Holocaust*, edited by Alan Milchman and Alan Rosenberg, 125–35. Amsterdam: Editions Rodopi B. V., 1998.

Marrati, Paola. "Derrida and Levinas: Ethics, Writing, Historicity." *Levinas Studies* 1 (2007): 19–35.

Mole, Gary D. *Lévinas, Blanchot, Jabès: Figures of Estrangement.* Gainesville: University Press of Florida, 1997.

Moore, F. C. T. *Bergson Thinking Backwards.* Cambridge: Cambridge University Press, 1996.

Moran, Dermot. *Introduction to Phenomenology.* New York: Routledge, 2000.

Morgan, Michael. *Discovering Levinas.* Cambridge: Cambridge University Press, 2007.

Moulard-Leonard, Valentine. *Bergson-Deleuze Encounters.* Albany: State University of New York Press, 2008.

Moyn, Samuel. *Origins of the Other: Emmanuel Levinas Between Revelation and Ethics.* Ithaca, NY: Cornell University Press, 2005.

Naas, Michael. "Lending Assistance Always to Itself: Levinas' Infinite Conversation with Platonic Dialogue." In *Levinas and the Ancients,* edited by Brian Schroeder and Silvia Benso, 79–102. Bloomington: Indiana University Press, 2008.

Nelson, Eric Sean, Antje Kapust, and Kent Still, eds. *Addressing Levinas.* Evanston, IL: Northwestern University Press, 2005.

Nemo, Philippe. *Job and the Excess of Evil.* Translated by Michael Kigel. Pittsburgh: Duquesne University Press, 1998.

Newman, Michael. "Sensibility, Trauma, and the Trace: Levinas from Phenomenology to the Immemorial." In *The Face of the Other and the Trace of God: Essays on the Philosophy of Emmanuel Levinas,* edited by Jeffrey Bloechl, 90–129. New York: Fordham University Press, 2000.

Olafson, Frederick. *Heidegger and the Ground of Ethics: A Study of Mitsein.* Cambridge: Cambridge University Press, 1998.

Olivier, Paul. "L'être et le temps chez Emmanuel Lévinas." *Recherches de science religieuse Paris* 71, no. 3 (1983): 337–80.

Ouaknin, Marc-Alain *The Burnt Book: Reading the Talmud.* Translated by Llewellyn Brown. Princeton, NJ: Princeton University Press, 1998.

Parmenides. *Fragments of Parmenides.* In *The Presocratics,* by Edward Hussey, 79–99. New York: Charles Scribner's Sons, 1973.

Pascal, Blaise. *Pensées and Other Writings.* Translated by Honor Levi. Oxford: Oxford University Press, 1999.

Peperzak, Adriaan T. *Beyond: The Philosophy of Emmanuel Levinas.* Evanston, IL: Northwestern University Press, 1997.

———. *To the Other: An Introduction to the Philosophy of Emmanuel Levinas.* West Lafayette: Purdue University Press, 1993.

Perpich, Diane. *The Ethics of Emmanuel Levinas.* Stanford: Stanford University Press, 2008.

Pinkard, Terry. *Hegel's Phenomenology: The Sociality of Reason.* Cambridge: Cambridge University Press, 1996.

Plato. *Phaedo.* In *Five Dialogues,* translated by G. M. A. Grube, 93–155. Indianapolis: Hackett, 1981.

———. *The Symposium.* Translated and edited by Christopher Gill. London: Penguin Books, 2003.

———. *Timaeus.* Translated by Benjamin Jowett. Upper Saddle River, NJ: Prentice Hall, 1997.

Plotinus. *The Enneads.* Abr. Translated by Stephen Mackenna, edited by John Dillon. London: Penguin Books, 1991.

Pyle, Andrew. *Malebranche.* London: Routledge, 2003.

Rajan, Tilottama. *Deconstruction and the Remainders of Phenomenology: Sartre, Derrida, Foucalt, Baudrillard.* Stanford: Stanford University Press, 2002.

Reinhard, Kenneth. "Kant with Sade, Lacan with Levinas." *Modern Language Notes* 110, no. 4 (Sept. 1995): 785–808.

Richardson, William. "The Irresponsible Subject." In *Ethics as First Philosophy: The Significance of Emmanuel Levinas for Philosophy, Literature and Religion,* edited by Adriaan T. Peperzak, 123–31. New York: Routledge, 1995.

Ricoeur, Paul. *Autrement: Lecture d'autrement qu'être.* Paris: Presses Universitaires de France, 1997.

———. *Oneself as Another.* Translated by Kathleen Blamey. Chicago: University of Chicago Press, 1992.

Riera, Gabriel. " 'The Possibility of the Poetic Said' in *Otherwise than Being* (Allusion, or Blanchot in Levinas)," *Diacritics,* Vol. 34, No. 2 (Summer 2004), 13–36.

Robbins, Jill. *Altered Reading: Levinas and Literature.* Chicago: University of Chicago Press, 1999.

Rodemeyer, Lanei. "Developments in the Theory of Time-Consciousness: An Analysis of Protention." In *The New Husserl: A Critical Reader,* edited by Don Welton, 125–54. Bloomington: Indiana University Press, 2003.

Rolland, Jacques. "Getting Out of Being by a New Path." In *On Escape,* translated by Bettina Bergo, 3–48. Stanford: Stanford University Press, 2003.

Rose, Gillian. *The Broken Middle.* Oxford: Blackwell, 1992.

Rosenzweig, Franz. *The Star of Redemption.* Translated by William W. Hallo. Notre Dame: University of Notre Dame Press, 1970.

Sallis, John, ed. *Chorology: On Beginning in Plato's Timaeus.* Bloomington: Indiana University Press, 1999.

Sandford, Stella. "Levinas, Feminism and the Feminine." In *The Cambridge Companion to Levinas,* edited by Simon Critchley and Robert Bernasconi, 139–60. Cambridge: Cambridge University Press, 2002.

———. *The Metaphysics of Love: Gender and Transcendence in Levinas.* London: Athlone Press, 2000.

Sartre, Jean-Paul. *Being and Nothingness: An Essay in Phenomenological Ontology.* Abr. Translated by Hazel E. Barnes. New York: Citadel Press, 2001.

———. *The Imaginary: A Phenomenological Psychology of the Imagination.* Translated by Jonathan Webber. London: Routledge, 2004.

Saussure, Ferdinand de. *Course in General Linguistics.* Translated by Roy Harris. Edited by Charles Bally and Albert Sechehaye. Chicago: Open Court, 1986.

Schroeder, Brian, and Silvia Benso, eds. *Levinas and the Ancients.* Bloomington: Indiana University Press, 2008.

Severson, Eric R., ed. *I More than Others: Responses to Evil and Suffering.* Newcastle upon Tyne: Cambridge Scholars Press, 2010.

Shakespeare, William. *The Tragedy of Macbeth.* Edited by E. K. Chambers. Boston: D. C. Heath, 1915.

Sikka, Sonia. "The Delightful Other: Portraits of the Feminine in Kierkegaard, Nietzsche, and Levinas." In *Feminist Interpretations of Emmanuel Levinas,* edited by Tina Chanter, 96–118. University Park, PA: Pennsylvania State University Press, 2001.

Smith, Colin. "The Philosophy of Vladimir Jankélévitch." *Philosophy* 32, no. 123 (Oct. 1957): 315–24.

Smith, Steven G. "Reason as One for Another: Moral and Theoretical Argument." In *Face to Face with Levinas,* edited by Richard A. Cohen, 53–71. Albany: State University of New York Press, 1986.

Sokolowski, Robert. *Husserlian Meditations: How Words Present Things.* Evanston, IL: Northwestern University Press, 1974.

Spiegelberg, Herbert. *The Phenomenological Movement: A Historical Introduction.* 3rd rev. ed. Dordrecht: Kluwer Academic, 1994.

Stähler, Tanja. "Getting under the Skin: Platonic Myths in Levinas." In *Levinas and the Ancients,* edited by Brian Schroeder and Silvia Benso, 62–78. Bloomington: Indiana University Press, 2008.

Taminiaux, Jacques. *Heidegger and the Project of Fundamental Ontology,* translated by Michael Gendre. Albany: State University of New York Press, 1991.

Tauber, Alfred. "Outside the Subject: Levinas's Jewish Perspective on Time." *Graduate Faculty Philosophy Journal* 20/21 (1998): 439–59.

Thomas, Elisabeth Louise. *Emmanuel Levinas: Ethics, Justice, and the Human Beyond Being.* New York: Routledge, 2004.

Vasey, Craig. "Faceless Women and Serious Others: Levinas, Misogyny, and Feminism." In *Ethics and Danger: Essays on Heidegger and Continental Thought,* edited by Arleen B. Dallery and Charles E. Scott, 317–30. Albany: State University of New York Press, 1992.

Vries, Hent de. *Religion and Violence: Philosophical Perspectives from Kant to Derrida.* Baltimore: Johns Hopkins University Press, 2002.

Waldenfels, Bernhard. "Levinas on the Saying and the Said." In *Addressing Levinas,* edited by Eric Sean Nelson, Antje Kapust, and Kent Still, 86–97. Evanston, IL: Northwestern University Press, 2005.

Westphal, Merold. "The Transparent Shadow: Kierkegaard and Levinas in Dialogue." In *Kierkegaard in Post/Modernity,* edited by Martin Joseph Matuštík and Merold Westphal, 265–82. Bloomington: Indiana University Press, 1995.

Wygoda, Shmuel. "The Phenomenology of Time in the Philosophy of Levinas: Temporality and Otherness in the Hebraic Tradition." *Analecta husserliana* 90 (2006): 283–301.

Wyschogrod, Edith. "Derrida, Levinas, and Violence." In *Derrida and Deconstruction,* edited by Hugh J. Silverman, 177–94. New York: Routledge, 1989.

———. *Emmanuel Levinas: The Problem of Ethical Metaphysics.* 2nd ed. New York: Fordham University Press, 2000.

Zahavi, Dan. *Husserl's Phenomenology.* Stanford: Stanford University Press, 2003.

Index

absence: and presence, 176, 186; of time, 28, 87, 93, 152
activity, 45, 135, 136, 256; and passivity, 35, 193
affectivity, 45, 46, 58, 62, 67, 83, 195–96; Husserl and, 32, 37–38; of the instant, 64, 65, 83; and passivity, 64, 194, 195–96, 242, 258, 333n31
Ainley, Alison, 165
Allen, Sarah, 27
Alliance Israélite Universelle, 138
alterity, 65, 100, 130, 147, 200, 286, 309n14, 324n39; and being, 196, 278; and death, 98, 102, 103, 118; Derrida on, 212, 215, 220, 224, 225, 226; and ethics, 232, 328n25; and evil, 165, 324n39; and exteriority, 150, 218, 321n3; and family, 173; and feminine, 170–71, 273; of God, 187–88, 199, 294, 327n10; in Hebrew texts, 200–01; Hegel on, 155; Kant on, 328n31; and knowledge, 98, 136; and language, 103–04, 132, 183, 190–91; Levinas's rethinking of, 173, 238, 260; and maternity, 275–76, 283; metaphors and images to evoke, 103–04, 142, 176, 181, 182, 217; of the other, 99, 101, 102–03, 114, 120–21, 124, 127, 137, 145, 185, 187, 197, 200, 210, 212, 227, 317n44; philosophy and, 200, 232, 338n27; spatialization of, 103–04, 132, 145, 176, 181, 182, 218; temporal nature of, 174, 175, 182; and time, 3, 20, 78, 101–02, 105, 108, 137, 161, 179, 185, 202–03, 226, 268–69; and violence, 121, 176. *See also* exteriority; other
Alweiss, Lilian, 196
ambush, 122
antecedence, 48, 310n27
anxiety, 27, 46–47, 81, 82, 86, 264, 315n15; and death, 85, 86, 316n26; and insomnia, 49–50, 81; Levinas on, 46, 47–48, 81, 86, 106; and nothingness, 82, 148, 315n14, 322n14
Aquinas, Thomas, 15
Arendt, Hannah, 289, 340n48
Aristotle, 79, 130, 132, 166, 198, 281, 328n38; Heidegger on, 12–13, 17, 215, 305–06n12; and time, 9, 13, 14–15, 60, 193, 194–95, 204–05, 305n6, 306n23
Aronowicz, Annette, 209
art, 254; and imagination, 71, 73; Levinas critique of, 70, 73, 74, 75; and representation, 69, 71, 72
assimilation, 202, 230
atheism, 128, 130, 131, 132, 136
Augustine, Saint, 15, 17
authenticity, 19, 28–29, 75, 82–83, 85, 149–50

Bass, Alan, 213
Beauvoir, Simone de, 165–66, 167, 324n39
being, 27, 67, 80, 98, 137, 178, 200, 231, 293; as barbarian, 30; and alterity, 196, 278; anonymity of, 39, 84; basic structure of, 112–13; and beings, 87, 136, 318n5; comprehension of, 113, 129; and creation, 132; Dasein and, 28–29, 129, 318n1; and death, 162, 246; and diachrony, 338n30; and dialogue, 256; economy of, 124, 167; and ego, 92; and evil, 63, 231, 279, 312n59, 338n30; and face, 162, 294–95; and feminine, 167; and freedom, 26, 313n76; fullness of, 79, 80; as game, 335n58; and Good, 240, 273, 277, 282; Heidegger influence on Levinas around, 21, 29, 37; and history, 178, 240; and human condition, 25–26; Husserl on, 32, 272; meaning of, 318nn1–2; and other, 150, 231, 239, 272, 289, 294, 333n29; philosophy and, 26, 28, 29, 30, 78, 90, 155–56, 160–61; Plato on, 102–03, 129, 147, 261; and psychic life, 323n24; solitary nature of, 79–80; temporality of, 12, 28, 30, 65, 87, 96, 305n18; and time, 14, 15, 18, 60, 98, 456; totality of, 90, 226; transcendence and, 29, 129; and violence, 145–46, 156. *See also* Dasein; existence; ontology

Being and Time (Heidegger), 17, 38, 81, 94, 129, 297, 318n1; critique of philosophy in, 12, 36–37, 55, 84, 139, 194; on "Dasein of Others," 113–16; enduring value of, 17–18; fundamental ontology of, 131, 315n23, 326n9; influence on Levinas of, 11, 305n12; Levinas on, 28, 37, 110–11, 128, 225
being-in-the-world, 16, 43–44, 85, 86, 106, 112–13
being-toward-death, 75, 83, 85, 86–87, 94, 316n25, 319n9; authentic and inauthentic, 82, 315n17; dwelling and, 162
Bennington, Geoffrey, 221
Bergo, Bettina, 194, 237, 332n19
Bergson, Henri, 41, 49, 56, 69, 139, 312n50; on conservation of past, 263–64; and *durée,* 50–55, 59–60, 311n36; on free will, 56, 60, 311n45, 312n55; and Heidegger, 17, 54–55, 57, 311nn40–41, 311n47; influence on Levinas by, 7, 8–9, 40, 305n5, 305n7, 310n32; Levinas's debt to, 2–3, 51, 55, 310n32; Levinas's turn away from, 59–60, 65; and science and mathematics, 237, 332n23; time conception of, 5, 9–10, 12, 51, 52–54, 59–60, 201, 211, 305n6
Bernasconi, Robert, 213, 214, 219, 226, 236, 331n86
Bernet, Rudolf, 1
Bible, 211, 226
Bloechl, Jeffrey, 63, 71, 73, 79, 303n9, 325n53; on God and religion, 197, 232; on paternity and ipseity, 275, 338n20; on women and gender, 166, 167
Blond, Phillip, 63
Blondel, Charles, 8, 9
bourgeois, 27–28, 329n41; egoism of, 153, 322–23n21; work ethic of, 26
Brody, Donna, 270, 273
Brothers Karamazov, The (Dostoyevsky), 250–52

Cain and Abel, 236
Caputo, John D., 17, 277, 305–06n18, 312n59, 340n61
caress, 64, 101–02, 329n42
Chalier, Catherine, 248, 329–30n53
Chanter, Tina, 168, 169, 307n39, 314n7; on feminine and gender, 163, 173, 269, 270, 285, 294, 299, 325n61; on Heidegger, 12, 47, 305n6, 318n5; on Levinas's philosophy of time, 1, 285, 303n6, 335n56
Chaplin, Charlie, 157, 323n28
children, 169, 173–74, 241, 242–43, 255
Chouchani, 206–08, 209
Christianity, 19–20, 25, 90, 124, 301, 307n39
Ciaramelli, Fabio, 236–37, 238, 240, 244, 245, 332n17
Cohen, Richard, 64, 78–79, 172, 308n52, 310n26, 334n46; on gendered themes, 164, 324n37; on Heidegger, 9, 311n41; on Husserl, 32, 195; on Levinas's theory of ethics, 209–10, 233; on Levinas's theory of time, 1, 3, 233; preface to *Otherwise than Being* by, 233, 291; on religious themes, 293, 329n40; on Rosenzweig, 77, 88, 90, 317n37
conatus essendi, 233, 290
consciousness, 71, 160, 233, 240, 257, 265, 283, 337n2; freedom of, 234, 238; Husserl on time and, 15–16, 32, 34, 193, 196, 237, 332n24; and hypostasis, 332n16; and language, 253, 254; of the other, 155, 231; and phenomenology, 194; and self, 238, 339n38; transcendental and intuitive, 195
Course in General Linguistics (Saussure), 189, 327n16
creation, 90, 122–23, 132, 275, 281; continuous, 58, 312n50; *ex nihilo,* 132, 246–47, 285–86, 334n44; time and temporality in, 132, 205
Cronos myth. *See* Saturn (Cronos) myth

Dalton, Drew, 281, 338–39n33
Dasein, 46, 130, 161, 297–98, 313n76; and authenticity, 19, 75, 82–83, 85, 149–50; and being, 28–29, 129, 318n1; and being-in-the-world, 94, 95, 113; and being-toward-death, 85, 86–87, 94, 319n9; as condition of care, 84, 315n21; and death, 47, 86, 205; and existence, 21, 94, 103, 111–12; and nothingness, 82–84, 315nn15–16; ontology of, 318nn3–4; and other, 113–16, 319n13; temporality of, 21, 96, 194, 305n18; thrownness of, 29, 57, 95, 311n48, 315n20; and time, 16, 98. *See also* being

Davidson, Arnold, 24
dead time, 154
death, 27, 96, 98, 100–01, 106, 119, 162, 252; and alterity, 98, 102, 103, 118; and anxiety, 85, 86, 316n26; and being, 162, 246; being-toward, 47, 75, 82, 83, 85, 86–87, 94, 162, 310n25, 315n17, 316n25; and existence, 99–100; and freedom, 95, 316n25; and future, 28, 96, 205; Heidegger on, 29–30, 46–47, 84–85, 86–87, 307n48, 316n26; inevitability of, 20, 28; Levinas-Heidegger differences over, 93–99; and passivity, 98, 100
Defoe, Daniel, 65
deformalization of time: by Heidegger, 201, 237–38, 332n24; by Levinas, 2, 10, 205, 238, 247, 264, 267–68, 336–37n2
Derrida, Jacques, 35, 106, 140, 168, 253, 259, 277; on alterity, 212, 215, 220, 224, 225, 226; critique of Levinas by, 143–44, 151, 180, 213, 322n18; deconstructive method of, 213–14, 217, 223; funeral oration for Levinas, 268; on Hegel, 151, 156, 222, 238, 322n18; on Heidegger, 326n9; on history, 221, 223, 330n77; as influenced by Levinas, 218, 219; influence on Levinas by, 212–13, 224; on interiority and exteriority, 217–18, 330n65; Levinas's response to criticisms by, 224, 330–31n78; relationship to Levinas, 221, 224–25; on time, 18, 25; on trace, 213, 219–20, 225; on violence, 220, 221–22, 223, 307n37; wave analogy of, 5–6, 108, 212; works: *Adieu to Emmanuel Levinas*, 225; *The Gift of Death*, 225; *Margins of Philosophy*, 219–20; "Violence and Metaphysics," 180, 212–27, 238
Descartes, René, 12, 272; on continued creation, 58, 312n50; Heidegger on, 148, 322n13; on infinite, 263, 264; on motion, 59, 60, 312n52; on senses, 147–48, 322n12
desire, 44–45, 154, 309n16, 309n18
Desmond, William, 175
determinism, 20, 230–31
diachrony, 173, 195, 210, 244, 248, 254, 294, 295; and debt, 302; denudes chronologies, 246, 257; and ethics, 210, 227, 283; and future, 264–65, 287, 289; God and, 210, 291, 292; and history of philosophy, 259–66; as hope, 301; and language, 188–89, 241, 253, 257; Levinas's appropriation of, 38, 123, 142, 172–73, 179–80, 189, 196–97, 212, 250; Levinas's usage of term, 180, 188, 189, 196, 327n13; and maternity, 249, 273, 285, 289, 294; matrix for, 281–86; and narrative, 236, 239, 243; and the other, 191, 229–30, 247, 272, 273, 286; and past, 286, 296; and responsibility, 256, 269; restoration and renewal of, 287, 299–302; and self, 229, 237, 240; signification of, 245–46; and synchrony, 190, 193; and time, 1, 4, 179–80, 189, 191, 206, 211, 227, 243, 264, 265–66, 269, 333n27; and transcendence, 190, 228, 260–61
dialogue, 199, 256, 257
différance, 219–20, 224
distance, 101, 127, 181, 226; and time, 151–54
divine time, 91
Dostoyevsky, Fyodor, 250–52
Drabinsky, John, 36–38
dualism, 20, 165, 278
duration, 10, 52–54, 58, 59
dwelling. *See* habitation and dwelling

economic and clock time, 61, 62, 63, 124, 151–52, 159, 161, 202; Heidegger on, 13, 14, 306n21
ecstasis, 47, 136, 140, 205
ego, 148, 162, 240, 248, 287; alter ego, 220; and being, 84, 92; captivity in instant of, 77; and diachrony, 240; and encounter with the face, 155, 233–34; and enjoyment, 233, 237, 279, 323n22; fecundity of, 163, 239; and freedom, 33, 34, 230, 231; guilt and shame of, 125, 281; and history, 243; Husserl on, 35, 237, 260, 332n24; and narration, 241, 242, 243, 244, 250; and the other, 63, 155, 234, 250; and pardon, 319–20n19; past and future of, 34, 35; as protagonist, 247, 324n35; reconfiguration of, 235–36, 340n54; and responsibility, 234–35; and separation, 152, 158–59; and time, 33, 34–35, 237, 238, 243, 246, 279, 285; and totality, 126. *See also* self; subject

egoism, 152, 257, 258, 292, 322n20; bourgeois, 153, 322–23n21; and freedom, 230, 231
emotion, 135
enjoyment, 153, 245, 279, 322n20, 332n19; and concupiscence, 244–45; and ego, 233, 237, 279, 323n22; and uterine existence, 152, 278–79, 289. *See also jouissance*
Enneads, The (Plotinus), 262
Epicurus, 95–96
epistemology, 48, 74, 198–99
eros, 102, 176, 325n66
eschatological time, 179
eschatology, 172, 177, 178, 179, 223; Derrida on, 223; *Totality and Infinity* on, 141, 202
essence, 243, 333n36
eternal time, 55
eternity, 21, 50, 71–72, 92; Bergson on, 51–52; and history, 280; and love, 262–63; philosophy and, 13, 14–15, 21, 55, 204; Rosenzweig on, 91, 92; and time, 13, 14–15, 16, 55, 61, 92, 262, 264
ethics, 41, 170, 215, 236, 264, 265, 340n61; and alterity, 232, 328n25; and art, 74; and diachrony, 210, 227, 283; as first philosophy, 1, 41; and freedom, 149; Hegel on, 156; Judaism and, 200–01; and language, 253; as masculine affair, 167, 171, 173; and maternity, 283, 284, 339n47; and responsibility, 2, 179; and time, 77, 201, 209–10
Euclid, 183
evil, 320n22; and being, 63, 231, 279, 312n59, 338n30
exegesis, 183, 209–10
existence, 58, 84, 91, 96, 99, 123, 130, 195; Dasein's relationship to, 21, 94, 103, 111–12; and death, 99–100; and existent, 87, 93, 94; individualized, 55; simple and complex, 126; and time, 57, 103; uterine, 152, 278–79, 289. *See also* being
Existence and Existents. See Levinas, Emmanuel—works
experience, 44, 135, 136, 137, 162, 183, 184; and language, 182–83, 190, 200, 252, 327n21; Merleau-Ponty on, 183–84, 326n5; of the other person, 190, 200, 327n21; and separation, 152, 158; of subject, 50, 310n30
exteriority, 126, 135, 138, 147, 160, 221; and alterity, 150, 218, 321n3; Derrida on, 217, 330n72; face's appeal to, 162, 216; God and, 132, 197–98; and interiority, 123, 140, 143–44, 277, 279; and language, 216, 217, 218; Levinas's use of, 104–05, 127, 139–40, 145, 150, 216, 217–18, 275, 330n65; of the other, 101, 104, 113, 118–19, 123, 168, 216, 321n3; philosophy and, 140, 215; spatial dimension of, 110, 113, 132, 143–44, 145, 148–49, 157, 217, 321n3; and transcendence, 144, 320n27; and violence, 145, 321n8. *See also* alterity

Fabian, Johannes, 189–90
face, 116, 134, 149, 152, 174, 176; and being, 162, 294–95; exteriority of, 162, 216; plastic form of, 282, 339n36; representation of, 229; spatial dimension of, 109, 185; temporal element to, 133
face-to-face encounter, 112, 127, 185, 190, 237, 239, 265, 319n15; and alterity, 100, 101, 102, 121; creation compared to, 122–23; and ego, 155, 233–34; as epiphany, 199; with feminine other, 272; and hospitality, 172, 232; and identity, 256; and infinite, 131, 133, 233; and language, 183; Levinas's emphasis on, 107, 216; as partnership, 118; presence and absence in, 271, 291; as rupture, 117, 301; and self, 110, 148, 237; and time, 172–73, 186, 301; and war, 120. *See also* other
family, 169, 173, 239
fate, 20, 91, 317n36
fecundity, 162–63, 169, 173–74, 239, 242, 333n32
feminine, 102, 170, 249, 269, 278; alterity of, 165, 170–71, 273, 294; and being, 167; and economy, 173; feminists on, 271, 337n13; and habitation, 163–67, 171, 174, 233, 270, 272, 279–81, 293; Levinas's problematic use of, 4, 168, 293; and maternity, 242–43, 272, 283; as metaphor and trope, 142, 165, 166–67, 271, 294, 333n34, 338n18; in past, present, future, 172, 174, 286; philosophy and, 167, 289; representation of, 166, 271, 274, 283; spatiality of, 157; and time, 269–70, 273, 285. *See also* women

forgiveness. *See* pardon and forgiveness
Franck, Didier, 63
freedom, 21, 63, 65, 130, 132, 154, 211, 230; and being, 26, 313n76; of consciousness, 234, 238; and death, 95, 316n25; and ego, 33, 34, 230, 231; from history, 22, 153; and the instant, 50, 55, 56–57, 63, 154; and language, 253–54; Levinas's primacy of, 22–23, 149, 307n39; of the other, 120, 121; and responsibility, 231; of the subject, 60, 117; and time, 33, 34, 36, 73, 195, 272, 328n28
free will, 56, 60, 231, 311n45, 312n55
future, 13, 24, 28, 61, 161, 172, 178, 218; and children, 169, 173–74; and death, 28, 96, 205; and diachrony, 264–65, 287, 289; and ego, 34, 35; and eschatology, 172, 177; God and, 292–93; and hope, 61–62, 286, 287; and hypostasis, 97; and intersubjectivity, 124; and the other, 172–73, 265; and pardon, 61, 125, 177; and present, 78, 261, 286; and time, 61, 152, 161; and women, 173, 174, 286

gender, 271; differences, 289, 339n4; Levinas on, 163–67, 169, 173, 269–73, 324n37
gift, 63, 287; time as, 60, 63, 65, 84, 152
gnosticism, 278, 338n29
God, 127, 188, 191, 200, 232–33; alterity of, 187–88, 199, 294, 327n10; Descartes on, 59, 60; and diachrony, 210, 291, 292; and exteriority, 132, 197–98; and future, 292–93; and history, 242, 292, 333n32; and infinitude, 148, 242; and justice, 291, 292; kingdom of, 232, 331–32n8; Levinas on, 127, 197–98, 290–91; and philosophy, 289–95, 301; Rosenzweig on, 90, 317n37; and time, 201, 205, 211; in *Totality and Infinity,* 188, 198, 327n10, 328n36; transcendence of, 112, 132, 182, 199, 317n37
Gold Rush, The, 157, 323n28
Good, 275, 277, 295; and being, 240, 273, 277, 282
grace, 20–21, 301–02
Gschwandtner, Christina, 290
Guenther, Lisa, 339n47, 340n48
guilt, 125, 129, 252, 289–90, 296

habitation and dwelling, 157–60, 265, 271–72, 284; as concept, 153, 169; feminine and, 163–67, 171, 174, 233, 270, 272, 279–81, 293; Heidegger and, 158, 323–24n30; interiority of, 157–58, 233; and self, 162, 237; and time, 159, 161–62
Hayat, Pierre, 260
Hegel, G. W. F., 89, 155, 156, 203, 263; conception of history, 144, 156–57, 196, 203, 223, 238, 240, 321n7; Derrida on, 151, 156, 222, 238, 322n18; Levinas and, 41, 144, 145, 151, 154–57, 238, 264, 322n18; and modern philosophy, 78, 155–56; on phenomenology, 154–56; on time, 12, 156, 196, 203, 238
Heidegger, Martin, 65, 102, 196, 211, 310n30; on anxiety, 46–47, 81, 82, 85, 86, 106, 148, 264, 315nn14–15, 322n14; on Aristotle, 12–13, 17, 215, 305–06n12; on authenticity, 19, 28–29, 75, 82–83, 85, 149–50; on being, 14–17, 28–29, 37, 160–61, 215, 272; and Bergson, 17, 54–55, 57, 311nn40–41, 311n47; on care, 55, 311n43; critique of philosophy by, 12, 28, 84, 128, 220; on death, 29–30, 46–47, 84–85, 86–87, 307n48, 316n26; deformalization of time by, 201, 237–38, 332n24; on Descartes, 148, 322n13; exegesis of, 17; and Husserl, 10–11, 15–16, 17, 32, 37, 337n2; influence on Levinas by, 7, 11–12, 28, 31, 305n12; on the instant, 47, 55, 57, 311n43; Levinas's relationship with, 6, 36–37, 139, 314n11; on mood, 297–98; Nazi affiliation of, 6, 11, 86, 139, 225, 320n22; neutral sense of persons in, 144, 321n4; on nothingness, 81–84, 314nn12–13; ontology of, 12, 18, 29, 30, 111, 113, 129, 131, 185, 200, 315–16n23, 318n1, 318n4, 320n25, 326n9; on plurality and unity, 93–94; on solitude, 47, 310n22; on time, 9, 13–14, 17–18, 60, 83, 84, 87, 139, 201, 205, 305n6; tool imagery of, 158, 323–24n30; on totality, 149–50; transitions in thinking by, 314n11
Heidegger, Martin—Levinas's critique, 2, 6, 19, 38, 77, 93–99, 113–16, 194, 225; in *On Escape,* 6, 28–31, 94, 127, 307–08n49; in *Existence*

and Existents on, 46–48, 57–58, 65, 80, 94, 127–28; in "Is Ontology Fundamental?," 109–13, 139; in "Meaning and Sense," 185; in "Philosophy and the Idea of Infinity," 127–31, 320n21; in "Some Meditations on the Philosophy of Hitlerism," 94; in *Time and the Other,* 76, 80–84, 86–88, 93–99, 315–16n23; in *Totality and Infinity,* 94
height, 180–82, 217
Heraclitus, 146, 147, 322n11
Hindenburg, Paul von, 24
historical time, 195, 328n28
history, 19, 21, 162, 177, 187, 210, 278, 285; and being, 178, 240; Derrida on, 221, 223, 330n77; and diachrony, 289; and eternity, 280, 284; freedom from, 22, 153; God and, 242, 292, 333n32; Hegel's conception of, 144, 156–57, 196, 203, 223, 238, 240, 321n7; Judaism and, 201–02, 203, 205–06; judging, 162, 200, 324n34; narration of, 239; and the other, 185–86, 278, 336n79; and religion, 205, 295; rupture of, 156, 223, 323n27; as a said, 241, 243; suffering in, 249–50; and time, 151, 153, 195, 202, 203, 204, 284, 328n28; and transcendence, 23, 221, 223, 243; universal, 202
Hitler, Adolf, 6, 11, 18, 22, 23, 24
Hitlerism, 6, 7, 18–25, 30, 124, 295, 301; and evil, 25; ideology and philosophy of, 23, 24; and violence, 23–24
Hobbes, Thomas, 222
Hodge, Joanna, 16
holiness, 88–89, 268, 291–92, 293, 294, 295
hope, 42, 61–65, 96, 265, 269, 299, 301; eschatological, 125, 177; and future, 61–62, 286, 287; and present, 63–64; and salvation, 63, 125; and time, 60, 63, 64; and transcendence, 62, 63
horizon, 116, 117, 118–19, 147, 185, 319n8
hospitality, 173, 340n61; maternal, 248–49, 270; toward the other, 172, 232
hostage, 179, 248, 258, 336n71; as concept, 295, 321n30; Levinas's use of the term, 226, 331n86; and the other, 257–58, 288, 297
How Natives Think (Lévy-Bruhl), 135
humanism, 229
hunting analogies, 113, 187, 318–19n7
Husserl, Edmund, 41, 56, 106, 196, 272, 310n30; and affectivity, 32, 37–38; on consciousness, 15–16, 32, 34, 193, 196, 237, 332n24; on ego, 35, 237, 260, 332n24; and Heidegger, 15–16, 17, 32, 37, 337n2; influence on Levinas, 5, 7, 9, 10–11, 31, 40, 192, 195; Levinas's philosophy and, 2–3, 6, 31–38, 139, 191–97; and phenomenology, 34, 36, 37, 184, 192, 194, 196, 326n7, 328n29; on philosophy, 34, 37, 43, 105, 211, 215; on presuppositions, 9, 10, 305n9; on time, 5, 9–10, 12, 15–16, 31, 35, 193, 194, 195, 196, 201, 205, 237, 332n24; and *Urimpression,* 34–35, 36, 38, 192, 196; works: *Cartesian Meditations,* 32; *Logical Investigations,* 15, 328n29; *The Phenomenology of Internal Time-Consciousness,* 9, 31, 193
hypostasis, 2, 79–80, 97, 158, 314n7, 332n16

idealism, 25, 29, 89, 111, 113, 134, 136, 195
il y a, 46, 47, 58, 83, 98, 107, 338n30; Bergson and, 310n32; *Existence and Existents* on, 39, 46, 83, 308n52; and hypostasis, 79, 80; importance for Levinas of, 48, 310n26; as metaphor, 39, 105–06; and time, 49, 152
images and imagery, 127, 269; Levinas critique of, 74–75; representation of, 72; Sartre on, 70–71; spatial determinations of, 71, 313n67. *See also* metaphors and tropes; spatial imagery and language
Imaginary, The (Sartre), 70, 312n63
imagination, 71, 73, 75, 313n66
incarnation, 282, 283, 285, 339nn38–39
infinity and infinite, 144, 149, 150, 223, 321n5; and alterity, 170; Derrida on, 219; and distance, 150; in face of the other, 131, 133, 233, 324n33; and finitude, 128; God and, 148, 242; Heidegger and, 129; and history, 284; and horizon, 117, 147; Levinas's metaphors for, 104, 144–45; as social relation, 131; spatial relation in, 117, 145; and time, 16, 148, 168, 169, 173–74, 211, 242; and totality, 128, 278; and transcendence, 263, 321n5

insomnia, 58, 80, 98; and anxiety, 49–50, 81; Levinas on phenomena of, 45–46, 47, 98, 159, 309–10n20; as loss of vision, 66–67, 100–01; as metaphor, 39, 43, 48–49, 105–07; and sense of time, 152
instant, 59, 61, 72, 80, 110, 117, 275, 279; affectivity of, 64, 65, 83; as captivity, 50, 60; death as, 96; and duration, 53–54, 58; and feminine, 272, 284; and freedom, 50, 55, 56–57, 63, 154; Heidegger on, 47, 55, 57, 311n43; Husserl on, 192; and interiority, 153, 279; Levinas's philosophy of, 40, 42, 48–50, 56, 57–58, 68, 151; ontological nature of, 153; passivity of, 64, 134; philosophy and, 49, 52, 62, 310n35; Sartre on, 73; and self, 60, 98–99, 100; as solitary and isolated, 50–51, 57–58, 98–99; solitude of, 79, 100; spatialization of, 153, 157; specificity and uniqueness of, 47, 57; substantiality of, 314n7; and time, 50, 60, 64, 72, 77, 152, 270
intellect, 69
intellectualism, 119, 326n7
intentionality, 33–34, 45–46, 48, 194, 195–96, 233–34, 337n2
interiority, 126, 215, 216, 218, 284; and dwelling, 157–58, 233, 323n23; and exteriority, 123, 140, 143–44, 277, 279; Heidegger and, 140, 161; and the instant, 153, 279; Levinas's use of, 104–05, 217–18, 330n65; and self, 140, 153, 174; spatial dimension of, 132, 143–44, 157; and time, 151–52, 154, 163; and transcendence, 149
internal time, 34–35, 36, 195, 237–38
intersubjectivity, 31, 40, 63, 76, 241, 249, 285; ethical, 40, 41; Heidegger and, 86, 321n4; and language, 213, 252–53; Levinas's turn to, 31, 41, 107, 308n52; and time, 5, 124, 154. *See also* subjectivity
intuition, 53, 54, 56, 311n44
ipseity, 275, 338n20
Israel, 169

Job, 211
jouissance, 47, 244–45. *See also* enjoyment
Judaism and Jews, 138, 180, 199; ethics of, 200–01; on forgiveness and pardon, 20, 25, 124, 301; and history, 201–02, 203, 205–06; as influence on Levinas, 7–8, 41, 88, 170, 201, 329n40; and messianism of, 201–02; Rosenzweig on, 89, 90, 205; and separation, 204–05; and suffering, 249–50; and time, 8, 88, 198, 201, 203, 212–13
justice, 133, 149, 231, 232, 301, 340n54; God and, 291, 292

Kagan, Shelly, 249
Kant, Immanuel, 40, 41, 222, 263, 264, 272, 317n34; on alterity, 328n31; Heidegger and, 15, 17; on time, 12, 196
Katz, Claire Elise, 166, 170, 330n55
khora, 276
Kierkegaard, Søren, 89, 144
Kigel, Michael, 269
knowledge, 43, 154, 158, 184, 195, 200, 240; and alterity, 98, 136; and light, 65–66; philosophy and, 106, 183, 198

labor, 120, 124–25, 159, 161, 162, 323n29; and the other, 121, 187; and self, 158, 159
Lacan, Jacques, 245
lamentation, 296–97, 299
language, 109–10, 112, 139, 142, 184–85, 199, 233, 323n27; and alterity, 103–04, 132, 183, 190–91; betrayal by, 241, 255–56, 257, 258; and consciousness, 253, 254; and diachrony, 188–89, 241, 253, 257; and experience, 182–83, 190, 200, 252, 327n21; and exteriority, 216, 217, 218; and freedom, 253–54; gendered, 164–67, 168; and intersubjectivity, 213, 252–53; and metaphor, 184; philosophy and, 143, 260; spatial, 103–04, 124, 132, 140, 163, 199, 221, 226; and speech, 253, 254, 335n61; temporal, 124; and writing, 68, 253
Legros, Robert, 1
Leiris, Michel, 65
Levinas, Emmanuel—biography: academic positions, 143, 321n2; childhood and schooling, 7–8, 41, 169, 206, 329n46; in French military, 38; murder of family members, 24, 38, 250, 307n42; in Nazi captivity, 7, 39, 41–42, 122, 139; university education, 8–9, 15, 51
Levinas, Emmanuel—evolution of thinking: Derrida's influence and,

212–13; on diachrony, 180, 188, 264, 327n13; on ego and self, 236–37, 238, 324n35; on face, 123–24; on Heidegger, 77, 127–28, 134; on Husserl, 197; on images and representation, 127; on love, 123–24; on narration, 233–34, 237, 238; on persecution and hostage, 179, 226, 331n86; on politics, 123; on proximity, 182; on responsibility, 2, 138–39; Sartre and, 41–42; on spatial imagery, 105, 117, 139, 168, 180, 181–82, 326n8; on time, 2, 3, 41, 108, 172, 174, 178, 180, 182, 200; in use of metaphors, 104, 109–10, 179
Levinas, Emmanuel—influences on: Bergson, 5, 8–9, 40, 51, 305n5, 310n32; Blondel, 8; Derrida, 212–13, 224; Heidegger, 7, 11–12, 28, 31, 40, 305n12; Husserl, 5, 7, 9, 10–11, 31, 40, 192, 195; Judaism, 7–8, 41, 88, 170, 198, 201, 329n40; Merleau-Ponty, 184, 326n6; Nazi holocaust, 39–40; Rosenzweig, 5, 7, 77, 88–89, 140, 198, 204, 316n31
Levinas, Emmanuel—works: "Between Two Worlds: The Way of Franz Rosenzweig," 88, 204–05, 316n30; *Carnets de captivité et autre inedits,* 40, 88, 304n14; "Diachrony and Representation," 2, 316n33; "The Ego and the Totality," 108, 120, 123–27, 319–20n19; "Enigma and Phenomenon," 180, 188, 191, 327n12; *On Escape,* 6, 25–31, 44, 94, 127, 151, 172, 304n1, 307–08n49, 309n17; *Existence and Existents,* 39–49, 53, 56–67, 72–74, 94, 101, 117, 126–28, 152, 165, 217, 279; "Freedom and Command," 108, 120–23; *God, Death, and Time,* 268; "God and Philosophy," 268; *Of God Who Comes to Mind,* 263–64; "Intentionality and Sensation," 188, 193; interview with Armengaud (1985), 338n27; interview with Poirié (1986), 305n7; interview with *Voix d'Israel* (1937), 8, 304–05n3; "Is Ontology Fundamental?," 29, 108–13, 117–20, 139, 147, 185, 318n1; "Jacques Derrida: Wholly Otherwise," 224; "Judaism and the Present," 203; "Lévy-Bruhl and Contemporary Philosophy," 108–09, 134–38; "Meaning and Sense," 180, 182–83, 185, 219, 252, 326n6; "The Name of God according to a Few Talmudic Texts," 210; "The Old and the New," 268; *Otherwise than Being,* 4, 39, 64, 68, 104, 138–39, 172, 178–80, 182, 188, 206, 228–61, 267, 269, 294, 296, 297, 299, 330n55, 340n51; "Philosophy and the Idea of Infinity," 108–09, 120, 127–34, 319n15, 320n21; "Philosophy and Transcendence," 260–62, 264, 268; "Questions and Answers," 265, 336n78; "Reality and Its Shadow," 40, 69–70, 74–75; "The Ruin of Representation," 191–92; "Signature," 199–200, 321n33; "Some Thoughts on the Philosophy of Hitlerism," 6, 7, 18–25, 91, 94; "Substitution," 331nn85–86; *The Theory of Intuition in Husserl's Phenomenology,* 32; "The Thinking of Being," 268; *Time and the Other,* 3, 41, 76–107, 118, 119, 139, 217, 313–18; *Totality and Infinity,* 44, 48, 64, 94, 103–04, 107–08, 123, 128, 132, 137, 140–78, 188, 199–200, 202, 212, 214, 216–18, 223, 226–27, 234, 236, 238, 240, 242, 246, 248, 260, 270–81, 284, 286, 288, 293, 321–25; "The Trace of the Other," 182, 185–86, 219, 293–94, 326n4; "Transcendence and Height," 180–82; "The Transcendence of Words," 40, 65, 66, 68, 116; *Unforeseen History,* 19–20; "Useless Suffering," 249–50; "The Work of Edmund Husserl," 6, 31–38
Levinas scholarship, 7, 197, 267; chronological studies by, 3, 43, 270, 303n2; possibilities for future research in, 269–70, 295; primary sources in, 3, 304n13; time neglected in, 1, 3, 7
Levy, Ze'ev, 197
Lévy-Bruhl, Lucien, 134–38
liberalism, 19, 20, 21–22, 307n39
light, 35, 43, 67, 93, 101, 216, 226; epistemology of, 48; phenomenology and, 35, 36, 216, 328n25; and truth, 65–66; and violence, 216, 221; and vision, 65, 306n34
Lingis, Alphonso, 2, 27, 40, 61, 283
Llewelyn, John, 2
Lotz, Christian, 38
love, 90, 124, 176, 262–63, 319n18

Macbeth, 97
Malebranche, Nicolas, 58–59
Malka, Salomon, 39, 212
Manning, Robert John Sheffler, 23
Marrati, Paola, 225
Marx, Karl, 222
maternity, 269, 273–81, 293, 299, 333n33; and alterity, 275–76, 283; bound to dwelling and instant, 284; and diachrony, 249, 273, 285, 289, 294; and ethics, 283, 284, 339n47; and feminine, 242–43, 272, 283; feminists on, 289, 337n16, 339n47; and hospitality, 248–49, 270; and incarnation, 282, 339n39; and passivity, 248, 282; and responsibility, 249, 283, 285; and time, 286; as trope and metaphor, 222, 271, 272–73, 278, 285, 286, 289
Mein Kampf (Hitler), 24
memory, 186, 239, 295–96, 340n51
Merleau-Ponty, Maurice, 106, 183–84, 211, 326nn5–7
messianic time, 91, 141, 175, 197, 203, 210, 329n42
messianism, 63, 127, 199; and eschatology, 223; in Judaism, 201–02; and time, 91, 141, 175, 197, 201–04, 210, 329n42
metaphors and tropes, 109–10, 119, 137, 269; for alterity, 103–04, 127, 142, 176, 181, 182, 217; for death, 106; for feminine, 142, 165, 166–67, 271, 294, 333n34, 338n18; hostage as, 179; *il y a* as, 39, 105–06; for infinite, 104, 144–45; insomnia as, 39, 43, 48–49, 105–07; and language, 184; for maternity, 222, 271, 272–73, 278, 285, 286, 289; and reality, 105–07; of totality, 144–45. *See also* images and imagery
Metaphysical Foundations of Logic, The (Heidegger), 15–17
metaphysics, 132, 136, 150, 221, 222–23; Heidegger and, 130–31, 320n25; ontology and, 132; philosophy and, 320n25, 320n27
Minkowski, Eugene, 181, 217
Miteinandersein, 85, 103, 111, 130
Mole, Gary, 30, 307–08n49
mood, 297–99
morality, 41, 146–47, 200, 215, 231, 239
Moran, Dermot, 43, 305n9, 309n14
More, Henry, 59
Morgan, Michael, 141
motion, 15; Descartes on, 59, 60, 312n52; Heidegger on, 46, 58, 82–83; and time, 52–53
Moyn, Samuel, 41, 303n2
murder, 118, 120, 137, 162, 176, 324n33
mythology, 136; Greek, 19, 20, 49, 54, 262–63, 307n36

narration, 290, 333n26; and diachrony, 236, 239, 243; and ego, 238, 241, 242, 243, 244, 250; Levinas and structure of, 233–34, 237, 238, 240; and representation, 240–41, 242, 294; as said, 241–42; and self, 240, 242; of suffering, 249; and synchrony, 238–39, 332n25
needs, 231; and desires, 44, 309n18
Nicomachean Ethics (Aristotle), 166
Nietzsche, Friedrich, 22, 23
nothingness, 81–84, 86, 152, 314–15nn12–20; and anxiety, 82, 148, 315n14, 322n14

obligation, 6, 179, 181, 182, 186, 243. *See also* responsibility
On Escape. *See* Levinas, Emmanuel—works
ontology, 117, 128, 134, 140, 150, 162; and alterity, 118, 121, 132, 187; and creation, 281; fundamental, 116, 127, 134, 185, 326n9; Heidegger and, 12, 18, 29, 30, 111, 113, 129, 131, 200, 315–16n23, 318n1, 318n4, 320n25; Levinas's attempt to go beyond, 112, 119, 139, 160, 202–03, 226, 283; Levinas's critique of Heidegger's, 109, 127, 128, 134, 225, 307n49; and metaphysics, 132, 320n25; and phenomena, 112, 318n4; and philosophy, 12, 110–11, 113, 116, 129, 147, 149, 253; Plato on, 263; and presence, 87–88, 200; and signs, 186; and time, 18, 31; and violence, 119–20, 146, 149, 162. *See also* being
other, 40, 71, 77, 98, 112, 154, 327n21; alterity of, 101, 102–03, 114, 120–21, 124, 127, 145, 185, 187, 197, 200, 210, 212, 227, 317n44; and being, 150, 231, 239, 272, 289, 294, 333n29; consciousness of, 155, 231; Dasein and, 113–16, 319n13; and diachrony, 191, 229–30, 247, 272, 273, 286; erotic relations with, 325n66; and ethics, 232; and

experience, 190, 200, 327n21; exteriority of, 101, 104, 113, 118–19, 123, 168, 216, 321n3; feminine, 164, 165, 166, 170–71, 174, 272; freedom of, 120, 121; future of, 172–73, 265; God and, 188, 191, 327n10; Heidegger and, 85–87, 102, 139; and history, 185–86, 278, 336n79; hostage to, 257–58, 288, 297; and labor, 121, 187; ontological role of, 84–85, 109, 316n24; and past, 172–73, 186, 191, 230; persecution of, 249, 258, 283; Plato and, 103; presence of, 185, 186; and present, 68, 176, 271–72; primacy of, 228, 248; proximity of, 149, 231, 254, 260, 265; representation of, 127, 136, 230, 273; responsibility to, 3, 138–39, 234, 237, 245, 258, 288, 336n69; and same, 126, 128, 129, 130, 134, 170, 181, 205, 320n23; and self, 63, 99, 100, 104, 115, 118–19, 125–26, 140, 150–51, 155, 165, 168–69, 182, 231, 319n12; and speech, 117–18, 248, 255, 265, 335n61; and subject, 68, 93; suffering of, 99–103, 230, 249, 250, 258, 290, 291, 297, 317–18n44; and time, 65, 84, 189, 190, 191, 228, 240, 243, 285; trace of, 185, 188, 191, 197, 294; transcendence of, 4, 148, 182, 190–91, 199, 229, 260, 317n37. *See also* alterity; face-to-face encounter

Otherwise than Being. *See* Levinas, Emmanuel—works

paganism, 91–92, 128, 131, 136

pain, 27, 28, 96

pardon and forgiveness: Christianity on, 19–20, 25, 124, 301; and ego, 319–20n19; and future, 61, 125, 177; as gift, 63, 65; God and, 127; Judaism on, 19–20, 25, 124, 301; and past, 62, 124; and time, 25, 124, 125

Parmenides, 147, 155, 313n2, 322n11; Levinas's break with, 77–81, 93, 145, 264; on unity of being, 78, 313n3

participation, 135, 137, 138

passivity, 122, 166, 184, 232, 237, 245, 255, 257, 281, 298; and activity, 35, 193; and affectivity, 64, 194, 195–96, 242, 258, 333n31; and death, 98, 100; and goodness, 331n5; Husserl and, 197; of the instant, 64, 134; Levinas's notion of, 79, 98, 194, 220; and light, 36, 328n25; and maternity, 248, 282; and the other, 244, 247, 249; and saying, 256, 335n64; and suffering, 103, 249, 250–51; and violence, 222

past, 21, 34, 61, 97, 137, 175, 187; Bergson on, 263–64; and diachrony, 286, 296; and intersubjectivity, 124; Levinas on, 172–75, 178, 182, 296; and the other, 172–73, 186, 191, 230; pardon for, 62, 124; philosophy and, 13, 299; and present, 45, 256–57, 261, 294; recuperation of, 230, 331n3; and responsibility, 186, 293

paternity, 162, 167–69, 239, 242, 274–75

peace, 146, 151, 177, 223, 232

Peperzak, Adriaan, 189, 267, 320n23, 327n10, 327n12, 336n2; on gendered language use, 164–65, 166–67; on Husserl, 32; on speech, 256

Perpich, Diane, 233–34, 236, 237, 238, 240, 244, 271, 324n35, 336–37n2

persecution, 244, 248, 295; Levinas's use of the term, 179, 226, 331n86; of the other, 249, 258, 283; and responsibility, 259, 336n71

phenomena, 40–41, 43, 184; Husserl on, 33, 308n56; ontology and, 112, 318n4

phenomenology: and alterity, 309n14; Derrida on, 35, 214; and experience, 183–84; failure of in field of language, 182; Hegel and, 154–56; Husserl and, 34, 36, 37, 184, 192, 194, 196, 326n7, 328n29; of the image, 312n63; Levinas and, 9, 31, 32–33, 36–37, 38, 42–43, 99, 184, 194; and light, 35, 36, 216; Merleau-Ponty and, 183–84, 326nn5–7; and philosophy's gaps, 10; and presuppositions, 40–41; and representation, 193; and science, 60; and time, 34, 59, 60

Phenomenology of Internal Time-Consciousness, The (Husserl), 9, 10, 31, 193

philosophy, 10, 22, 44, 54, 58, 104, 127, 132, 136–37, 146; and alterity, 200, 232, 338n27; and atheism, 130, 131; and being, 26, 28, 29, 30, 78, 90, 155–56, 160–61; birth of, 328–29n38; and contamination, 214–15; and diachrony, 249, 259–66; and eternity, 13, 14–15, 21, 55, 204; and ethics, 1, 41, 129; and exteriority, 140, 215; and feminine, 167, 289; formalism in, 336–37n2; God and, 289–95, 301;

Hegel and, 78, 155–56; Heidegger's critique of, 12, 28, 84, 128, 220; Husserl on, 34, 37, 43, 105, 211, 215; and the instant, 49, 52, 62, 310n35; intellectualism of, 33, 337n2; and knowledge, 106, 183, 198; and language, 143, 260; Levinas's challenge to, 143, 147, 149, 215, 226, 227, 232, 259, 260, 268; Levinas's place in history of, 31, 142–43; and metaphysics, 320n25, 320n27; and ontology, 12, 110–11, 113, 116, 129, 147, 149, 253; and paganism, 91–92; and past, 13, 299; Plato and, 73, 129, 277, 338n27; and present, 13, 16, 61; and reason, 215, 263, 264; Rosenzweig and, 79, 89–90, 211; and separation, 93, 198; and theology, 291; and time, 7, 10, 23, 25, 34, 55, 57, 119, 137, 205, 211; and transcendence, 198–99, 211, 232, 260, 261, 262, 263, 338n27; and violence, 19, 120, 121, 151, 222, 223, 319n14
Plato, 73, 129, 132, 281; on being and good, 102–03, 129, 147; *eidos* of, 52, 103, 112, 130, 276–77; Heidegger on, 12, 17, 129, 215, 264, 305–06n12; on *khora,* 276; Levinas and, 130, 261, 262, 277, 336n77, 338n27; on love, 262–63; on time and eternity, 13, 14–15, 193, 264; on transcendence, 132, 198, 262; on unity of the One, 260, 264; works: *Phaedo,* 130; *Republic,* 276; *Symposium,* 262–63; *Timaeus,* 276–77
Plotinus, 262
pluralism, 96, 175, 200; and unity, 78, 86, 313n3
plurality, 78–79, 85, 93–94
poetry, 74, 313n74
politics, 19, 167, 232, 291, 301, 340n61
possession, 66, 70, 71, 112, 159–60, 162
prayer, 116–17, 199, 231–32
presence, 177, 185, 191, 199, 220, 226; and absence, 176, 186; Levinas's critique of, 185; ontology of, 87–88; and trace, 187
present, 47, 70, 75, 124, 126, 177, 193, 230; controllability of, 27–28, 99, 117; and dwelling, 161–62, 271–72; and future, 78, 261, 286; Husserl on, 34; irreparablity of, 61; and the other, 68, 176, 271–72; and past, 45, 256–57, 261, 294; philosophy and, 13, 16, 61; redemption of, 45, 63–64, 99, 100; and time, 36, 99, 100, 161–62; and women, 171, 174–75
presuppositions, 9, 10, 40–41, 305n9
primitive societies, 134–36, 137
Prometheus, 161
prophecy, 264, 265
Proust, Marcel, 41
proximity, 182, 230, 260, 265, 326n3, 331n4, 336n26; and the other, 149, 231, 254, 260, 265
psyche, 284, 339n41

racism, 23, 24
Rajan, Tilottama, 36
reality, 51, 74, 310n34, 340n48; and metaphors, 105–07; and representation, 72, 193
reason, 22, 116, 130, 232; domination of, 113, 318–19n7; philosophical, 215, 263, 264
redemption: of ego, 287; Judaism and Christianity on, 90; of present, 45, 63–64, 99, 100; of time, 72, 91, 124
Reinhard, Kenneth, 245
religion, 112, 123, 132, 200, 299; and history, 205, 295; Levinas's confessional writings on, 48, 88, 138, 202–03, 304n13; in Levinas's thought, 88, 123, 126–27, 197, 199; and responsibility, 231, 232; and time, 21, 201–04
representation, 137, 138, 160, 173; and art, 69, 71, 72; of the face, 127, 229; feminine, 166, 271, 274, 283; Levinas's treatment of, 35, 134, 320–21n29; and narration, 240–41, 242, 294; of the other, 127, 136, 230, 273; privileging of, 134, 135; and reality, 72, 193; and time, 69
responsibility, 179, 231, 236, 257–58, 266, 335n65; and diachrony, 256, 269; and ethics, 2, 179; Levinas's emphasis on, 3, 109, 138, 139, 230, 233, 259, 272, 301; and maternity, 249, 283, 285; to the other, 3, 138–39, 234, 237, 245, 258, 288, 336n69; and past, 186, 293; and persecution, 259, 336n71; and religion, 231, 232; and self, 181, 234–35, 240, 288, 290; and singularity, 244, 245; and suffering, 250, 251; and time, 6, 74, 269, 289. *See also* obligation
resurrection, 62, 63, 312n565

revelation, 90, 205
Ricoeur, Paul, 247–48, 267, 334n47, 337n3
Riera, Gabriel, 74
Robbins, Jill, 74
Robinson Crusoe, 65, 67, 68–69, 125
Rodin, Auguste, 72–73
Rolland, Jacques, 7, 31, 304n1, 307n46, 308n52
Rose, Gillian, 2
Rosenzweig, Franz, 89–90, 91–92, 204–06, 313n2, 317n37; influence on Levinas of, 5, 7, 77, 88–89, 140, 198, 204, 316n31; on Judaism, 89, 90, 205; Levinas's borrowing from, 2–3, 76, 77, 88–92, 110, 180, 317n37; Levinas's differences with, 92, 317n37; on philosophy, 79, 89–90, 211; on separation, 88–89, 92, 110, 211; on time, 2–3, 5, 180, 211
Rousseau, Jean-Jacques, 41, 222

salvation, 63, 72, 92, 125, 202, 289–90
Sandford, Stella, 166–67, 333n34
Sartre, Jean-Paul, 50, 70, 106, 310n30, 312n63; Levinas's engagement with, 38, 41–42, 69–74, 76, 138, 320–21n29
Saturn (Cronos) myth, 49, 54, 61, 154, 243, 262–63, 264
Saussure, Ferdinand de, 188–89, 190, 327n16
saying and said, 247, 255, 267, 282, 294; betrayal of each other, 257, 335–36n68; diachrony and, 68, 191; distinction between, 74, 240–41, 255–56, 333n29, 335n56; language and, 253; maternity and, 284–85; mood and, 298; passivity and, 256, 335n64; speech and, 254–55, 335n62; time and, 335n56; trace of saying in said, 239, 273, 294
science, 49, 53, 54, 60, 130, 204
secrecy, 154, 323n24
self, 20, 96, 200, 210, 248, 290, 298; and diachrony, 229, 237, 240; and exteriority, 138, 140; and face-to-face encounter, 110, 115; and habitation, 162, 237; horizon of, 116, 301; and *il y a*, 83; and instant, 60, 98–99, 100; interiority of, 140, 152, 153, 174; and the other, 63, 99, 100, 104, 115, 118–19, 125–26, 140, 150–51, 155, 165, 168–69, 182, 231, 319n12; renarration of, 242; and responsibility, 181, 234–35, 240, 288, 290; and spatiality, 125–26, 182; and time, 26, 58, 61, 78, 96, 169, 189, 205, 237, 238; and world, 44, 98. *See also* ego, subject
self-consciousness, 238
sensation, 135, 183, 184, 194, 230; affectivity and passivity in, 195–96, 328n25; Descartes on, 147–48, 322n12; and language, 184–85
separation, 145, 150, 154, 175–76, 204–05, 322n20; between ego and world, 152, 158–59; and experience, 152, 158; Levinas language of, 139–40; philosophy and, 93, 198; Rosenzweig view of, 88–89, 92, 110, 211; spatial, 145, 211; and time, 119, 206
shame, 281, 290, 338–39n33
sight. *See* vision
signs and signals, 186, 188, 189, 191, 313n67, 335n65
Sikka, Sonia, 167, 271
sin, 19, 20, 289–90
singularity, 129, 244, 245, 333–34n37
skepticism, 236
Smith, Steven, 236
Socrates, 129, 132, 263
Sokolowski, Robert, 9–10
solitude, 47, 78, 79, 87, 93, 100, 310n22
sound, 66, 67, 69, 117–18
space and spatiality, 52, 103–05, 117, 119, 142; alterity and, 105, 125–26, 145, 218; distance and, 101, 226; exteriority and, 105, 321n3; separation and, 145, 211; time and, 54, 55; as visual, 199
spatial imagery and language, 113, 118, 120, 199; Derrida on, 216, 221, 226; to describe alterity, 103–04, 132, 181; in "Is Ontology Fundamental,"? 119; Levinas's shift in usage of, 105, 139, 168, 180, 181–82, 326n8; in "Meaning and Sense," 185; temporal imagery and, 110, 124; in "The Ego and the Totality," 127; in "The Trace of the Other," 185–86; time and, 177; in *Totality and Infinity*, 104, 137, 140, 141, 143, 163, 168, 174, 175, 217. *See also* images and imagery
speech, 253; and language, 253, 254, 335n61; and the other, 117–18, 248, 265, 335n61; saying and said in, 254–55, 335n62

Spinza, Baruch, 233
spirituality, 195, 290
Star of Redemption, The (Rosenzweig), 79, 89–90, 92, 316n31
structuralism, 189, 190
subject, 58, 83, 98, 152, 194, 237; and death, 96, 340n48; and existence, 92, 93; freedom of, 60, 117; identity of, 335n65; and time, 16, 50, 61, 100; unity of, 92, 106–07. *See also* ego, self
subjectivity, 60, 129, 130, 200, 212, 333n32. *See also* intersubjectivity
substance, 136–37
substitution, 236, 245, 272–73; and suffering, 249, 250, 258–59
suffering, 27, 95, 96, 97, 98, 149, 229, 245; Dostoyevsky on, 250–52; and guilt, 252; in history, 249–50; Levinas's fixation on, 296; maternal, 272–73; of the other, 99–103, 230, 249, 250, 258, 290, 291, 297, 317–18n44; and pain, 27, 323n22; and passivity, 103, 249, 250–51; and responsibility, 250, 251; and substitution, 249, 250, 258–59; and vision, 100–01
suicide, 106, 246
supernatural, 135, 136
symmetry, 221, 330n73
synchronization, 15, 246
synchrony, 189, 229, 241, 249, 250, 292; and diachrony, 190, 193; narration as, 238–39, 332n25

Talmud, 204; Levinas's study of, 8, 200, 206–08, 329nn46–50; Levinas's utilization of, 180, 198, 201, 209, 226; time in, 180, 211, 212–13
Tauber, Alfred, 1, 201
temporality, 16, 27, 61, 125, 194; and alterity, 174, 175, 182; Aristotle and, 17, 205; of being, 12, 21, 28, 30, 85, 87, 96, 305n18; and creation, 132, 205; and ethics, 209–10; evolution of Levinas's thinking on, 2, 11, 173, 174, 192; and face, 133; Heidegger's reconfiguration of, 12, 29, 87, 238, 332n23; history as field of, 202; imagery and language of, 110, 124; Judeo-Christian, 20; modern philosophy and, 29; and time, 176–77, 202, 220
Ten Commandments, 132
Thales of Miletos, 198, 328–29n38
theology, 89, 291
Thomas, Elisabeth Louise, 56
Time and the Other. See Levinas, Emmanuel—works
totalitarianism, 19
totality, 126, 144–45, 147, 162, 175–76; and being, 155, 178; as concept, 123; Hegel on, 155, 203; Heidegger on, 149–50; and infinity, 128, 278; and time, 177; and war, 120, 146, 321n8
Totality and Infinity. See Levinas, Emmanuel—works
totalization, 147, 154, 182, 203; and violence, 145, 321n8
trace, 182, 185, 186, 187, 210, 294; Derrida on, 213, 219–20, 225; of the other, 185, 188, 191, 197, 294; as sign, 186, 191; Talmud and, 329–30n53
tragedy, 72, 99
transcendence, 42, 60, 127, 144, 147, 180–82, 325n65; and being, 29, 129; and diachrony, 190, 228, 260–61; and exteriority, 144, 320n27; in the face, 4, 277; God and, 112, 132, 182, 199, 317n37; and history, 23, 221, 223, 243; and hope, 62, 63; and infinitude, 174, 263, 321n5; and labor, 159; of the other, 4, 148, 190–91, 199, 229, 260, 317n37; philosophy and, 198–99, 211, 232, 261, 262; Plato on, 132, 198, 262; and religion, 112; as social relation, 149; of the son, 275, 276, 278; spatialization of, 132, 143, 145, 174, 181, 182, 199; and time, 143, 175
trauma, 244, 245, 340n51
truth, 137–38, 320n27; and history, 156; and light, 65–66; universal, 21–22; and violence, 22, 23, 307n37

unity, 136, 176, 225, 262; of being, 78; and pluralism, 78, 86, 313n3; plurality and, 78–79, 93–94; and separation, 204–05; of subject, 92, 106–07
universal time, 50, 60, 151, 156, 176–77, 195

Vasey, Craig, 270–71, 272, 293
violence, 19, 120, 125, 154, 177; and alterity, 121, 176; and being, 145–46, 156; Derrida on, 220, 221–22, 223, 307n37; economy of, 221; and exteriority, 145, 321n8; Hegel on,

155, 156; Hitlerism and, 23–24; and light, 216, 221; and morality, 146–47; ontology and, 119–20, 146, 149, 162; and peace, 151, 223; philosophy and, 19, 120, 151, 221, 222, 223, 319n14; and time, 25, 202; and totalization, 145, 321n8; and truth, 22, 23, 307n37
"Violence and Metaphysics" (Derrida), 180, 212–27, 238; alter ego argument of, 220–23; Levinas on, 214; on light and space, 216–20
virility, 83, 98, 166
vision, 69, 75, 93, 122, 127, 326n5; and alterity, 100, 105; Dasein and, 57, 75; insomnia as loss of, 66–67, 100–01; and light, 65, 306n34; moment of, 83, 311n49; and possession, 66, 70, 71; and sound, 117–18; and time, 60, 306n34

Waldenfels, Bernhard, 254
war, 122, 155, 162, 176, 177, 319n14; and alterity of the other, 120–21; and totality, 120, 146, 321n8
Weil, Simone, 127
Westphal, Merold, 289
"What Is Metaphysics?" (Heidegger), 81–82
Wiesel, Elie, 207
Wild, John, 143, 324n34
women, 270, 272, 324n39; excluded from politics and ethics, 166, 167, 173; as metaphor, 166–67, 242–43; as other, 164, 165, 166, 170–71; and past, present, future, 171, 174–75; relegated to home, 171, 285, 325n61. *See also* feminine
writing, 68, 253
Writing in General Linguistics (Saussure), 189
Wygoda, Shmuel, 1
Wyschogrod, Edith, 203, 224

Zahavi, Dan, 192